IN SEARCH OF MEMORY

ALSO BY ERIC R. KANDEL

Psychiatry, Psychoanalysis, and the New Biology of Mind

Principles of Neural Science, Fourth Edition (with James H. Schwartz and Thomas M. Jessell)

Memory: From Mind to Molecules (with Larry Squire)

Essentials of Neural Science and Behavior (with James H. Schwartz and Thomas M. Jessell)

Behavioral Biology of Aplysia

Cellular Basis of Behavior

IN SEARCH OF MEMORY

The Emergence of a New Science of Mind

ERIC R. KANDEL

W. W. NORTON & COMPANY
NEW YORK • LONDON

Printed in the United States of America
First Edition

Manufacturing by Maple-Vail Book Manufacturing Group
Book design by Chris Welch
Production manager: Julia Druskin

Library of Congress Cataloging-in-Publication Data

Kandel, Eric R.
In search of memory : the emergence of a new science of mind / Eric R. Kandel.— 1st ed.
p. cm.
Includes bibliographical references and index.
ISBN 0-393-05863-8 (hardcover)
1. Kandel, Eric R. 2. Neurologists—United States—Biography.
3. Medical scientists—United States—Biography. 4. Nobel Prizes. 5. Memory.
6. Neurobiology. 7. Cellular signal transduction. I. Title.
RC339.52.K362A3 2006
616.80092—dc22

2005028565

W. W. Norton & Company, Inc., 500 Fifth Avenue, New York, N.Y. 10110
www.wwnorton.com

W. W. Norton & Company Ltd., Castle House, 75/76 Wells Street, London W1T 3QT

1 2 3 4 5 6 7 8 9 0

POUR DENISE

CONTENTS

Preface *xi*

ONE

1. Personal Memory and the Biology of Memory Storage 3
2. A Childhood in Vienna 12
3. An American Education 33

TWO

4. One Cell at a Time 53
5. The Nerve Cell Speaks 74
6. Conversation Between Nerve Cells 90
7. Simple and Complex Neuronal Systems 103
8. Different Memories, Different Brain Regions 116
9. Searching for an Ideal System to Study Memory 135
10. Neural Analogs of Learning 150

THREE

11. Strengthening Synaptic Connections 165
12. A Center for Neurobiology and Behavior 180
13. Even a Simple Behavior Can Be Modified by Learning 187
14. Synapses Change with Experience 198
15. The Biological Basis of Individuality 208
16. Molecules and Short-Term Memory 221
17. Long-Term Memory 240
18. Memory Genes 247
19. A Dialogue Between Genes and Synapses 261

FOUR

20. A Return to Complex Memory 279
21. Synapses Also Hold Our Fondest Memories 286
22. The Brain's Picture of the External World 295
23. Attention Must Be Paid! 307

FIVE

24. A Little Red Pill 319
25. Mice, Men, and Mental Illness 335
26. A New Way to Treat Mental Illness 352
27. Biology and the Renaissance of Psychoanalytic Thought 363
28. Consciousness 376

SIX

29. Rediscovering Vienna via Stockholm 393

30. Learning from Memory: Prospects 416

Glossary 431

Notes and Sources 453

Acknowledgments 485

Index 489

PREFACE

Understanding the human mind in biological terms has emerged as the central challenge for science in the twenty-first century. We want to understand the biological nature of perception, learning, memory, thought, consciousness, and the limits of free will. That biologists would be in a position to explore these mental processes was unthinkable even a few decades ago. Until the middle of the twentieth century, the idea that mind, the most complex set of processes in the universe, might yield its deepest secrets to biological analysis, and perhaps do this on the molecular level, could not be entertained seriously.

The dramatic achievements of biology during the last fifty years have now made this possible. The discovery of the structure of DNA by James Watson and Francis Crick in 1953 revolutionized biology, giving it an intellectual framework for understanding how information from the genes controls the functioning of the cell. That discovery led to a basic understanding of how genes are regulated, how they give rise to the proteins that determine the functioning of cells, and how development turns genes and proteins on and off to determine the body plan of an organism. With these extraordinary accomplishments behind it, biology assumed a central position in the constellation of sciences, one in parallel with physics and chemistry.

Imbued with new knowledge and confidence, biology turned its attention to its loftiest goal, understanding the biological nature of the human mind. This effort, long considered to be prescientific, is already in full swing. Indeed, when intellectual historians look back on the last two decades of the twentieth century, they are likely to comment on the surprising fact that the most valuable insights into the human mind to emerge during this period did not come from the disciplines traditionally concerned with mind—from philosophy, psychology, or psychoanalysis. Instead, they came from a merger of these disciplines with the biology of the brain, a new synthesis energized recently by the dramatic achievements in molecular biology. The result has been a new science of mind, a science that uses the power of molecular biology to examine the great remaining mysteries of life.

This new science is based on five principles. First, mind and brain are inseparable. The brain is a complex biological organ of great computational capability that constructs our sensory experiences, regulates our thoughts and emotions, and controls our actions. The brain is responsible not only for relatively simple motor behaviors, such as running and eating, but also for the complex acts that we consider quintessentially human, such as thinking, speaking, and creating works of art. Looked at from this perspective, mind is a set of operations carried out by the brain, much as walking is a set of operations carried out by the legs, except dramatically more complex.

Second, each mental function in the brain—from the simplest reflex to the most creative acts in language, music, and art—is carried out by specialized neural circuits in different regions of the brain. This is why it is preferable to use the term "biology of mind" to refer to the set of mental operations carried out by these specialized neural circuits rather than "biology of *the* mind," which connotes a place and implies a single brain location that carries out all mental operations.

Third, all of these circuits are made up of the same elementary signaling units, the nerve cells. Fourth, the neural circuits use specific molecules to generate signals within and between nerve cells. Finally, these specific signaling molecules have been conserved—retained as it were—through millions of years of evolution. Some of them were present in the cells of our most ancient ancestors and can be found today in our most distant and primitive evolutionary relatives: single-celled organisms such as bacteria and yeast and simple multicellular organisms such as worms, flies, and snails. These creatures use the same molecules to organize their maneuvering through their environment that we use to govern our daily lives and adjust to our environment.

Thus, we gain from the new science of mind not only insights into ourselves—how we perceive, learn, remember, feel, and act—but also a new perspective of ourselves in the context of biological evolution. It makes us appreciate that the human mind evolved from molecules used by our lowly ancestors and that the extraordinary conservation of the molecular mechanisms that regulate life's various processes also applies to our mental life.

Because of its broad implications for individual and social well-being, there is now a consensus in the scientific community that the biology of mind will be to the twenty-first century what the biology of the gene was to the twentieth century.

In addition to addressing the central issues that have occupied Western thought since Socrates and Plato first speculated about the nature of mental processes more than two thousand years ago, the new science of mind gives us the practical insights to understand and cope with important issues about mind that affect our everyday lives. Science is no longer the exclusive domain of scientists. It has become an integral part of modern life and contemporary culture. Almost daily, the media report technical information that the general public cannot be expected to understand. People read about the memory loss caused by Alzheimer's disease and about age-related memory loss and try, often unsuccessfully, to understand the difference between these two disorders of memory—one progressive and devastating, the other comparatively benign. They hear about cognitive enhancers but do not quite know what to expect from them. They are told that genes affect behavior and that disorders of those genes cause mental illness and neurological disease, but they are not told how this occurs. And finally, people read that gender differences in aptitude influence the academic and career paths that men and women follow. Does this mean there are differences between the brains of women and of men? Do men and women learn differently?

In the course of our lives, most of us will have to make important private and public decisions that involve a biological understanding of mind. Some of these decisions will arise in the attempt to understand variations in normal human behavior, while others will concern more serious mental and neurological disorders. It is essential, therefore, that everyone have access to the best available scientific information presented in clear, understandable form. I share the view now current in the scientific community that we have a responsibility to provide the public with such information.

Early in my career as a neuroscientist I realized that people without a background in science are as eager to learn about the new science of mind as we scientists are to explain it. In this spirit, one of my colleagues at Columbia University, James H. Schwartz, and I wrote *Principles of Neural Science*, an introductory college and medical school textbook that is now entering its fifth edition. The publication of that textbook led to invitations to give talks about brain science to general audiences. That experience convinced me that nonscientists are willing to work to understand the key issues of brain science *if* scientists are willing to work at explaining them. I have therefore written this book as an introduction to the new science of mind for the general reader who has no background in science. My purpose is to explain in simple terms how the new science of mind emerged from the theories and observations of earlier scientists into the experimental science that biology is today.

A further impetus for writing this book came in the fall of 2000, when I was privileged to receive the Nobel Prize in Physiology or Medicine for my contributions to the study of memory storage in the brain. All Nobel laureates are invited to write an autobiographical essay. In the course of writing mine, I saw more clearly than before how my interest in the nature of memory was rooted in my childhood experiences in Vienna. I also saw more vividly, and with great wonder and gratitude, that my research has allowed me to participate in a historic period of science and to be part of an extraordinary international community of biological scientists. In the course of my work

I have come to know several outstanding scientists in the front ranks of the recent revolution in biology and neuroscience, and my own research has been greatly influenced by my interactions with them.

Thus, I interweave two stories in this book. The first is an intellectual history of the extraordinary scientific accomplishments in the study of mind that have taken place in the last fifty years. The second is the story of my life and scientific career over those five decades. It traces how my early experiences in Vienna gave rise to a fascination with memory, a fascination that focused first on history and psychoanalysis, then on the biology of the brain, and finally on the cellular and molecular processes of memory. *In Search of Memory* is thus an account of how my personal quest to understand memory has intersected with one of the greatest scientific endeavors—the attempt to understand mind in cellular and molecular biological terms.

ONE

It is not the literal past that rules us, save, possibly, in a biological sense. It is images of the past. These are often as highly structured and selective as myths. Images and symbolic constructs of the past are imprinted, almost in the manner of genetic information, on our sensibility. Each new historical era mirrors itself in the picture and active mythology of its past.

—George Steiner, *In Bluebeard's Castle* (1971)

1

PERSONAL MEMORY AND THE BIOLOGY OF MEMORY STORAGE

Memory has always fascinated me. Think of it. You can recall at will your first day in high school, your first date, your first love. In doing so you are not only recalling the event, you are also experiencing the atmosphere in which it occurred—the sights, sounds, and smells, the social setting, the time of day, the conversations, the emotional tone. Remembering the past is a form of mental time travel; it frees us from the constraints of time and space and allows us to move freely along completely different dimensions.

Mental time travel allows me to leave the writing of this sentence in my study at home overlooking the Hudson River and project myself backward sixty-seven years and eastward across the Atlantic Ocean to Vienna, Austria, where I was born and where my parents owned a small toy store.

It is November 7, 1938, my ninth birthday. My parents have just given me a birthday gift that I have craved endlessly: a battery-operated, remote-controlled model car. This is a beautiful, shiny blue car. It has a long cable that connects its motor to a steering wheel with which I can control the car's movement, its destiny. For the next two days, I drive that little car everywhere in our small apartment—through the living room, into the dining area, under the legs of the dining room table where my parents, my older brother, and I sit down for dinner each evening, into the bedroom and out again—steering with great pleasure and growing confidence.

But my pleasure is short-lived. Two days later, in the early evening, we are startled by heavy banging on our apartment door. I remember that banging even today. My father has not yet returned from working at the store. My mother opens the door. Two men enter. They identify themselves as Nazi policemen and order us to pack something and leave our apartment. They give us an address and tell us that we are to be lodged there until further notice. My mother and I pack only a change of clothes and toiletries, but my brother, Ludwig, has the good sense to bring with him his two most valued possessions—his stamp and coin collections.

Carrying these few things, we walk several blocks to the home of an elderly, more affluent Jewish couple whom we have never seen before. Their large, well-furnished apartment seems very elegant to me, and I am impressed with the man of the house. He wears an elaborately ornamented nightgown when he goes to bed, unlike the pajamas my father wears, and he sleeps with a nightcap to protect his hair and a guard over his upper lip to maintain the shape of his moustache. Even though we have invaded their privacy, our appointed hosts are thoughtful and decent. With all their affluence, they also are frightened and uneasy about the events that brought us to them. My mother is embarrassed to be imposing on our hosts, conscious that they are probably as uncomfortable to have three strangers suddenly thrust upon them as we are to be there. I am bewildered and frightened during the days we live in this couple's carefully arranged apartment. But the greatest source of anxiety for the three of us is not being in a stranger's apartment; it is my father—he disappeared abruptly and we have no idea where he is.

After several days we are finally allowed to return home. But the apartment we now find is not the one we left. It has been ransacked and everything of value taken—my mother's fur coat, her jewelry, our silver tableware, the lace tablecloths, some of my father's suits, and all of my birthday gifts, including my beautiful, shiny, remote-controlled blue car. To our very great relief, however, on November 19, a few days after we have returned to our apartment, my father comes back to us. He tells us that he had been rounded up, together with hundreds of other Jewish men, and incarcerated in an army barracks. He won his release because he was able to prove that he had been a soldier in the Austro-Hungarian army, fighting on the side of Germany during World War I.

The memories of those days—steering my car around the apartment with increasing assurance, hearing the bangs on the door, being ordered by the Nazi policemen to go to a stranger's apartment, finding ourselves robbed of our belongings, the disappearance and reappearance of my father—are the most powerful memories of my early life. Later, I would come to understand that these events coincided with Kristallnacht, the calamitous night that shattered not just the windows of our synagogues and my parents' store in Vienna, but also the lives of countless Jews all over the German-speaking world.

In retrospect, my family was fortunate. Our suffering was trivial compared with that of millions of other Jews who had no choice but to remain in Europe under the Nazis. After one humiliating and frightening year, Ludwig, then age fourteen, and I were able to leave Vienna for the United States to live with our grandparents in New York. Our parents joined us six months later. Although my family and I lived under the Nazi regime for only a year, the bewilderment, poverty, humiliation, and fear I experienced that last year in Vienna made it a defining period of my life.

IT IS DIFFICULT TO TRACE THE COMPLEX INTERESTS AND actions of one's adult life to specific experiences in childhood and youth. Yet I cannot help but link my later interest in mind—in how people behave, the unpredictability of motivation, and the persistence of memory—to my last year in Vienna. One theme of post-Holocaust Jewry has been "Never forget," an exhortation to future generations to be vigilant against anti-Semitism, racism, and hatred, the mind-sets that allowed the Nazi atrocities to occur. My scientific work investigates the biological basis of that motto: the processes in the brain that enable us to remember.

My remembrances of that year in Vienna first found expression even before I became interested in science, when I was a college student in the United States. I had an insatiable interest in contemporary Austrian and German history and planned to become an intellectual historian. I struggled to understand the political and cultural context in which those calamitous events had occurred, how a people who loved art and music at one moment could in the very next moment commit the most barbaric and cruel acts. I wrote several term papers on Austrian and German history, including an honors thesis on the response of German writers to the rise of Nazism.

Then, in my last year in college, 1951–52, I developed a fascination with psychoanalysis, a discipline focused on peeling back the layers of personal memory and experience to understand the often irrational roots of human motivation, thoughts, and behavior. In the early 1950s most practicing psychoanalysts were also physicians. I therefore decided to go to medical school. There, I was exposed to the revolution occurring in biology, to the likelihood that fundamental mysteries of the nature of living things were about to be revealed.

Less than a year after I entered medical school in 1952, the structure of DNA was being elucidated. As a result, the genetic and molecular workings of the cell were beginning to open up under scientific scrutiny. With time, that investigation would extend to the cells that make up the human brain, the most complex organ in the universe. It was then that I began to think about exploring the mystery of learning and memory in biological terms. How did the Viennese past leave its lasting traces in the nerve cells of my brain? How was the complex three-dimensional space of the apartment where I steered my toy car woven into my brain's internal representation of the spatial world around me? How did terror sear the banging on the door of our apartment into the molecular and cellular fabric of my brain with such permanence that I can relive the experience in vivid visual and emotional detail more than a half century later? These questions, unanswerable a generation ago, are yielding to the new biology of mind.

The revolution that captured my imagination as a medical student transformed biology from a largely descriptive field into a coherent science firmly grounded in genetics and biochemistry. Prior to the advent of molecular biology, three disparate ideas held sway: Darwinian evolution, the idea that human beings and other animals evolved gradually from simpler animal ancestors quite unlike themselves; the genetic basis of the inheritance of bodily form and mental traits; and the theory that the cell is the basic unit of all living things. Molecular biology united those three ideas by focusing on the actions of genes and proteins in individual cells. It recognized the gene as the unit of heredity, the driving force for evolutionary change, and it recognized the products of the gene, the proteins, as the elements of cellular function. By examining the fundamental elements of life processes, molecular biology revealed what all life-forms have in common. Even more than quantum mechanics or cosmology, the other fields of science that saw great revolutions in the twentieth century, molecular biology commands our attention because it directly affects our everyday lives. It goes to the core of our identity, of who we are.

The new biology of mind has emerged gradually over the five decades of my career. The first steps were taken in the 1960s, when the philosophy of mind, behaviorist psychology (the study of simple behavior in experimental animals), and cognitive psychology (the study of complex mental phenomena in people) merged, giving rise to modern cognitive psychology. This new disipline attempted to find common elements in the complex mental processes of animals ranging from mice to monkeys to people. The approach was later extended to simpler invertebrate animals, such as snails, honeybees, and flies. Modern cognitive psychology was at once experimentally rigorous and broadly based. It focused on a range of behavior, from simple reflexes in invertebrate animals to the highest mental processes in people, such as the nature of attention, of consciousness, and of free will, traditionally the concern of psychoanalysis.

In the 1970s cognitive psychology, the science of mind, merged with neuroscience, the science of the brain. The result was cognitive neuroscience, a discipline that introduced biological methods of exploring mental processes into modern cognitive psychology. In the 1980s cognitive neuroscience received an enormous boost from brain imaging, a technology that enabled brain scientists to realize their dream of peering inside the human brain and watching the activity in various regions as people engage in higher mental functions—perceiving a visual image, thinking about a spatial route, or initiating a voluntary action. Brain imaging works by measuring indices of neural activity: positron-emission tomography (PET) measures the brain's consumption of energy, and functional magnetic resonance imaging (fMRI) measures its use of oxygen. In the early 1980s cognitive neuroscience incorporated molecular biology, resulting in a new science of mind—a molecular biology of cognition—that has allowed us to explore on the molecular level such mental processes as how we think, feel, learn, and remember.

EVERY REVOLUTION HAS ITS ORIGINS IN THE PAST, AND THE revolution that culminated in the new science of mind is no exception. Although the central role of biology in the study of mental processes was new, the ability of biology to influence the way we see ourselves was not. In the mid-nineteenth century, Charles Darwin argued that we are not uniquely created, but rather evolved gradually from lower animal ancestors; moreover, he held, all life can be traced back to a common ancestor—all the way back to the creation of life itself. He proposed the even more daring idea that evolution's driving force is not a conscious, intelligent, or divine purpose, but a "blind" process of natural selection, a completely mechanistic sorting process of random trial and error based on hereditary variations.

Darwin's ideas directly challenged the teaching of most religions. Since biology's original purpose had been to explain the divine design of nature, his ideas rent the historic bond between religion and biology. Eventually, modern biology would ask us to believe that living beings, in all their beauty and infinite variety, are merely the products of ever new combinations of nucleotide bases, the building blocks of DNA's genetic code. These combinations have been selected for over millions of years by organisms' struggle for survival and reproductive success.

The new biology of mind is potentially more disturbing because it suggests that not only the body, but also mind and the specific molecules that underlie our highest mental processes—consciousness of self and of others, consciousness of the past and the future—have evolved from our animal ancestors. Furthermore, the new biology posits that consciousness is a biological process that will eventually be explained in terms of molecular signaling pathways used by interacting populations of nerve cells.

Most of us freely accept the fruits of experimental scientific research as they apply to other parts of the body: for instance, we are comfortable with the knowledge that the heart is not the seat of emotions, that it is a muscular organ that pumps blood through the circulatory system. Yet the idea that the human mind and spirituality originate in a physical organ, the brain, is new and startling for some people. They find it hard to believe that the brain is an information-processing computational organ made marvelously powerful not by its mystery, but by its complexity—by the enormous number, variety, and interactions of its nerve cells.

For biologists working on the brain, mind loses none of its power or beauty when experimental methods are applied to human behavior. Likewise, biologists do not fear that mind will be trivialized by a reductionist analysis, which delineates the component parts and activities of the brain. On the contrary, most scientists believe that biological analysis is likely to increase our respect for the power and complexity of mind.

Indeed, by unifying behaviorist and cognitive psychology, neural science and molecular biology, the new science of mind can address philosophical questions that serious thinkers have struggled with for millennia: How does mind acquire knowledge of the world? How much of mind is inherited? Do innate mental functions impose on us a fixed way of experiencing the world? What physical changes occur in the brain as we learn and remember? How is an experience lasting minutes converted to a lifelong memory? Such questions are no longer the province of speculative metaphysics; they are now fertile areas of experimental research.

THE INSIGHTS PROVIDED BY THE NEW SCIENCE OF MIND ARE most evident in our understanding of the molecular mechanisms the brain uses to store memories. Memory—the ability to acquire and store information as simple as the routine details of daily life and as complex as abstract knowledge of geography or algebra—is one of the most remarkable aspects of human behavior. Memory enables us to solve the problems we confront in everyday life by marshaling several facts at once, an ability that is vital to problem solving. In a larger sense, memory provides our lives with continuity. It gives us a coherent picture of the past that puts current experience in perspective. The picture may not be rational or accurate, but it persists. Without the binding force of memory, experience would be splintered into as many fragments as there are moments in life. Without the mental time travel provided by memory, we would have no awareness of our personal history, no way of remembering the joys that serve as the luminous milestones of our life. We are who we are because of what we learn and what we remember.

Our memory processes serve us best when we can easily recall the joyful events of our lives and dilute the emotional impact of traumatic events and disappointments. But sometimes, horrific memories persist and damage people's lives, as happens in post-traumatic stress disorder, a condition suffered by some people who have experienced at first hand the terrible events of the Holocaust, of war, rape, or natural disaster.

Memory is essential not only for the continuity of individual identity, but also for the transmission of culture and for the evolution and continuity of societies over centuries. Although the size and structure of the human brain have not changed since *Homo sapiens* first appeared in East Africa some 150,000 years ago, the learning capability of individual human beings and their historical memory have grown over the centuries through shared learning—that is, through the transmission of culture. Cultural evolution, a nonbiological mode of adaptation, acts in parallel with biological evolution as the means of transmitting knowledge of the past and adaptive behavior across generations. All human accomplishments, from antiquity to modern times, are products of a shared memory accumulated over centuries, whether through written records or through a carefully protected oral tradition.

Much as shared memory enriches our lives as individuals, loss of memory destroys our sense of self. It severs the connection with the past and with other people, and it can afflict the developing infant as well as the mature adult. Down's syndrome, Alzheimer's disease, and age-related memory loss are familiar examples of the many diseases that affect memory. We now know that defects in memory contribute to psychiatric disorders as well: schizophrenia, depression, and anxiety states carry with them the added burden of defective memory function.

The new science of mind holds out the hope that greater understanding of the biology of memory will lead to better treatments for both memory loss and persistent painful memories. Indeed, the new science is likely to have practical implications for many areas of health. Yet it goes beyond a search for solutions to devastating illnesses. The new science of mind attempts to penetrate the mystery of consciousness, including the ultimate mystery: how each person's brain creates the consciousness of a unique self and the sense of free will.

2

A CHILDHOOD IN VIENNA

At the time of my birth, Vienna was the most important cultural center in the German-speaking world, rivaled only by Berlin, capital of the Weimar Republic. Vienna was renowned for great music and art, and it was the birthplace of scientific medicine, psychoanalysis, and modern philosophy. In addition, the city's great tradition of scholarship provided a foundation for experiments in literature, science, music, architecture, philosophy, and art, experiments from which many modern ideas were derived. It was home to a diverse collection of thinkers, including Sigmund Freud, the founder of psychoanalysis; outstanding writers, such as Robert Musil and Elias Canetti; and the originators of modern philosophy, including Ludwig Wittgenstein and Karl Popper.

Vienna's culture was one of extraordinary power, and it had been created and nourished in good part by Jews. My life has been profoundly shaped by the collapse of Viennese culture in 1938—both by the events I experienced that year and by what I have learned since about the city and its history. This understanding has deepened my appreciation of Vienna's greatness and sharpened my sense of loss at its demise. That sense of loss is heightened by the fact that Vienna was my birthplace, my home.

2-1 My parents, Charlotte and Hermann Kandel, at the time of their wedding in 1923. (From Eric Kandel's personal collection.)

My parents met in Vienna and married in 1923 (figure 2-1), shortly after my father had established his toy store in the Eighteenth District on the Kutschkergasse (figure 2-2), a lively street that also contained a produce market, the Kutschker Market. My broth7er, Ludwig, was born in 1924 and I five years later (figure 2-3). We lived in a small apartment at Severingasse in the Ninth District, a middle-class neighborhood near the medical school and not far from Berggasse 19, the apartment of Sigmund Freud. As both my parents worked in the store, we had a series of full-time housekeepers at home.

I went to a school on a street appropriately named Schulgasse (School Street), located halfway between our apartment and my parents' store. Like most elementary schools, or *Volksschulen*, in Vienna, it had a traditional, academically rigorous curriculum. I followed in the footsteps of my exceptionally gifted brother, who had had the same teachers as I. Throughout my childhood in Vienna I felt that Ludwig had an intellectual virtuosity I would never match. By the time I began reading and writing, he was starting to master Greek, to become proficient at piano, and to construct radio sets.

Ludwig had just finished building his first short-wave radio receiver a few days before Hitler's triumphal march into Vienna in

2-2 My parents' toy and luggage store on the Kutschkergasse. My mother with me, or perhaps my brother. (From Eric Kandel's personal collection.)

March 1938. On the evening of March 13, Ludwig and I were listening with earphones as the broadcaster described the advance of German troops into Austria on the morning of March 12. Hitler had followed in the afternoon, crossing the border first at his native village, Braunau am Inn, and then moving on to Linz. Of the 120,000 citizens of Linz, almost 100,000 turned out to greet him, screaming "Heil Hitler" in unison. In the background, the "Horst Wessel song," a hypnotic Nazi marching song that even I found captivating, blared forth on the radio. On the afternoon of March 14, Hitler's entourage reached Vienna, where he was greeted by a wildly enthusiastic crowd of 200,000 people in the Heldenplatz, the great central square, and hailed as the hero who had unified the German-speaking people (figure 2-4). For my brother and me, this overwhelming support for the man who had destroyed the Jewish community of Germany was terrifying.

Hitler had expected the Austrians to oppose Germany's annexation of their country and to demand a relatively independent German protectorate instead. But the extraordinary reception he received, even from those who had opposed him forty-eight hours earlier, convinced

2-3 My brother and I in 1933. I was three years old and Ludwig was eight. (From Eric Kandel's personal collection.)

him that Austria would readily accept—would indeed welcome—annexation. It seemed as if everyone, from modest shopkeepers to the most elevated members of the academic community, now openly embraced Hitler. Theodor Cardinal Innitzer, the influential archbishop of Vienna, once a sympathetic defender of the Jewish community, ordered all the Catholic churches in the city to fly the Nazi flag and ring their bells in honor of Hitler's arrival. Greeting Hitler in person, the cardinal pledged his own loyalty and that of all Austrian Catholics, the majority of the population. He promised that Austria's Catholics would become "the truest sons of the great Reich into whose arms they had been brought back on this momentous day." The archbishop's only request was that the liberties of the Church be respected and its role in the education of the young guaranteed.

That night and for days to come, all hell broke loose. Viennese mobs, both adults and young people, inspired by Austrian Nazis and screaming "Down with Jews! Heil Hitler! Destroy the Jews!" erupted in a nationalistic frenzy, beating up Jews and destroying their property. They humiliated Jews by forcing them to get on their knees and scrub the streets to eliminate every vestige of anti-annexation political

2-4 Hitler enters Vienna in March of 1938. He is greeted with great enthusiasm by the crowds, including groups of girls waving Nazi flags emblazoned with swastikas (above). Hitler speaks to the Viennese public in the Heldenplatz (below). The largest turnout in the history of Vienna, 200,000 people, came to hear him. (Photos courtesy of Dokumentationsarchiv des Österreichischer Widerstands and Hoover Institute Archives.)

2-5 Jews forced to scrub the streets of Vienna to remove political graffiti advocating a free Austria. (Courtesy of Yad Vashem Photo Archives.)

graffiti (figure 2-5). In my father's case, he was forced to use a toothbrush to rid Vienna of the last semblance of Austrian independence—the word "yes" scrawled by Viennese patriots encouraging the citizenry to vote for Austria's freedom and to oppose annexation. Other Jews were forced to carry paint buckets and to demarcate stores owned by Jews with the Star of David or with the word *Jude* (Jew). Foreign commentators, long accustomed to Nazi tactics in Germany, were astonished by the brutality of the Austrians. In *Vienna and Its Jews*, George Berkley quotes a German storm trooper: "the Viennese have managed to do overnight what we Germans have failed to achieve . . . up to this day. In Austria, a boycott of the Jews does not need organizing—the people themselves have initiated it."

In his autobiography, German playwright Carl Zuckmayer, who

had moved to Austria in 1933 to escape Hitler, described Vienna during the days following the annexation as a city transformed "into a nightmare painting of Hieronymus Bosch." It was as if:

> Hades had opened its gates and vomited forth the basest, most despicable, most horrible demons. In the course of my life I had seen something of untrammeled human insights of horror or panic. I had taken part in a dozen battles in the First World War, had experienced barrages, gassings, going over the top. I had witnessed the turmoil of the postwar era, the crushing uprisings, street battles, meeting hall brawls. I was present among the bystanders during the Hitler Putsch in 1923 in Munich. I saw the early period of Nazi rule in Berlin. But none of this was comparable to those days in Vienna. What was unleashed upon Vienna had nothing to do with [the] seizure of power in Germany. . . . What was unleashed upon Vienna was a torrent of envy, jealousy, bitterness, blind, malignant craving for revenge. All better instincts were silenced . . . only the torpid masses had been unchained. . . . It was the witch's Sabbath of the mob. All that makes for human dignity was buried.

The day after Hitler marched into Vienna, I was shunned by all of my classmates except one—a girl, the only other Jew in the class. In the park where I played, I was taunted, humiliated, and roughed up. At the end of April 1938, all the Jewish children in my elementary school were expelled and transferred to a special school run by Jewish teachers on Pantzergasse in the Nineteenth District, quite far from where we lived. At the University of Vienna, almost all Jews—more than 40 percent of the student body and 50 percent of the faculty—were dismissed. This malevolence toward Jews, of which my treatment was but a mild example, culminated in the horrors of Kristallnacht.

MY FATHER AND MOTHER HAD EACH COME TO VIENNA BEFORE World War I, when they were very young and the city was a very different, more tolerant place. My mother, Charlotte Zimels, was born in

1897 in Kolomyya, a town of about 43,000 inhabitants on the Prut River in Galicia. This region of the Austro-Hungarian Empire near Romania was then part of Poland and is now part of Ukraine. Almost half the population of Kolomyya was Jewish, and the Jewish community had a lively culture. My mother came from a well-educated middle-class family. Although she spent only one year at the University of Vienna, she spoke and wrote English in addition to German and Polish. My father, Hermann Kandel—to whom my mother was immediately attracted because she found him handsome, energetic, and filled with humor—was born in 1898 into a poor family in Olesko, a town of about 25,000 near Lvov (Lemberg), also now part of Ukraine. He moved to Vienna with his family in 1903, when he was five. He was drafted directly from high school into the Austro-Hungarian army, fought in the First World War, and sustained a shrapnel wound in battle. After the war, he worked to support himself and never finished high school.

I was born eleven years after the Austro-Hungarian Empire collapsed following its defeat in World War I. Before the war, it was the second largest country in Europe, surpassed in area only by Russia. The empire extended in the northeast to what is now Ukraine, its eastern provinces included what are now the Czech and Slovak republics, and its southern provinces contained Hungary, Croatia, and Bosnia. After the war, Austria was drastically reduced in size, having lost all of its foreign-speaking provinces and retaining only the German-speaking core. Consequently, it was greatly reduced in population (from 54 million inhabitants to 7 million) and in political significance.

Still, the Vienna of my youth, a city of almost 2 million people, remained intellectually vibrant. My parents and their friends were pleased when the municipal government, under the leadership of the Social Democrats, initiated a highly successful and widely admired program of social, economic, and health care reforms. Vienna was a thriving cultural center. The music of Gustav Mahler and Arnold Schönberg, as well as that of Mozart, Beethoven, and Haydn, resonated throughout the city, as did the bold expressionist images of Gustav Klimt, Oskar Kokoschka, and Egon Schiele.

Even as it thrived culturally, however, Vienna in the 1930s was the capital city of an oppressive, authoritarian political system. As a child,

I was too young to understand this. It was only later, from the perspective of a more carefree adolescence in the United States, that I understood just how oppressive the conditions that formed my first impressions of the world actually were.

Although Jews had lived in Vienna for over a thousand years and had been instrumental in developing the city's culture, anti-Semitism was chronic. At the beginning of the twentieth century, Vienna was the only major city in Europe where anti-Semitism formed the basis of the political platform of the party in power. Karl Lueger, the anti-Semitic populist mayor of Vienna from 1897 to 1910, focused his spellbinding orations specifically on "the wealthy Jews" of the middle class. That middle class had emerged following the adoption of a new constitution in 1867, which extended equal civil rights to Jews and other minority groups and gave them the freedom to practice their religion openly.

Despite the provisions of the new constitution, the Jews, who made up about 10 percent of the city's overall population and almost 20 percent of its vital core (the nine inner districts), were discriminated against everywhere: in the civil service, in the army, in the diplomatic corps, and in many aspects of social life. Most social clubs and athletic organizations had an Aryan clause that prevented Jews from joining. From 1924 until 1934, when it was outlawed, there existed in Austria a Nazi party with a strongly anti-Semitic platform. The party protested, for example, the performance of an opera by Ernst Krenek, a Jewish composer, at the Vienna Opera House in 1928 (figure 2-6).

Nonetheless, the Jews of Vienna, my parents included, were entranced by the city. Berkley, the historian of Jewish life in Vienna, has commented aptly: "The fierce attachment of so many Jews to a city that throughout the years demonstrated its deep-rooted hate for them remains the greatest grim irony of all." In later years, I learned from my parents why the city exerted such a powerful hold. To begin with, Vienna is beautiful: the museums, the opera house, the university, the Ringstrasse (Vienna's main boulevard), the parks, and the Hapsburg Palace in the city center are all architecturally interesting. The renowned Vienna Woods outside the city are easily accessible, as is the Prater, the almost magical amusement park with its giant Ferris wheel

Wiener und Wienerinnen!

Die Zersetzung und Vergiftung unserer bodenständigen Bevölkerung durch das östliche Gesindel nimmt einen gefahrdrohenden Umfang an. Nicht genug, daß unser Volk durch die Geldentwertung einer durchgreifenden Ausplünderung zugeführt wurde, sollen nun auch alle sittlich-kulturellen Grundfesten unseres Volkstumes zerstört werden.

Unsere Staatsoper,

die erste Kunst- und Bildungsstätte der Welt, der Stolz aller Wiener,

ist einer frechen jüdisch-negerischen Besudelung zum Opfer gefallen.

Das Schandwerk eines tschechischen Halbjuden

„Jonny spielt auf!"

in welchem Volk und Heimat, Sitte, Moral und Kultur brutal zertreten werden soll, wurde der Staatsoper aufgezwungen. Eine volksfremde Meute von Geschäftsjuden und Freimaurern setzt alles daran, unsere Staatsoper zu einer Bedürfnisanstalt ihrer jüdisch-negerischen Perversitäten herabzuwürdigen. **Der Kunst-Bolschewismus erhebt frech** sein Haupt. Die Schamröte muß jedem anständigen Wiener ins Gesicht steigen, wenn er hört, welch ungeheuerliche Schmach und Demütigung der berühmten Musikstadt Wien durch volksfremdes Gesindel angetan wurde.

Da die christlich-großdeutsche Regierung diesem schamlosen Treiben untätig zusieht und von keiner Seite eine Abwehr versucht wird, so rufen wir alle Wiener zu einer

Riesen-Protest-Kundgebung

auf, in welcher über die Wahrheit der jüdischen Verseuchung unseres Kunstlebens und über die der Staatsoper angetane Schmach gesprochen werden wird.

Christliche Wiener und Wienerinnen, Künstler, Musiker, Sänger und Antisemiten erscheint in Massen und protestiert mit uns gegen diese unerhörten Schandzustände in Oesterreich.

Ort: Lembachers Saal, Wien, III., Landstraße Hauptstraße.
Zeitpunkt: Freitag, den 13. Jänner 1928, 8 Uhr abends.
Kostenbeitrag: 20 Groschen. / Juden haben keinen Zutritt!

Nationalsozialistische deutsche Arbeiterpartei Großdeutschlands

2-6 An Austrian Nazi party poster from 1928, a decade before Hitler entered Vienna, protests the performance at the Vienna Opera House of an opera by the Jewish composer Ernst Krenek: "Our opera house, the foremost arts and educational institution in the world, the pride of all Viennese, has fallen victim to an insolent Jewish-Negro defilement . . . protest with us against this unheard-of shame in Austria." (Courtesy of Wiener Stadt-und Landesbibliothek.)

later made famous in the movie *The Third Man*. "After an evening at the theater or a May Day in the Prater, a Viennese might with equanimity regard his city as the pivot of the universe. Where else did appearance so beguilingly sweeten reality?" wrote the historian William Johnston. Although my parents were not deeply cultivated people, they felt themselves to be connected to the intellectual values of Vienna, especially to the theatre, the opera, and the city's melodic dialect, a dialect I still speak.

My parents shared the values of most other Viennese parents: they wanted their children to achieve something professionally—ideally, something intellectual. Their aspirations reflected typical Jewish values. Ever since the destruction of the Second Temple in Jerusalem in 70 A.D., when Yohanan ben Zakkai left for the coastal town of Yabneh and established there the first academy for the study of the Torah, Jews have been a people of the book. Every man, irrespective of financial position or social class, was expected to be literate in order to read the prayer book and the Torah. By the end of the nineteenth century, upwardly mobile Jewish parents were encouraging their daughters as well as their sons to become well educated. Beyond that, the goal of life was not simply to achieve economic security, but rather to use economic security to rise to a higher cultural plane. What was most important was *Bildung*—the pursuit of education and culture. It meant a great deal, even to a poor Jewish family in Vienna, that at least one son succeed in becoming a musician, a lawyer, a doctor, or, better still, a university professor.

Vienna was one of the few cities in Europe where the cultural aspirations of the Jewish community coincided fully with the aspirations of most non-Jewish citizens. After the repeated defeat of Austria's armies by Prussia, first in the War of the Austrian Succession from 1740 to 1748, and then in the Austro-Prussian War in 1866, the Hapsburgs—Austria's ruling family—lost all hope of military predominance among the German-speaking states. As their political and military power waned, they replaced their desire for territorial preeminence with a desire for cultural preeminence. The lifting of restrictions under the new constitution led to a major emigration of Jews and other minority groups from all over the empire to Vienna in the last quarter of the nineteenth century. Vienna became home to people from Germany, Slovenia, Croatia, Bosnia, Hungary, northern Italy, the Balkans, and Turkey. Between 1860 and 1880, its population increased from 500,000 to 700,000. The middle-class citizens of Vienna began to see themselves as citizens of the world, and they exposed their children to culture early in life. Being reared "in museums, theaters, and concert halls of the new Ringstrasse, the middle-class Viennese acquired culture not as an ornament of life, or a badge of status, but as the air they breathed," wrote Carl Schorske, the cultural historian of Vienna. Karl Kraus, the great satirical social and literary critic, said of Vienna that "its streets are not paved with asphalt but with culture."

In addition to being culturally vibrant, Vienna was also alive sensually. My fondest early memories are typically Viennese: one, a modest but sustained bourgeois contentment that came from being raised within a close-knit and supportive family that shared holidays in a regular, prescribed manner, and the other, a moment of erotic happiness that came naturally from our seductive housekeeper, Mitzi.

That erotic experience was right out of one of Arthur Schnitzler's short stories, wherein a young, middle-class Viennese adolescent is introduced to sexuality by *ein süsses Mädchen*, a sweet young maiden, either a servant in the house or a working girl outside the house. Andrea Lee, writing in *The New Yorker*, has said that one of the criteria bourgeois families in Austria-Hungary used in selecting girls for housework was that they be suitable to relieve the family's adolescent boys of their virginity, in part to entice them away from any possible attraction to homosexuality. I find it interesting to look back and realize that an encounter that easily could have become, or could have been perceived by others as being exploitative, never had that connotation for me.

My encounter with Mitzi, an attractive, sensual young woman of about twenty-five, began one afternoon as I was recovering from a cold at age eight. She sat down at the edge of my bed and touched my face. When I responded with pleasure, she opened her blouse, exposing her ample bosom, and asked me whether I would like to touch her. I barely grasped what she was talking about, but her attempt at seduction had its effect on me, and I suddenly felt different than I ever had before.

As I began with some guidance to explore her body, she suddenly became uncomfortable and said we had better stop or I'd become pregnant. How could I become pregnant? I knew full well that only women have babies. Where can a baby come from in boys?

"From the belly button," she answered. "The doctor puts some powder on it, and the belly button opens up to allow the baby to come out."

Part of me knew this was impossible. But part of me was not certain—and even if it seemed improbable, I became slightly anxious at the potential consequences of that event. My worry was, What would my mother think if ever I were to become pregnant? That worry and Mitzi's change of mood ended my first sexual encounter. But Mitzi continued thereafter to speak freely to me about her sexual yearnings and said that she might have realized them with me were I older.

Mitzi did not, as it turned out, remain celibate until I reached her age qualifications. Several weeks after our brief rendezvous in my bed, she took up with a gas repairman who came by to fix our stove. A month or two later, she ran off with him to Czechoslovakia. For many years thereafter, I thought that running off to Czechoslovakia was the equivalent of devoting one's life to the happy pursuit of sensuality.

Our bourgeois familial happiness was typified by the weekly card game at my parents' house, family gatherings on the occasion of Jewish holidays, and our summer vacations. On Sunday afternoons my Aunt Minna, my mother's younger sister, and her husband, Uncle Srul, would come for tea. My father and Srul would spend most of the time playing pinochle, a card game at which my father excelled and which he played with great animation and humor.

Passover was a festive occasion that brought our family together at the home of my grandparents, Hersch and Dora Zimels; we read the Haggadah, an account of the escape of the Jews from slavery in Egypt, and then enjoyed one of my grandmother's carefully prepared seder meals, the high point of which was her gefilte fish, which to my mind still has no equal. I particularly remember the Passover of 1936. A few months earlier, Aunt Minna had married Uncle Srul, and I was an attendant at her wedding—I helped manage the train of her beautiful gown. Srul was quite wealthy. He had developed a successful leather business, and his wedding to Minna was elaborate in a way I had not previously experienced. I was therefore very pleased with my role in it.

On the first night of Passover, I recalled fondly for Minna how much I had enjoyed their wedding with everyone dressed so nicely and food served in an elegant way. The wedding was so beautiful, I said, that I hoped she would have another soon so I could experience a special moment like that again. Minna, as I learned later, felt somewhat ambivalent about Srul. She considered him her intellectual and social inferior and therefore immediately assumed that I was referring not to the event but to her choice of partner. She inferred that I would like to see her remarried to someone else—someone perhaps more appropriately matched to her intellect and breeding. Minna became enraged and lectured me at length on the sanctity of marriage. How dare I suggest that she would want another wedding so soon, to marry someone else? As I was to learn later, in reading Freud's *Psychopathology of Everyday Life*, a fundamental principle of dynamic psychology is that the unconscious never lies.

Every August my parents, Ludwig, and I spent our summer holidays in Mönichkirchen, a small farming village fifty miles south of Vienna. Just as we were about to depart for Mönichkirchen in July 1934, the Austrian chancellor, Engelbert Dollfuss, was assassinated by a band of Austrian Nazis disguised as policemen—the first storm to register on my emerging political consciousness.

Modeling himself on Mussolini, Dollfuss, who had been elected chancellor in 1932, had absorbed the Christian Socialists into the Fatherland Front and established an authoritarian regime, choosing as an emblem a traditional form of a cross rather than the swastika, to express Christian rather than Nazi values. To ensure his control of the government, he had abolished Austria's constitution and outlawed all opposition parties, including the Nazis. Although Dollfuss opposed the efforts of the Austrian National Socialist movement to form a state consisting of all German-speaking people—a pan-German state—his abolition of the old constitution and competing political parties helped open the door for Hitler. Following Dollfuss's assassination and during the early years of the chancellorship of his successor, Kurt von Schuschnigg, the Austrian Nazi party was driven further underground. It nonetheless continued to gain new adherents, especially among teachers and other civil servants.

HITLER WAS AUSTRIAN AND HAD LIVED IN VIENNA. HE HAD LEFT his childhood home in Braunau am Inn for the capital in 1908, at age nineteen, hoping to become an artist. Despite a reasonable talent for painting, he failed repeatedly to gain entrance to the Art Academy of Vienna. While in Vienna, he came under the influence of Karl Lueger. It was from Lueger that Hitler first learned the power of demagogic oratory and the political benefits of anti-Semitism.

Hitler had dreamed of a union of Austria and Germany since his youth. Consequently, from its very beginning in the 1920s, the agenda of the Nazi party, which was framed in part by Austrian Nazis, included the merger of all German-speaking people into a Greater Germany. In the fall of 1936 Hitler began to act on this agenda. In full control of Germany since 1933, he had reinstated conscription in 1935, and the next year he had ordered his troops to reoccupy the Rhineland, a German-speaking region that had been demilitarized and placed under French supervision by the Treaty of Versailles. He then intensified his rhetoric, threatening to move against Austria. Schuschnigg was eager to appease Hitler while ensuring Austria's independence, and he responded to the threats by requesting a meeting with Hitler. On February 12, 1938, they met in Berchtesgaden, the private retreat Hitler had selected, for sentimental reasons, to be close to the Austrian border.

In a show of power, Hitler arrived at the meeting with two of his generals and threatened to invade Austria unless Schuschnigg lifted the restrictions on the Austrian Nazi party and appointed three Austrian Nazis to key ministerial posts in the cabinet. Schuschnigg refused. As the day wore on, however, Hitler stepped up the pressure. Finally, the exhausted chancellor gave in, agreeing to legalize the Nazi party, free Nazis held as political prisoners, and grant the Nazi party two cabinet positions. But the agreement between Schuschnigg and Hitler only whetted the Austrian Nazis' appetite for power. Now a sizable group, they emerged into public view and challenged Schuschnigg's government in a series of insurgencies that the police had difficulty controlling. Faced with Hitler's threatened aggression from without and the rebellion of the Austrian Nazis from within, Schuschnigg took the offensive and boldly called for a plebiscite to be held on March 13, a mere month after his meeting with Hitler. The question for the voters was simple: Should Austria remain free and independent, yes or no?

This courageous move by Schuschnigg, much admired by my parents, unsettled Hitler, as it seemed almost certain that the vote would favor an independent Austria. Hitler responded by mobilizing troops and threatening to invade the country unless Schuschnigg postponed the plebiscite, resigned as chancellor, and formed a new government with an Austrian Nazi, Arthur Seyss-Inquart, as chancellor. Schuschnigg turned for help to Britian and Italy, two countries that had formerly supported Austrian independence. To the dismay of Viennese liberals like my family, neither responded. Abandoned by potential allies and concerned about needless bloodshed, Schuschnigg resigned on the evening of March 11.

Even though the president of Austria acquiesced to all of Germany's demands, Hitler invaded the country the next day.

Now came a surprise. Rather than being met by angry crowds of Austrians, Hitler was welcomed enthusiastically by a substantial majority of the population. As George Berkley has pointed out, this dramatic turnabout from people who screamed loyalty to Austria and supported Schuschnigg one day to people who greeted Hitler's troops as "German brothers" the next cannot be explained simply by the emergence from the underground of tens of thousands of Nazis. Rather, what happened was one of history's "fastest and fullest mass conversions." Hans Ruzicka was to write, "These are the people who cheered the Emperor and then cursed him, who welcomed democracy after the Emperor was dethroned and then cheered [Dollfuss's] fascism when the system came to power. Today he is a Nazi, tomorrow he will be something else."

The Austrian press was no exception. On Friday, March 11, the *Reichspost*, one of the country's major newspapers, endorsed Schuschnigg. Two days later, the same newspaper printed a front-page editorial entitled "Toward Fulfillment," which stated: "Thanks to the genius and determination of Adolf Hitler, the hour of all-German unity has arrived."

The attacks on Jews that had begun in mid-March 1938 reached a peak of viciousness eight months later in Kristallnacht. When I later read about Kristallnacht, I learned that it had originated in part from the events of October 28, 1938. On that day seventeen thousand German Jews who were originally from Eastern Europe were rounded up by the Nazis and dumped near the town of Zbszyn, which lies on the border between Germany and Poland. At that time, the Nazis still considered emigration—voluntary or forced—to be the solution to "the Jewish question." On the morning of November 7, a seventeen-year-old Jewish boy, Herschel Grynszpan, distraught over the deportation of his parents from their home in Germany to Zbszyn, shot and killed Ernst vom Rath, a third secretary in the German embassy in Paris, mistaking him for the German ambassador. Two days later, using this one act as a pretext for acting against the Jews, organized mobs set almost every synagogue in Germany and Austria on fire.

Of all the cities under Nazi control, Vienna was the most debased on Kristallnacht. Jews were taunted and brutally beaten, expelled from their businesses, and temporarily evicted from their homes. Their businesses and homes were then looted by avaricious neighbors. Our beautiful synagogue on Schopenhauerstrasse was completely destroyed. Simon Wiesenthal, the leading Nazi hunter after World War II, was later to say that "compared to Vienna, the Kristallnacht in Berlin was a pleasant Christmas festival."

On the day of Kristallnacht, as my father was rounded up, his store was taken away from him and turned over to a non-Jew. This was part of the so-called Aryanization (*Arisierung*) of property, a purportedly legal form of theft. From the time of my father's release from prison in the middle of November 1938 until he and my mother left Vienna in August 1939, they were destitute. As I was to learn much later, my parents received provisions and an occasional opportunity for my father to work at jobs such as moving furniture, from the Israelitische Kultusgemeinde der Stadt Wien, the Jewish Community Council of Vienna.

Aware of the anti-Jewish laws instituted in Germany following Hitler's rise to power, my parents understood that the violence in Vienna was not likely to fade away. They knew that we had to leave—and to leave as soon as possible. My mother's brother, Berman Zimels, had left Austria for New York a decade earlier and established himself as an accountant. My mother wrote him on March 15, 1938, just three days after Hitler's invasion, and he quickly sent us affidavits assuring the U.S. authorities that he would support us upon our arrival in the United States. However, Congress had passed an immigration act in 1924 that set a quota on the number of people who could enter the United States from the countries of Eastern and Southern Europe. Since my parents were born in territory that was at that time Poland, it took about a year for our quota number to come up, despite our having the necessary affidavits. When the number was finally called, we had to emigrate in stages, also because of the immigration laws, which specified the sequence with which family members could enter the United States. According to this sequence, my mother's parents could leave first, which they did in February 1939; my brother and I next, in April; and finally my parents, in late August, just days before World War II broke out.

Because my parents' only source of income had been taken from them, they had no money to pay for our voyage to the United States. They therefore applied to the Kultusgemeinde for one and a half tickets on the Holland America Line, one ticket for my brother and a half for me. A few months later, they applied for two tickets for their own voyage. Fortunately, both requests were granted. My father was a scrupulous, honest person who always paid his bills on time. I have in my possession today all the documents supporting his request, which show that he religiously paid his membership dues to the Kultusgemeinde. This view of him as an upstanding man of integrity and character is specifically mentioned by an officer of the Kultusgemeinde in his evaluation of my father's request for assistance.

MY LAST YEAR IN VIENNA WAS A DEFINING ONE. CERTAINLY, IT fostered a profound, lasting gratitude for the life I found in the United States. But without a doubt, the spectacle of Vienna under the Nazis also presented me for the first time with the darker, sadistic side of human behavior. How is one to understand the sudden, vicious brutality of so many people? How could a highly educated society so quickly embrace punitive policies and actions rooted in contempt for an entire people?

Such questions are difficult to answer. Many scholars have struggled to come up with partial and inconsistent explanations. One conclusion, which is troubling to my sensibilities, is that the quality of a society's culture is not a reliable indicator of its respect for human life. Culture is simply incapable of enlightening people's biases and modifying their thinking. The desire to destroy people outside the group to which one belongs may be an innate response and may thus be capable of being aroused in almost any cohesive group.

I doubt very much that any such quasi-genetic predisposition would operate in a vacuum. The Germans as a whole did not share the vicious anti-Semitism of the Austrians. How, then, did Vienna's cultural values become so radically dissociated from its moral values? Certainly one important reason for the actions of the Viennese in 1938 was sheer opportunism. The successes of the Jewish community—economic, political, cultural, and academic—generated envy and a desire for revenge among non-Jews, especially those in the university. Nazi party membership among university professors greatly exceeded that in the population at large. As a result, the non-Jewish Viennese were eager to advance themselves by replacing Jews in the professions: Jewish university professors, lawyers, and doctors quickly found themselves without jobs. Many Viennese simply took possession of Jewish homes and belongings. Thus, as Tina Walzer and Stephen Templ's systematic study of the period has revealed, a "large number of lawyers, judges, and physicians improved their living standards in 1938 by plundering their Jewish neighbors. The success of many Austrians today is based on the money and property stolen sixty years ago."

Another reason for the dissociation of cultural and moral values was the move from a cultural to a racial form of anti-Semitism. Cultural anti-Semitism is based on the idea of "Jewishness" as a religious or cultural tradition that is acquired through learning, through distinctive traditions and education. This form of anti-Semitism attributes to Jews certain unattractive psychological and social characteristics that are acquired through acculturation, such as a profound interest in making money. However, it also holds that as long as Jewish identity is acquired through upbringing in a Jewish home, these characteristics can be undone by education or religious conversion, in which case the Jew overcomes the Jew in himself or herself. A Jew who converts to Catholicism can, in principle, be as good as any other Catholic.

Racial anti-Semitism, on the other hand, is thought to have its origins in the belief that Jews as a race are genetically different from other races. This idea derives from the Doctrine of Deicide, which was long taught by the Roman Catholic Church. As Frederick Schweitzer, a Catholic historian of Jews, has argued, this doctrine gave rise to the popular belief that the Jews killed Christ, a view not renounced by the Catholic Church until recently. According to Schweitzer, this doctrine argued that the Jewish perpetrators of deicide were a race so innately lacking in humanity that they must be genetically different, subhuman. One therefore could remove them from the other human races without compunction. Racial anti-Semitism was evidenced in the Spanish Inquisition of the 1400s and was adopted in the 1870s by some of Austria's (and Germany's) intellectuals, including Georg von Schönerer, leader of the Pan-German nationalists in Austria, and by Karl Lueger, the mayor of Vienna. Although racial anti-Semitism had not been a dominant force in Vienna before 1938, it became official public policy after March of that year.

Once racial anti-Semitism replaced cultural anti-Semitism, no Jew could ever become a "true" Austrian. Conversion—that is to say, religious conversion—was no longer possible. The only solution to the Jewish question was expulsion or elimination of the Jews.

MY BROTHER AND I LEFT FOR BRUSSELS BY TRAIN IN APRIL 1939. Leaving my parents behind when I was only nine years old was deeply distressing, despite our father's persistent optimism and our mother's calm reassurances. As we reached the border between Germany and Belgium, the train stopped for a brief time and German customs officers came on board. They demanded to see any jewelry or other valuables we might have. Ludwig and I had been forewarned of this request by a young woman who was traveling with us. I had therefore hidden in my pocket a small gold ring with my initials on it, which I had been given as a present on my seventh birthday. My normal anxiety in the presence of Nazi officers reached almost unbearable heights as they boarded the train, and I feared that they would discover the ring. Fortunately, they paid little attention to me and allowed me to quake undisturbed.

In Brussels, we stayed with Aunt Minna and Uncle Srul. With their substantial financial resources, they had succeeded in purchasing a visa that allowed them to enter Belgium and settle in Brussels. They were to join us in New York a few months later. From Brussels, Ludwig and I went by train to Antwerp, where we boarded the S.S. *Geroldstein* of the Holland-American Line for the ten-day journey that took us to Hoboken, New Jersey—directly past the welcoming Statue of Liberty.

3

AN AMERICAN EDUCATION

Arriving in the United States was like starting life anew. Although I lacked both the prescience and the language to say "Free at last," I felt it then and have felt it ever since. Gerald Holton, a historian of science at Harvard University, has pointed out that for many Viennese émigrés of my generation, the solid education we obtained in Vienna, combined with the sense of liberation we experienced on arriving in America, released boundless energy and inspired us to think in new ways. That certainly proved true for me. One of the many gifts I was to receive in this country was a superb liberal arts education in three highly distinctive institutions: the Yeshivah of Flatbush, Erasmus Hall High School, and Harvard College.

My brother and I moved in with my mother's parents, Hersch and Dora Zimels, who had arrived in Brooklyn in February 1939, two months ahead of us. I spoke no English and felt I had to fit in. I therefore dropped the last letter from my name, Erich, and assumed the current spelling. Ludwig underwent an even more dramatic metamorphosis, to Lewis. My Aunt Paula and Uncle Berman, who had lived in Brooklyn since coming to the United States in the 1920s, enrolled me in a public elementary school, P.S. 217, located in the Flatbush section not far from where we lived. I attended that school for only twelve weeks, but by the time I left for the summer break, I spoke English well enough to make myself understood. That summer I reread Erich Kästner's *Emil and the Detectives*, one of my childhood favorites, this time in English, an accomplishment that gave me a sense of pride.

I was not very comfortable at P.S. 217. Although many Jewish children attended the school, I was not aware of it. On the contrary, because so many students were blond and blue-eyed, I was convinced they were non-Jews and I was afraid they would in the long term be hostile toward me. I was therefore receptive to the urgings of my grandfather that I attend a Hebrew parochial school. My grandfather was a religious and very scholarly man, although somewhat unworldly. My brother has said that our grandfather was the only man he knew who could speak seven languages but could not make himself understood in any of them. My grandfather and I liked each other a great deal, and he readily convinced me that he could tutor me in Hebrew during the summer so that I might be eligible for a scholarship at the Yeshivah of Flatbush in the autumn. This well-known Hebrew day school offered secular classes in English and religious studies in Hebrew, both on a highly demanding level.

Thanks to my grandfather's tutelage, I entered the yeshivah in the fall of 1939. By the time I graduated in 1944, I spoke Hebrew almost as well as English. I had read in Hebrew the five books of Moses, the books of Kings, the Prophets, and some of the Talmud. It gave me both pleasure and pride to learn later that Baruch S. Blumberg, who won the Nobel Prize in Physiology or Medicine in 1976, had also benefited from the extraordinary educational experience provided by the Yeshivah of Flatbush.

MY PARENTS LEFT VIENNA IN LATE AUGUST 1939. BEFORE THEY left, my father was arrested for a second time and taken to the Vienna Soccer Stadium, where he was interrogated and intimidated by the brown-shirted troops of the Sturm Abteilung, the SA. The fact that he had obtained a visa for the United States and was about to depart led to his release and probably saved his life.

When my parents arrived in New York, my father, who spoke not a word of English, found a job in a toothbrush factory. While the toothbrush had been the emblem of his humiliation in Vienna, in New York it started him on the path to a better life. Even though he was not fond of the work, he threw himself into it with his usual energy and was soon reprimanded by the union steward for producing too many toothbrushes too quickly and thus making the other workers appear slow. My father was undeterred. He loved America. Like many other immigrants, he often referred to it as the *goldene Medina*, the land of gold that promised Jews safety and democracy. In Vienna he had read the novels of Karl May, which mythologized the conquest of the American West and the bravery of American Indians, and my father was in his own way possessed of the frontier spirit.

In time, my parents saved enough money to rent and outfit a modest clothing store. My father and mother worked together and sold simple women's dresses and aprons, as well as men's shirts, ties, underwear, and pajamas. We rented the apartment above the store at 411 Church Avenue in Brooklyn. My parents earned enough not only to support us, but after a while to buy the building in which the store and apartment were located. In addition, they were able to help send me to college and medical school.

My parents were so preoccupied with the store—the key to financial stability for them and their children—that they did not share in the cultural life of New York, which Lewis and I were beginning to enjoy. Despite their constant labors, however, they were always optimistic and supportive of us, and they never tried to dictate decisions about our work or play. My father was an obsessively honest person who felt compelled to pay immediately the bills for the merchandise he received from his suppliers, and he often counted the change he gave his customers one more time. He expected Lewis and me to behave similarly with regard to financial matters. But other than a general expectation of reasonable and correct behavior, I never felt any pressure from him to follow one academic track or another. In turn, I never thought him in a position to advise me on those issues, given his limited social and educational experiences. For advice, I typically turned to my mother or, more often, to my brother, my teachers, and most frequently, my friends.

My father worked in his store until the week before he died, at seventy-nine, in 1977. Soon thereafter, my mother sold both the store and the building in which it was located and moved into a more comfortable and somewhat more elegant apartment around the corner, on Ocean Parkway. She died in 1991 at the age of ninety-four.

WHEN I GRADUATED FROM THE YESHIVAH OF FLATBUSH IN 1944, there was no affiliated high school, as there is today, so I went to Erasmus Hall High School, a local public school that was very strong academically. There, I became interested in history, in writing, and in girls. I worked on the school newspaper, *The Dutchman*, and became sports editor. I also played soccer and was one of the captains of the track team (figure 3-1). My co-captain, Ronald Berman, one of my closest friends in high school, was an extraordinary runner who went on to win the half-mile race in the city championship; I placed fifth. Ron later became a Shakespearean scholar and professor of English literature at the University of California, San Diego. He served as the first head of the National Endowment for the Humanities, in the Nixon administration.

At the urging of my history teacher, John Campagna, a Harvard alumnus, I applied to Harvard College. When I first discussed apply-

John Rucker Eric Kandel John Bartel Ronald Berman Peter Mannus

3-1 The winning team at the Pennsylvania Relays, 1948. The Pennsylvania Relays is an annual national event for high school and college track athletes. We won one of the one-mile events for high schools. (Courtesy of Ron Berman.)

ing to Harvard with my parents, my father (who, like me, was not familiar with the distinctions among various American universities) discouraged me because of the cost of submitting another college application. I had already applied for admission to Brooklyn College, an excellent school that my brother had attended. Upon hearing of my father's concerns, Mr. Campagna volunteered to cover from his own pocket the fifteen dollars required for my application. I was one of two students (Ron Berman was the other) in our class of about 1,150 to be admitted to Harvard, both of us with scholarships. After receiving the scholarships, Ron and I appreciated the true meaning of Harvard's alma mater, "Fair Harvard." Fair Harvard, indeed!

Even though I was thrilled by my good fortune and immensely grateful to Mr. Campagna, I was apprehensive about leaving Erasmus Hall, convinced that I would never again feel the sheer joy of social acceptance and academic and athletic achievement that I had experienced there. At the yeshivah, I had been a scholarship student. At Erasmus, I was a scholar-athlete. The difference, for me, was enormous. It was at Erasmus that I first sensed myself emerging from the shadow of my brother, a shadow that I had found so imposing while in school in Vienna. For the first time, I had interests of my own.

At Harvard, I majored in modern European history and literature. This was a selective major that required its students to commit to writing an honors thesis in their senior year. Those accepted had the opportunity, unique to this major, of having tutorials from the beginning of their sophomore year onward, first in small groups and then individually. My honors thesis was on the attitude toward National Socialism of three German writers: Carl Zuckmayer, Hans Carossa, and Ernst Jünger. Each writer represented a different position along a spectrum of intellectual responses. Zuckmayer, a courageous liberal and lifelong critic of National Socialism, left Germany early and went first to Austria and then to the United States. Carossa, a physician-poet, took a neutral position and remained physically in Germany, although his spirit, he claimed, escaped elsewhere. Jünger, a dashing German military officer in the First World War, extolled the spiritual virtues of war and of the warrior and was an intellectual precursor of the Nazis.

I came to the depressing conclusion that many German artists and intellectuals—including such apparently fine minds as Jünger, the great philosopher Martin Heidegger, and the conductor Herbert von Karajan—had succumbed all too eagerly to the nationalistic fervor and racist propaganda of National Socialism. Subsequent historical studies by Fritz Stern and others have found that Hitler did not have widespread popular support in his first year in office. Had intellectuals mobilized effectively and been able to bring along segments of the general population, Hitler's aspirations for complete control of the government might well have been prevented, or at least severely curtailed.

I began working on my honors thesis during my junior year, at a time when I was thinking of doing graduate work in European intellectual history. However, toward the end of my junior year I met and fell in love with Anna Kris, a student at Radcliffe College who had also emigrated from Vienna. At the time, I was taking two superb seminars with Karl Vietor, one on Goethe, the great German poet, the other on modern German literature. Vietor was one of the most inspired German scholars in the United States as well as an insightful and charismatic teacher, and he encouraged me to continue in German history and literature. He had written two books on Goethe—one on the young man, the other on the mature poet—and a groundbreaking study of Georg Büchner, a relatively unknown dramatist whom Vietor helped rediscover. In Büchner's brief life, he pioneered realist and expressionist writing in his unfinished play *Woyzeck*, the first drama to portray a relatively inarticulate common person in heroic dimensions. Published as a fragment after Büchner's death from typhoid fever in 1837 (at the age of twenty-four), *Woyzeck* was later converted into an opera (*Wozzeck*) and set to music by Alban Berg.

Anna took great pleasure in my knowledge of German literature, and in the early days of our friendship we would spend evenings together reading German poetry: Novalis, Rilke, and Stefan George. I was planning to take two further seminars with Vietor in my senior year. But suddenly, at the end of my junior year, he died of cancer. Vietor's death was a personal loss; it also created a large void in the curriculum I had planned. A few months before Vietor's death I had met

Anna's parents, Ernst and Marianne Kris, both prominent psychoanalysts from Freud's circle. The Krises fired my interest in psychoanalysis and changed my ideas about what I might want to do with my now open schedule.

IT IS DIFFICULT TO CAPTURE TODAY THE FASCINATION THAT psychoanalysis held for young people in the 1950s. Psychoanalysis had developed a theory of mind that gave me my first appreciation of the complexity of human behavior and of the motivations that underlie it. In Vietor's course on contemporary German literature, I had read Freud's *Psychopathology of Everyday Life*, as well as works of three other writers concerned with the inner workings of the human mind—Arthur Schnitzler, Franz Kafka, and Thomas Mann. Even by these daunting literary standards Freud's prose was a joy to read. His German—for which he had received the Goethe Prize in 1930—was simple, beautifully clear, humorous, and unendingly self-referential. The book opened a new world.

The Psychopathology of Everyday Life contains a series of anecdotes that have entered our culture to such an extent that today they could serve as the script for a Woody Allen movie or a stand-up comic routine. Freud recounts the most ordinary, apparently insignificant events—slips of the tongue, unaccountable accidents, misplacements of objects, misspellings, failures to remember—and uses them to show that the human mind is governed by a precise set of rules, most of which are unconscious. These oversights seem on the surface to be routine errors, little accidents that happen to everyone; they certainly had happened to me. But what Freud made me see was that none of these slips is accidental. Each has a coherent and meaningful relationship to the rest of one's psychic life. I found it particularly amazing that Freud could have written all this without ever having met my Aunt Minna!

Freud argued further that psychological determinacy—the idea that little, if anything, in one's psychic life occurs by chance, that every psychological event is determined by an event that precedes it—is central not only to normal mental life, but also to mental illness. A neurotic symptom, no matter how strange it may seem, is not strange to the unconscious mind; it is related to other, preceding mental processes. The connection between a slip of the tongue and its cause or between a symptom and the underlying cognitive process is obscured by the operation of defenses—ubiquitous, dynamic, unconscious mental processes—resulting in a constant struggle between self-revealing and self-protective mental events. Psychoanalysis held the promise of self-understanding and even of therapeutic change based on an analysis of the unconscious motivations and defenses underlying individual actions.

What made psychoanalysis so compelling to me while I was in college was that it was at once imaginative, comprehensive, and empirically grounded—or so it appeared to my naïve mind. No other views of mental life approached psychoanalysis in scope or subtlety. Earlier psychologies were either highly speculative or very narrow.

INDEED, UNTIL THE END OF THE NINETEENTH CENTURY, THE only approaches to the mysteries of the human mind were introspective philosophical inquiries (the reflections of specially trained observers on the nature of their own patterns of thought) or the insights of great novelists, such as Jane Austen, Charles Dickens, Fyodor Dostoevsky, and Leo Tolstoy. Those are the readings that inspired my first years at Harvard. But, as I learned from Ernst Kris, neither trained introspection nor creative insights would lead to the systematic accretion of knowledge needed for the foundation of a science of mind. That sort of foundation requires more than insight, it requires experimentation. Thus, it was the remarkable successes of experimental science in astronomy, physics, and chemistry that spurred students of mind to devise experimental methods for studying behavior.

This search began with Charles Darwin's idea that human behavior evolved from the behavioral repertory of our animal ancestors. That idea gave rise to the notion that experimental animals could be used as models to study human behavior. The Russian physiologist Ivan Pavlov and the American psychologist Edward Thorndike tested in animals an extension of the philosophical notion, first enunciated by Aristotle and later elaborated by John Locke, that we learn by associating ideas. Pavlov discovered classical conditioning, a form of learning in which an animal is taught to associate two stimuli. Thorndike discovered instrumental conditioning, a form of learning in which an animal is taught to associate a behavioral response with its consequences. These two learning processes provided the foundation for the scientific study of learning and memory not only in simple animals, but also in people. Aristotle's and Locke's suggestion that learning involves the association of ideas was replaced by the empirical fact that learning occurs through the association of two stimuli or a stimulus and a response.

In the course of studying classical conditioning, Pavlov discovered two nonassociative forms of learning: habituation and sensitization. In habituation and sensitization an animal learns only about the features of a single stimulus; it does not learn to associate two stimuli with each other. In habituation the animal learns to ignore a stimulus because it is trivial, whereas in sensitization it learns to attend to a stimulus because it is important.

The discoveries of Thorndike and Pavlov had an extraordinary impact on psychology, giving rise to behaviorism, the first empirical school of learning. Behaviorism held out the promise that behavior could be studied with the same rigor as the natural sciences. By the time I was at Harvard, the leading proponent of behaviorism was B. F. Skinner. I was exposed to his thinking through discussions with friends taking his courses. Skinner followed the philosophical path outlined by the founders of behaviorism. Together, they narrowed the view of behavior by insisting that a truly scientific psychology had to be restricted only to those aspects of behavior that could be publicly observed and objectively quantified. There was no room for introspection.

Consequently, Skinner and the behaviorists focused exclusively on observable behavior and excluded from their work all references to mental life and all efforts at introspection, because such things could not be observed, measured, or used to develop general rules about how people behave. Feelings, thoughts, plans, desires, motivations, and values—the internal states and personal experiences that make us human and that psychoanalysis brought to the fore—were considered inaccessible to experimental science and unnecessary for a science of behavior. The behaviorists were convinced that all of our psychological activities can be adequately explained without recourse to such mental processes.

The psychoanalysis I encountered through the Krises was worlds apart from Skinner's behaviorism. In fact, Ernst Kris went to great pains to discuss the differences and to bridge them. He argued that part of the appeal of psychoanalysis was that, like behaviorism, it attempts to be objective, to reject conclusions drawn from introspection. Freud argued that one cannot understand one's own unconscious processes by looking into oneself; only a trained neutral outside observer, the psychoanalyst, can discern the content of the unconscious in another person. Freud also favored observable experimental evidence, but he considered overt behavior to be simply one of several means of examining internal states, whether they be conscious or unconscious. Freud was just as interested in the internal processes that determined a person's responses to particular stimuli as he was in the responses per se. The psychoanalysts who followed Freud argued that, by limiting the study of behavior to observable, measurable actions, behaviorists ignored the most important questions about mental processes.

My attraction to psychoanalysis was further enhanced by the facts that Freud was Viennese and Jewish and had been forced to leave Vienna. Reading his work in German awakened in me a yearning for the intellectual life I had heard about but never experienced. More important even than reading Freud were my conversations about psychoanalysis with Anna's parents, who were extraordinarily interesting people filled with enthusiasm. Ernst Kris was already an established art historian and curator of applied art and sculpture at the Kunsthistorisches Museum in Vienna before marrying Marianne and taking up psychoanalysis. He trained, among others, the great art historian Ernst Gombrich, with whom he later collaborated, and they each contributed importantly to the development of a modern psychology of art. Marianne Kris was a distinguished psychoanalyst and teacher, as well as a wonderfully warm person. Her father was Oskar Rie, an outstanding pediatrician, Freud's best friend, and the physician to his children. Marianne was a close friend of Freud's highly accomplished daughter, Anna. Indeed, Marianne Kris named her daughter after Anna Freud.

Ernst and Marianne Kris were generous and encouraging to me, as they were to all of their daughter's friends. Through my frequent interactions with them I also had occasional interactions with their colleagues the psychoanalysts Heinz Hartmann and Rudolph Lowenstein. Together, the three men had forged a new direction in psychoanalysis.

When Hartmann, Ernst Kris, and Lowenstein immigrated to the United States, they joined forces to write a series of groundbreaking papers in which they pointed out that psychoanalytic theory had placed too much emphasis on frustration and anxiety in the development of the ego, the component of the psychic apparatus that, according to Freud's theory, is in contact with the outside world. More emphasis should be placed on normal cognitive development. To test their ideas, Ernst Kris urged empirical observations of normal child development. By bridging in this way the gap between psychoanalysis and cognitive psychology, which was just beginning to emerge in the 1950s and 1960s, he encouraged American psychoanalysis to become more empirical. Kris himself joined the faculty of the Child Study Center at Yale University and participated in their observational studies.

Listening in on these exciting discussions, I was converted to their view that psychoanalysis offered a fascinating approach, perhaps the only approach, to understanding mind. Psychoanalysis opened an unsurpassed view not only into the rational and irrational aspects of motivation and unconscious and conscious memory but also into the orderly nature of cognitive development, the development of perception and thought. This area of study began to seem much more exciting to me than European literature and intellectual history.

TO BECOME A PRACTICING PSYCHOANALYST IN THE 1950s, IT WAS generally considered best to go to medical school, become a physician, and then train as a psychiatrist, a course of study I had not previously considered. But Karl Vietor's death had left an opening for two full-year courses in my schedule. So in the summer of 1951 I took, almost on impulse, the introductory course in chemistry, which was required for medical school. The idea was that I would take physics and biology in my senior year, while writing my thesis, and then, if I continued with the plan, take organic chemistry, the final requirement for medical school, after graduating from Harvard.

That summer of 1951 I shared a house with four men who became lifelong friends: Henry Nunberg, Anna's cousin and the son of another great psychoanalyst, Herman Nunberg; Robert Goldberger; James Schwartz; and Robert Spitzer. A few months later, based on that single chemistry course and my overall college record, I was accepted at New York University Medical School, with the proviso that I complete the remaining course requirements before enrolling in the fall of 1952.

I entered medical school dedicated to becoming a psychoanalyst and stayed with that career plan through my internship and residency in psychiatry. By my senior year in medical school, however, I had become very interested in the biological basis of medical practice. I decided I had to learn something about the biology of the brain. One reason was that I had greatly enjoyed the course on the anatomy of the brain that I had taken during my second year in medical school. Louis Hausman, who taught the course, had each of us build out of colored clays a large-scale model that was four times the size of the human brain. As my classmates later described it in our yearbook, "The clay model stirred the dormant germ of creativity, and even the least sensitive among us begat a multihued brain."

Building this model gave me my first three-dimensional view of how the spinal cord and the brain come together to make up the central nervous system (figure 3-2). I saw that the central nervous system is a bilateral, essentially symmetrical structure with distinct parts, each bearing an intriguing name, such as hypothalamus, thalamus, cerebellum, or amygdala. The spinal cord contains the machinery needed for simple reflex behaviors. Hausman pointed out that by examining the spinal cord, one can understand in microcosm the overall purpose of the central nervous system. That purpose is to receive sensory information from the skin through bundles of long nerve fibers, called axons, and to transform it into coordinated motor commands that are relayed to the muscles for action through other bundles of axons.

As the spinal cord extends upward toward the brain, it becomes the brain stem (figure 3-3), a structure that conveys sensory information to higher regions of the brain and motor commands from those regions downward to the spinal cord. The brain stem also regulates attentive-

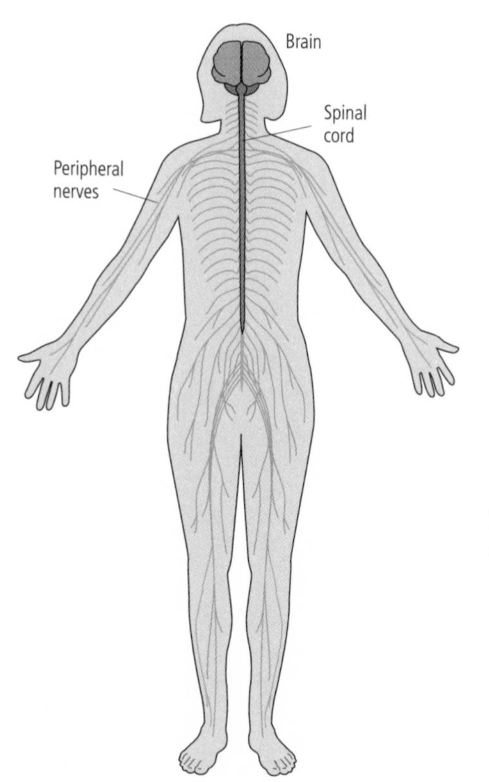

3-2 The central and peripheral nervous systems. The central nervous system, which consists of the brain and the spinal cord, is bilaterally symmetrical. The spinal cord receives sensory information from the skin through bundles of long axons that innervate the skin. These bundles are called peripheral nerves. The spinal cord also sends motor commands to muscles through the axons of the motor neurons. These sensory receptors and motor axons are part of the peripheral nervous system.

ness. Above the brain stem lie the hypothalamus, the thalamus, and the cerebral hemispheres, whose surfaces are covered by a heavily wrinkled outer layer, the cerebral cortex. The cerebral cortex is concerned with higher mental functions: perception, action, language, and planning. Three structures lie deep within it: the basal ganglia, the hippocampus, and the amygdala (figure 3-3). The basal ganglia help regulate motor performance, the hippocampus is involved with aspects of memory storage, and the amygdala coordinates autonomic and endocrine responses in the context of emotional states.

It was hard to look at the brain, even a clay model of it, without wondering where Freud's ego, id, and superego were located. A keen student of the anatomy of the brain, Freud had written repeatedly about the relevance of the biology of the brain to psychoanalysis. For example, in 1914 he wrote in his essay "On Narcissism": "We must

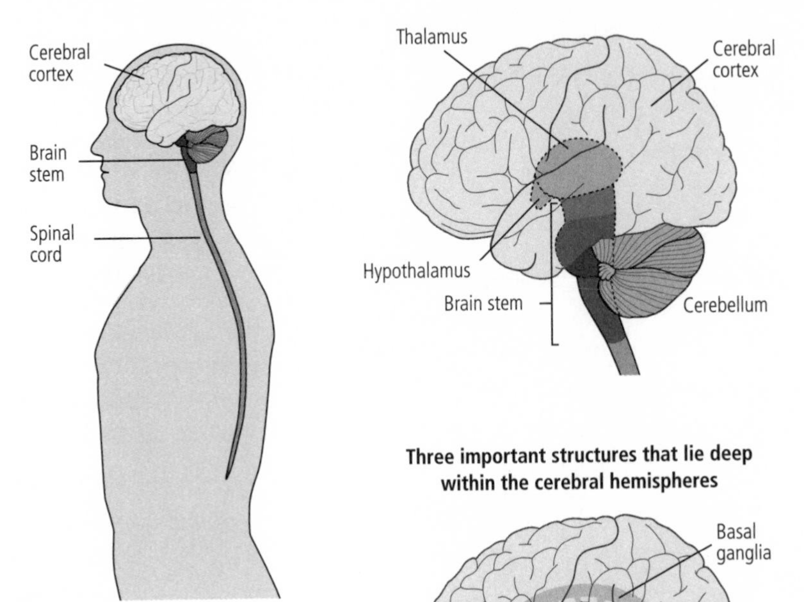

3-3 The central nervous system.

recollect that all of our provisional ideas in psychology will presumably one day be based on an organic substructure." In 1920 Freud again noted, in *Beyond the Pleasure Principle*:

> The deficiencies in our description would probably vanish if we were already in a position to replace the psychological terms by physiological or chemical ones. . . .

Although most psychoanalysts in the 1950s thought of mind in nonbiological terms, a few had begun to discuss the biology of the brain and its potential importance for psychoanalysis. Through the

Krises, I met three such psychoanalysts: Lawrence Kubie, Sidney Margolin, and Mortimer Ostow. After some discussion with each of them, I decided in the fall of 1955 to take an elective at Columbia University with the neurophysiologist Harry Grundfest. At the time, the study of brain science was not an important discipline at many medical schools in the United States, and no one on the NYU faculty was teaching basic neural science.

I WAS STRONGLY SUPPORTED IN THIS DECISION BY DENISE Bystryn, an extremely attractive and intellectually stimulating Frenchwoman I had recently started to date. While I had been taking Hausman's anatomy course, Anna and I had started to drift apart. A relationship that had been very special for both of us when we were together in Cambridge did not work as well with her in Cambridge and me in New York. In addition, our interests were beginning to diverge. So we parted ways in September 1953, soon after Anna graduated from Radcliffe. Anna now has a highly successful practice of psychoanalysis in Cambridge.

I subsequently had two serious but brief relationships, each of which lasted only a year. As the second relationship was breaking up, I met Denise. I had heard about her from a mutual friend, and I called her to ask her out. As the conversation progressed, she made it clear that she was busy and not particularly interested in meeting me. I nevertheless persisted, pulling out one stop after another. All to no avail. Finally, I dropped the fact that I came from Vienna. Suddenly, the tone of her voice changed. Realizing that I was European, she must have thought that I might not be a complete waste of her time and she agreed to meet with me.

When I picked her up at her apartment on West End Avenue, I asked her whether she wanted to go to a movie or to the best bar in town. She said she would like to go to the best bar, so I took her to my apartment on Thirty-first Street near the medical school, which I shared with my friend Robert Goldberger. When we moved into the apartment, Bob and I renovated it and built a very nicely functioning bar, certainly the best among our circle of acquaintances. Bob, a connoisseur of scotch, had a fine collection, even including some single-malt scotches.

Denise was impressed with our woodworking skill (mostly Bob's), but she did not drink scotch. So I uncorked a chardonnay and we spent a delightful evening in which I told her about life in medical school and she talked about her graduate work in sociology at Columbia. Denise's specific interest was in using quantitative methods to study how people's behavior changed over time. Many years later, she applied this methodology to the study of how adolescents become involved in drug abuse. Her epidemiological work was of landmark proportions: it became the basis for the gateway hypothesis, which holds that particular developmental sequences underlie progressively more severe drug use.

Our courtship was amazingly smooth. Denise combined intelligence and curiosity with a wonderful capability for beautifying everyday life. She was a fine cook, had excellent taste in clothing—some of which she made herself—and liked to surround herself with vases, lamps, and art that enlivened the space in which she lived. Much as Anna influenced my thinking about psychoanalysis, Denise influenced my thinking about both empirical science and the quality of life.

She also strengthened in me the sense of being a Jew and a Holocaust survivor. Denise's father, a gifted mechanical engineer, came from a long line of rabbis and scholars and had trained as a rabbi in Poland. He left Poland when he was twenty-one years old and went to Caen in Normandy, France, where he studied mathematics and engineering. Although he became an agnostic and stopped going to synagogue, he kept an impressive collection of Hebrew religious texts in his large library, including the Mishnah and a Vilna edition of the Talmud.

The Bystryns stayed in France throughout the war. Denise's mother helped her husband escape from a French concentration camp, and they both survived the war by hiding from the Nazis in the small town of St.-Céré, located in the southwest. During a good part of that time, Denise was separated from her parents, hidden in a Catholic convent in Cahors, about fifty miles away. Denise's experiences, though much more difficult, paralleled mine in a number of ways. Over the years, our memories of our individual experiences in a Europe dominated by Hitler proved to be enduring for each of us and brought us closer together.

One incident in Denise's life made an indelible impression on me. In the few years she spent in the convent, no one but the Mother Superior knew that she was Jewish and no one put any pressure on her to convert to Catholicism. But Denise felt awkward in relation to her classmates because she was different. She did not go to confession, nor did she take holy communion at mass every Sunday. Denise's mother, Sara, became uncomfortable about her daughter's standing out in this way and was afraid that her true identity might be uncovered, which could endanger her. Sara discussed this dilemma with Denise's father, Iser, and they decided to have Denise baptized.

Sara traveled on foot and by bus the almost fifty miles from their hiding place to the convent in Cahors. When she arrived at the convent, she stood in front of the heavy, dark wooden door and was about to knock on it and announce her presence, when at the last moment she could not bring herself to convey the fateful decision. She turned around without entering the convent and walked back home, certain that her husband would be furious that she had not lessened the danger to their daughter. When she entered the house in St.-Céré, she found that Iser was immensely relieved. All the time that Sara had been gone, he had obsessed about the error he had made in agreeing to allow Denise to be converted. Despite the fact that Iser did not believe in God, he and Sara were very proud to be Jews.

In 1949 Denise, her brother, and her parents immigrated to the United States. Denise attended the Lycée Français de New York for one year and was admitted to Bryn Mawr College as a junior at age seventeen. On graduating from Bryn Mawr at nineteen, she enrolled as a graduate student in sociology at Columbia University. When we met in 1955, she had started research for her Ph.D. thesis in medical sociology with Robert K. Merton, one of the great contributors to modern sociology and a founder of the sociology of science. Her thesis was a study of the career decisions of medical students based on an empirical longitudinal survey.

A few days after I graduated from medical school, in June 1956, Denise and I married (figure 3-4). After a brief honeymoon in Tanglewood, Massachusetts, where I spent some time studying for the

3-4 Denise at our wedding in 1956. She was twenty-three and a graduate student in sociology at Columbia University. (From Eric Kandel's personal collection.)

national boards in medicine—a point that Denise has never allowed me to forget—I started a one-year internship at Montefiore Hospital in New York City while Denise continued her doctoral research at Columbia.

Denise sensed, perhaps more than I did, that my idea of examining the biological basis of mental function was original and bold, and she urged me to explore it. I was concerned, however. Neither of us had any financial resources, and I thought it essential to have a private practice in order to support us. Denise simply gave the issue of money short shrift. It was of no importance, she insisted. Her father, who had died a year before I met her, had advised his daughter to marry a poor intellectual because such a man would value scholarship above all and would strive to pursue exciting academic goals. Denise believed she was following that advice (she certainly married someone who was poor), and she always encouraged me to make bold decisions that favored my doing something genuinely new and original.

TWO

Biology is truly a land of unlimited possibilities. We may expect it to give us the most surprising information, and we cannot guess what answers it will return in a few dozen years. . . . They may be of a kind which will blow away the whole of our artificial structure of hypotheses.

—Sigmund Freud, *Beyond the Pleasure Principle* (1920)

4

ONE CELL AT A TIME

I entered Harry Grundfest's laboratory at Columbia University for a six-month elective period in the fall of 1955, hoping to learn something about higher brain functions. I did not anticipate embarking on a new career, a new way of life. But my very first conversation with Grundfest gave me reason to reflect. In that conversation I described my interest in psychoanalysis and my hope of learning something about where in the brain the ego, the id, and the superego might be located.

My desire to find these three psychic agencies had been sparked by a diagram Freud published in the course of summarizing his new structural theory of mind, which he developed in the decade 1923 to 1933 (figure 4-1). That new theory maintained his earlier distinction between conscious and unconscious mental functions, but it added three interacting psychic agencies: the ego, the id, and the superego. Freud saw consciousness as the *surface* of the mental apparatus. Much of our mental function is submerged below that surface, Freud argued, just as the bulk of an iceberg is submerged below the surface of the ocean. The deeper a mental function lies below the surface, the less accessible it is to consciousness. Psychoanalysis provided a way of digging down to the buried men-

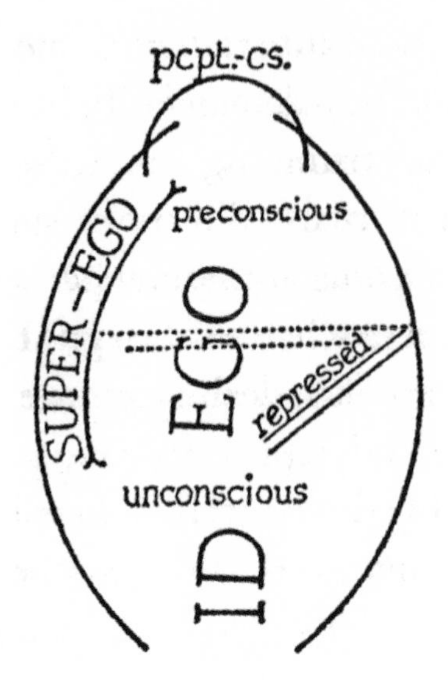

4-1 Freud's structural theory. Freud conceived of three main psychic structures—the ego, the id, and the superego. The ego has a conscious component (perceptual consciousness, or *pcpt.-cs.*) that receives sensory information and is in direct contact with the outside world, as well as a preconscious component, an aspect of unconscious processing that has ready access to consciousness. The ego's unconscious components act through repression and other defenses to inhibit the instinctual urges of the id, the generator of sexual and aggressive instincts. The ego also responds to the pressures of the superego, the largely unconscious carrier of moral values. The dotted lines indicate the divisions between those processes that are accessible to consciousness and those that are completely unconscious. (From *New Introductory Lectures on Psychoanalysis* [1933]).

tal strata, the preconscious and the unconscious components of the personality.

What gave Freud's new model a dramatic turn was the three interacting psychic agencies. Freud did not define the ego, the id, and the superego as either conscious or unconscious, but as differing in cognitive style, goal, and function.

According to Freud's structural theory, the ego (the "I," or autobiographical self) is the executive agency, and it has both a conscious and an unconscious component. The conscious component is in direct contact with the external world through the sensory apparatus for sight, sound, and touch; it is concerned with perception, reasoning, the planning of action, and the experiencing of pleasure and pain. In their work, Hartmann, Kris, and Lowenstein emphasized that this conflict-free component of the ego operates logically and is guided in its actions by the reality principle. The unconscious component of the ego is concerned with psychological defenses (repression, denial, sublimation), the mechanisms whereby the ego inhibits, channels, and redirects both the sexual and the aggressive instinctual drives of the id, the second psychic agency.

The id (the "it"), a term that Freud borrowed from Friedrich Nietszche, is totally unconscious. It is not governed by logic or by reality but by the hedonistic principle of seeking pleasure and avoiding pain. The id, according to Freud, represents the primitive mind of the infant and is the only mental structure present at birth. The superego, the third governor, is the unconscious moral agency, the embodiment of our aspirations.

Although Freud did not intend his diagram to be a neuroanatomical map of mind, it stimulated me to wonder where in the elaborate folds of the human brain these psychic agencies might live, as it had earlier stimulated the curiosity of Kubie and Ostow. As I mentioned, these two psychoanalysts with a keen interest in biology had encouraged me to study with Grundfest.

Grundfest listened patiently as I told him of my rather grandiose ideas. Another biologist might well have dismissed me, wondering what to do with this naïve and misguided medical student. But not Grundfest. He explained that my hope of understanding the biological basis of Freud's structural theory of mind was far beyond the grasp of contemporary brain science. Rather, he told me, to understand mind we needed to look at the brain one cell at a time.

One cell at a time! I initially found those words demoralizing. How could one address psychoanalytic questions about the unconscious motivation of behavior, or the action of our conscious life, by studying the brain on the level of single nerve cells? But as we talked I suddenly remembered that in 1887, when Freud began his own career, he had sought to solve the hidden riddles of mental life by studying the brain one nerve cell at a time. Freud started out as an anatomist, studying single nerve cells, and had anticipated a key point of what later came to be called the neuron doctrine, the view that nerve cells are the building blocks of the brain. It was only later, after he began treating mentally ill patients in Vienna, that Freud made his monumental discoveries about unconscious mental processes.

I found it ironic and remarkable that I was now being encouraged to take that journey in reverse, to move from an interest in the top-down structural theory of mind to the bottom-up study of the signaling elements of the nervous system, the intricate inner worlds of nerve cells. Harry Grundfest offered to guide me into this new world.

I HAD SPECIFICALLY SOUGHT TO WORK WITH GRUNDFEST because he was the most knowledgeable and intellectually interesting neurophysiologist in New York City—indeed, one of the best in the country. At the age of fifty-one, he was at the peak of his very considerable intellectual powers (figure 4-2).

Grundfest had obtained a Ph.D. in zoology and physiology at Columbia in 1930 and continued there as a postdoctoral fellow. In 1935 he joined the Rockefeller Institute (now Rockefeller University) to work in the laboratory of Herbert Gasser, a pioneer in the study of electrical signaling in nerve cells, a process at the very heart of how nervous systems function. At the time that Grundfest joined him, Gasser was at the

4-2 Harry Grundfest (1904–1983), professor of neurology at Columbia University, introduced me to neuroscience, letting me work in his laboratory for six months in 1955–56, at the beginning of my senior year in medical school. (From Eric Kandel's personal collection.)

high point of his career, having just been appointed president of the Rockefeller Institute. In 1944, while Grundfest was still in his laboratory, Gasser was awarded the Nobel Prize in Physiology or Medicine.

By the time Grundfest had finished his training with Gasser, he combined a broad biological outlook with a solid grounding in electrical engineering. Moreover, he had gained a good grasp of the comparative biology of the nervous system in animals ranging from simple invertebrates (crayfish, lobsters, squid, and the like) to mammals. Few people at the time had a comparable background. As a result, Grundfest was recruited back to his alma mater in 1945 to head the new neurophysiology laboratory at the Neurological Institute of the College of Physicians and Surgeons. Soon after his arrival, he began an important collaboration with David Nachmansohn, a well-known biochemist. Together, they studied the biochemical changes associated with nerve cell signaling. Grundfest's future seemed assured, but his career soon ran into trouble.

In 1953 Grundfest was summoned to testify before the Senate Permanent Subcommittee on Investigations, chaired by Senator Joseph McCarthy. During World War II, Grundfest, an outspoken radical, had worked on wound healing and nerve regeneration in the Climatic Research Unit of the Signal Laboratories at Fort Monmouth, New Jersey. McCarthy implied that Grundfest had been a Communist sympathizer and that he or his friends had conveyed technical knowledge to the Soviet Union during the war. At the McCarthy hearings, Grundfest testified that he was not a Communist. Invoking his rights under the Fifth Amendment, he refused to discuss his own political views further or to discuss those of his colleagues.

Not a shred of evidence was ever produced by McCarthy to support his allegations. Nevertheless, Grundfest lost his funding from the National Institutes of Health (NIH) for a number of years. Nachmansohn, fearing that his own governmental funding might be compromised, shut Grundfest out of their shared laboratory and broke off their collaboration. Grundfest had to reduce his research group to two persons, and his career would have been damaged even more severely had it not been for the strong support he received from the academic leadership at Columbia.

For Grundfest, the reduction of his research capability at what proved to be the peak of his scientific career was devastating. Paradoxically, the circumstances proved beneficial for me. Grundfest had more time available than he otherwise would have, and he devoted a substantial amount of it to teaching me what brain science was actually about and how it was soon to be transformed from a descriptive and unstructured field into a coherent discipline based on cell biology. I knew next to nothing about modern cell biology, yet the new direction in brain research, as outlined by Grundfest, fascinated me and stirred my imagination. The mysteries of brain function were beginning to unravel as a result of examining the brain one cell at a time.

AFTER BUILDING THE CLAY MODEL IN MY NEUROANATOMY course, I thought of the brain as an organ apart, one that functions in ways radically different from other parts of the body. This is true, of course: the kidney and the liver cannot receive and process the stimuli that impinge on our sensory organs, nor can their cells store and recall memory or give rise to conscious thought. However, as Grundfest pointed out, all cells share a number of common features. In 1839, the anatomists Mattias Jakob Schleiden and Theodor Schwann formulated the cell theory, which holds that all living entities, from the simplest plants to complex human beings, are made up of the same basic units called cells. Although the cells of different plants and animals differ importantly in detail, they all share a number of common features.

As Grundfest explained, every cell in a multicellular organism is surrounded by an oily membrane that separates it from other cells and from the extracellular fluid that bathes all cells. The cell surface membrane is permeable to certain substances, thereby allowing an exchange of nutrients and gases to take place between the interior of the cell and the fluid surrounding it. Inside the cell is the nucleus, which has a membrane of its own and is surrounded by an intracellular fluid called the cytoplasm. The nucleus contains the chromosomes, long thin structures made of DNA that carry genes like beads on a string. In addition to controlling the cell's ability to reproduce itself, genes tell the cell what proteins to make to carry out its activities. The actual machinery for making proteins is located in the cytoplasm. Seen from this shared perspective, the cell is the fundamental unit of life, the structural and functional basis of all tissues and organs in all animals and plants.

Besides their common biological features, all cells have specialized functions. Liver cells, for instance, carry out digestive activities, while brain cells have particular ways of processing information and communicating with one another. These interactions allow nerve cells in the brain to form complete circuits that carry and transform information. Specialized functions, Grundfest emphasized, make a liver cell uniquely suited to metabolism and a brain cell uniquely suited to processing information.

All of this knowledge I had encountered in my basic science courses at New York University and in the assigned textbook readings, but none of it excited my curiosity or even meant much to me until Grundfest put it into context. The nerve cell is not simply a marvelous piece of biology. It is the key to understanding how the brain works. As Grundfest's teachings began to have an impact on me, so did his insights into psychoanalysis. I came to realize that before we could understand how the ego operates in biological terms, we needed to understand how the nerve cell operates.

Grundfest's emphasis on the importance of understanding how nerve cells function was fundamental to my later studies of learning and memory, and his insistence on a cellular approach to brain function was critical to the emergence of the new science of mind. In retrospect, considering that the human brain is made up of about 100 billion nerve cells, it is remarkable how much scientists have learned about mental activity in the last half century by examining individual cells in the brain. Cellular studies have provided the first glimpse into the biological basis of perception, voluntary movement, attention, learning, and memory storage.

THE BIOLOGY OF NERVE CELLS IS GROUNDED IN THREE PRINCIPLES that emerged for the most part during the first half of the twentieth century and that form to this day the core of our understanding of the brain's functional organization. The *neuron doctrine* (the cell theory as it applies to the brain) states that the nerve cell, or neuron, is the fundamental building block and elementary signaling unit of the brain.

The *ionic hypothesis* focuses on the transmission of information within the nerve cell. It describes the mechanisms whereby individual nerve cells generate electrical signals, called action potentials, that can propagate over a considerable distance within a given nerve cell. The *chemical theory of synaptic transmission* focuses on the transmission of information between nerve cells. It describes how one nerve cell communicates with another by releasing a chemical signal called a neurotransmitter; the second cell recognizes the signal and responds by means of a specific molecule in its surface membrane called a receptor. All three concepts focus on individual nerve cells.

4-3 Santiago Ramón y Cajal (1852–1934), the great Spanish anatomist, formulated the neuron doctrine, the basis for all modern thinking about the nervous system. (Courtesy of the Cajal Institute.)

The person who made this cellular study of mental life possible was Santiago Ramón y Cajal, a neuroanatomist who was a contemporary of Freud (figure 4-3). Cajal laid the foundation for the modern study of the nervous system and is arguably the most important brain scientist who ever lived. He had originally aspired to be a painter. To become familiar with the human body, he studied anatomy with his father, a surgeon, who taught him by using bones unearthed from an ancient cemetery. A fascination with these skeletal remains ultimately led Cajal from painting to anatomy, and then specifically to the anatomy of the brain. In turning to the brain, he was driven by the same curiosity that drove Freud and that many years later drove me. Cajal wanted to develop a "rational psychology." He thought the first step was to have detailed knowledge of the cellular anatomy of the brain.

He brought to his task an uncanny ability to infer the properties of living nerve cells from static images of dead nerve cells. This leap of the imagination, perhaps derived from his artistic bent, enabled him to capture and describe in vivid terms and in beautiful drawings the essential nature of any observation he made. The noted British physiologist Charles Sherrington would later write of him, "in describing what the microscope showed, [Cajal] spoke habitually as though it were a living scene. This was perhaps the more striking because . . . his preparations [were] all dead and fixed." Sherrington went on to say:

> The intense anthropomorphic descriptions of what Cajal saw in stained fixed sections of the brain were at first too startling to accept. He treated the microscopic scene as though it were alive and were inhabited by beings which felt and did and hoped and tried as we do. . . . A nerve cell by its emergent fiber "groped to find another"! . . . Listening to him, I asked myself how far this capacity for anthropomorphizing might not contribute to his success as an investigator. I never met anyone else in whom it was so marked.

Prior to Cajal's entry into the field, biologists were thoroughly confused by the shape of nerve cells. Unlike most other cells of the body which have a simple shape, nerve cells have highly irregular shapes and are surrounded by a multitude of exceedingly fine extensions known at that time as processes. Biologists did not know whether those processes were part of the nerve cell or not, because there was no way of tracing them back to one cell body or forward to another and thus no way of knowing where they came from or where they led. In addition, because the processes are extremely thin (about one-hundredth the thickness of a human hair), no one could see and resolve their surface membrane. This led many biologists, including the great Italian anatomist Camillo Golgi, to conclude that the processes lack a surface membrane. Moreover, because the processes surrounding one nerve cell come in close apposition to the processes surrounding other nerve cells, it appeared to Golgi that the cytoplasm inside the processes intermingles freely, creating a continuously connected nerve net much like the web of a spider, in which signals can be sent in all directions at once. Therefore, Golgi argued, the fundamental unit of the nervous system must be the freely communicating nerve net, not the single nerve cell.

In the 1890s Cajal tried to find a better way to visualize the nerve cell in its entirety. He did so by combining two research strategies. The first was to study the brain in newborn rather than adult animals. In newborns, the number of nerve cells is small, the cells are packed less densely, and the processes are shorter. This enabled Cajal to see single trees in the cellular forest of the brain. The second strategy was to use a specialized silver staining method developed by Golgi. The method is quite capricious and marks, on a fairly random basis, only an occasional neuron—less than 1 percent of the total number. But each neuron that is labeled is labeled in its entirety, permitting the viewer to see the nerve cell body and all the processes. In the newborn brain, the occasionally labeled cell stood out in the unlabeled forest like a lighted Christmas tree. Thus Cajal wrote:

> Since the full grown forest turns out to be impenetrable and indefinable, why not revert to the study of the young wood, in the nursery stage, as we might say? . . . If the stage of development is well chosen . . . the nerve cells, which are still relatively small, stand out complete in each section; the terminal ramifications . . . are depicted with the utmost clearness.

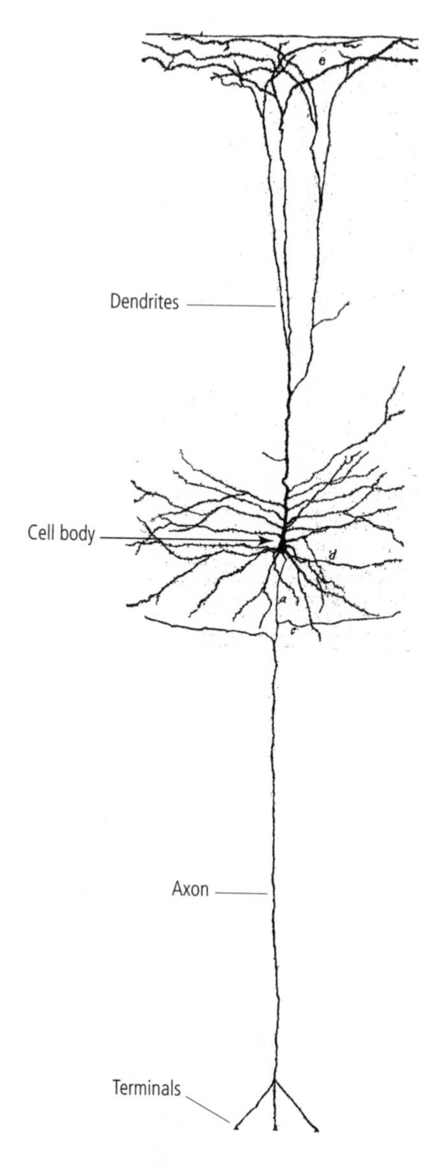

4-4 A neuron in the hippocampus, as drawn by Cajal. Cajal realized that both the dendrites (top) and the axon (bottom) of a cell extend from the cell body and that information flows from the dendrites to the axon. This drawing is modified from Cajal. (Adapted from "Figure 23," *Cajal on the Cerebral Cortex*, edited by Javier DeFelipe and Edward Jones, translated by Javier DeFelipe and Edward Jones, © 1988 by Oxford University Press, Inc. Used by permission of Oxford University Press, Inc.)

These two strategies revealed that, despite their complex shape, nerve cells are single, coherent entities (figure 4-4). The fine processes surrounding them are not independent but emanate directly from the cell body. Moreover, the entire nerve cell, including the processes, is fully enclosed by a surface membrane, consistent with the cell theory. Cajal went on to distinguish two sorts of processes, axons and dendrites. He named this three-component view of the nerve cell the neu-

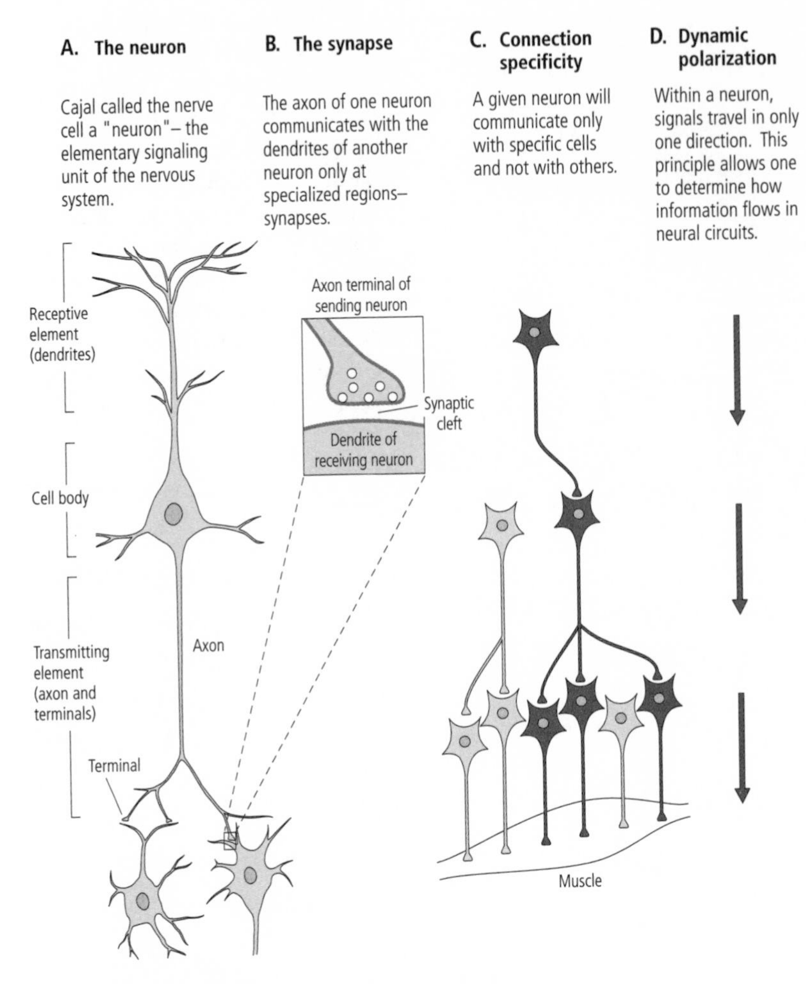

4-5 Cajal's four principles of neural organization.

ron. With rare exceptions, all nerve cells in the brain have a cell body that contains a nucleus, a single axon, and many fine dendrites.

The axon of a typical neuron emerges at one end of the cell body and can extend up to several feet. The axon often splits into one or more branches along its length; at the end of each of these branches are many tiny axon terminals. The several dendrites usually emerge on the opposite side of the cell body (figure 4-5-A). They branch extensively, forming a treelike structure that grows out from the cell body and spreads over a large area. Some neurons in the human brain have as many as forty dendritic branches.

In the 1890s Cajal pulled his observations together and formulated the four principles that make up the neuron doctrine, the theory of neural organization that has governed our understanding of the brain ever since.

The first principle is that the neuron is the fundamental structural and functional element of the brain—that is, both the basic building block and the elementary signaling unit of the brain. Moreover, Cajal inferred that the axons and dendrites play quite different roles in this signaling process. A neuron uses its dendrites to receive signals from other nerve cells and its axon to send information to other cells.

Second, he inferred that the terminals of one neuron's axon communicate with the dendrites of another neuron only at specialized sites, later named synapses by Sherrington (from the Greek *synaptein*, meaning to bind together). Cajal further inferred that the synapse between two neurons is characterized by a small gap, now called the synaptic cleft, where the axon terminals of one nerve cell—which Cajal called the presynaptic terminals—reach out to, but do not quite touch, the dendrites of another nerve cell (figure 4-5-B). Thus, like lips whispering very close to an ear, synaptic communication between neurons has three basic components: the presynaptic terminal of the axon, which sends signals (corresponding to the lips in our analogy); the synaptic cleft (the space between lips and ear); and the postsynaptic site on the dendrite that receives signals (the ear).

Third, Cajal inferred the principle of connection specificity, which holds that neurons do not form connections indiscriminately. Rather, each nerve cell forms synapses and communicates with certain nerve cells and not with others (figure 4-5-C). He used the principle of connection specificity to show that nerve cells are linked in specific pathways he called neural circuits; signals travel along these circuits in a predictable pattern.

Typically, a single neuron makes contact through its many presynaptic terminals with the dendrites of many target cells. In this way, a single neuron can disseminate the information it receives widely to different target neurons, sometimes located in different regions of the brain. Conversely, the dendrites of a target nerve cell can receive information from the presynaptic terminals of a number of different neurons. In this way a neuron can integrate information from a number of different neurons, even those located in different areas of the brain.

Based on his analysis of signaling, Cajal conceived of the brain as an organ constructed of specific, predictable circuits, unlike the prevailing view, which saw the brain as a diffuse nerve net in which every imaginable type of interaction occurred everywhere.

With an amazing leap of intuition, Cajal arrived at the fourth principle, dynamic polarization. This principle holds that signals in a neural circuit travel in only one direction (figure 4-5-D). Information flows, from the dendrites of a given cell to the cell body along the axon to the presynaptic terminals and then across the synaptic cleft to the dendrites of the next cell, and so on. The principle of the one-way flow of signals was enormously important because it related all components of the nerve cell to a single function—signaling.

The principles of connection specificity and the one-way flow of signals gave rise to a logical set of rules that has been used ever since to map the flow of signals between nerve cells. Efforts to delineate neural circuits received a further boost when Cajal showed that such circuits in the brain and spinal cord contain three major classes of neurons, each with a specialized function. *Sensory neurons*, which are located in the skin and in various sense organs, respond to a specific type of stimulus from the outside world—mechanical pressure (touch), light (vision), sound waves (hearing), or specific chemicals (smell and taste)—and send this information to the brain. *Motor neurons* send their axons out of the brain stem and spinal cord to effector cells, such as muscle and gland cells, and control the activity of those cells. *Interneurons*, the most numerous class of neurons in the brain, serve as relays between sensory and motor neurons. Thus Cajal was able to trace the flow of information from sensory neurons in the skin to the spinal cord and from there to interneurons and to motor neurons that signal muscle cells to move (figure 4-6). Cajal derived these insights from work on rats, monkeys, and people.

In time, it became clear that each cell type is biochemically distinct

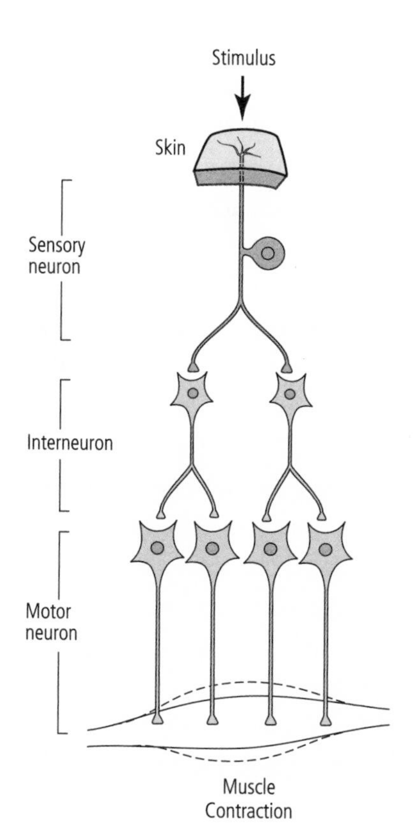

4-6 Three major classes of neurons, as identified by Cajal. Each class of neurons in the brain and spinal cord has a specialized function. Sensory neurons respond to stimuli from the outside world. Motor neurons control the activity of muscle or gland cells. Interneurons serve as relays between sensory and motor neurons.

and can be affected by distinct disease states. Thus, for example, sensory neurons from the skin and joints are compromised by a late stage of syphilis; Parkinson's disease attacks a certain class of interneurons; and motor neurons are selectively destroyed by amyotrophic lateral sclerosis and poliomyelitis. Indeed, some diseases are so selective that they affect only specific parts of the neuron: multiple sclerosis affects certain classes of axons; Gaucher's disease affects the cell body; fragile X syndrome affects dendrites; botulism toxin affects synapses.

For his revolutionary insights, Cajal received the Nobel Prize in Physiology or Medicine in 1906, together with Golgi, whose silver stain made Cajal's discoveries possible.

It is one of the strange twists of the history of science that Golgi, whose technical developments paved the way for Cajal's brilliant discoveries, continued to disagree vehemently with Cajal's interpretations and never subscribed to any aspect of the neuron doctrine.

Indeed, Golgi used the occasion of his Nobel Prize lecture to renew his attack on the neuron doctrine. He began by asserting once again that he had always been opposed to the neuron doctrine and that "this doctrine is generally recognized as going out of favor." He went on to say, "In my opinion, we cannot draw any conclusion, one way or the other, from all that has been said . . . in being for or against the neuron doctrine." He argued further that the principle of dynamic polarization was wrong and that it was incorrect to think that the elements of a neural circuit connected in precise ways or that different neural circuits had different behavioral functions.

Until his death in 1926, Golgi continued to think, quite erroneously, that nerve cells are not self-contained units. For his part, Cajal later wrote of the shared Nobel Prize, "What a cruel irony of fate to pair, like Siamese twins, united by the shoulders, scientific adversaries of such contrasting character."

This disagreement reveals several interesting things about the sociology of science that I was to observe repeatedly during my own career. To begin with, there are scientists, like Golgi, who are very strong technically but who do not necessarily have the deepest insights into the biological questions they are studying. Second, even the best scientists can disagree with one another, especially in the early stages of discovery.

Occasionally, disputes that start out as disagreements about science take on a personal, almost vindictive quality, as they did with Golgi. Such disputes reveal that the qualities that characterize competition—ambition, pride, and vindictiveness—are just as evident among scientists as are acts of generosity and sharing. The reason for this is clear. The aim of science is to discover new truths about the world, and discovery means priority, being there first. As Alan Hodgkin, the formulator of the ionic hypothesis, wrote in an autobiographical essay, "If pure scientists were motivated by curiosity alone, they should be delighted when someone else solves the problem they are working on—but that is not the usual reaction." Recognition by their peers and esteem come only to those who have made original contributions to the common stock of knowledge. This caused Darwin to point out that his "love of natural science . . . has been much aided by the ambition to be esteemed by my fellow scientists."

Finally, great controversies often originate when available methodologies are insufficient to provide an unambiguous answer to a key question. It was not until 1955 that Cajal's intuitions were borne out conclusively. Sanford Palay and George Palade at the Rockefeller Institute used the electron microscope to demonstrate that in the vast majority of cases, a slight space—the synaptic cleft—separates the presynaptic terminal of one cell from the dendrite of another cell. Those new images also revealed that the synapse is asymmetrical, and that the machinery for releasing chemical transmitters, discovered much later, is located only in the presynaptic cell. This explains why information in a neural circuit flows in just one direction.

PHYSIOLOGISTS WERE QUICK TO SEE THE IMPORTANCE OF Cajal's contributions. Charles Sherrington (figure 4-7) became one of Cajal's greatest supporters and invited him to England in 1894 to give the Croonian Lecture to the Royal Society in London, one of the most distinguished honors Great Britain can bestow on a biologist.

In his memorial to Cajal in 1949, Sherrington wrote:

> Is it too much to say of him that he is the greatest anatomist the nervous system has ever known? The subject had long been a favorite with some of the best investigators, previous to Cajal there were discoveries, discoveries which often left the physician more mystified than before, adding mystification without enlightenment. Cajal made it possible even for a tyro to recognize at a glance the direction taken by the nerve-current in the living cell, and in a whole chain of nerve cells.
>
> He solved at a stroke the great question of the direction of the nerve-currents in their travel through brain and spinal cord. He showed, for instance, that each nerve-path is always a line of one-way traffic only, and that the direction of that traffic is at all times irreversibly the same.

In his own influential book, *The Integrative Action of the Nervous System*, Sherrington built on Cajal's findings about the structure of nerve cells and succeeded in linking structure to physiology and to behavior.

4-7 Charles Sherrington (1857–1952) studied the neural basis of reflex behavior. He discovered that neurons can be inhibited as well as excited and that integration of these signals determines the actions of the nervous system. (Reprinted from *The Integrative Action of the Nervous System*, Cambridge University Press, 1947.)

He did this by examining the spinal cord of cats. The spinal cord receives and processes sensory information from the skin, joints, and muscles of the limbs and trunk. It contains within itself much of the basic neuronal machinery for controlling the movement of the limbs and the trunk, including the movements involved in walking and running. Trying to understand simple neural circuits, Sherrington studied two reflex behaviors—the cat's equivalent of the human knee jerk and the withdrawal response of the cat's paw when exposed to a stimulus that causes an unpleasant sensation. Such innate reflexes require no learning. Moreover, they are intrinsic to the spinal cord and do not require that messages be sent to the brain. Instead, they are elicited instantly by an appropriate stimulus, such as a tap on the knee or exposure of the paw to a shock or a hot surface.

In the course of his research on reflexes, Sherrington discovered something that Cajal could not have anticipated from anatomical studies alone—namely, that not all nervous action is excitatory—that is, not all nerve cells use their presynaptic terminals to stimulate the next receiving cells in line to transmit information onward. Some cells are inhibitory; they use their terminals to stop the receiving cells from relaying information. Sherrington made this discovery while studying how different reflexes are coordinated to yield a coherent behavioral response. He found that when a particular site is stimulated so as to elicit a specific reflex response, only that reflex is elicited; other, opposing reflexes are inhibited. Thus a tap on the tendon of the kneecap elicits one reflex action—an extension of the leg, a kick. That tap simultaneously inhibits the opposing reflex action—flexion, the drawing backward of the leg.

Sherrington then explored what was happening to the motor neurons during this coordinated reflex response. He found that when he tapped on the tendon of the kneecap, the motor neurons that extend the limb (the extensors) were actively excited, while the motor neurons that flex the limbs (the flexors) were actively inhibited. Sherrington called the cells that inhibit the flexors *inhibitory neurons*. Later work found that almost all inhibitory neurons are interneurons.

Sherrington immediately appreciated the importance of inhibition not only for coordinating reflex responses but also for increasing the stability of a response. Animals are often exposed to stimuli that may elicit contradictory reflexes. Inhibitory neurons bring about a stable, predictable, coordinated response to a particular stimulus by inhibiting all but one of those competing reflexes, a mechanism called reciprocal control. For example, extension of the leg is invariably accompanied by inhibition of flexion, and flexion of the leg is invariably accompanied by inhibition of extension. Through reciprocal control, inhibitory neurons select among competing reflexes and ensure that only one of two or even several possible responses is expressed as behavior.

Integration of reflexes and the decision-making capabilities of the spinal cord and brain derive from the integrative features of individual motor neurons. A motor neuron totals up all the excitatory and inhibitory signals it receives from the other neurons that converge upon it and then carries out an appropriate course of action based on that calculation. If and only if the sum of excitation exceeds that of inhibition by a critical minimum will the motor neuron signal the target muscle to contract.

Sherrington saw reciprocal control as a general means of coordinating priorities to achieve the singleness of action and purpose required for behavior. His work on the spinal cord revealed principles of neuronal integration that were likely to underlie some of the brain's higher cognitive decision making as well. Each perception and thought we have, each movement we make, is the outcome of a vast multitude of basically similar neural calculations.

Some of the details of the neuron doctrine and its implications for physiology had yet to be established in the mid-1880s, when Freud abandoned his basic research studies of nerve cells and their connections. However, he kept abreast of neurobiology and tried to incorporate some of Cajal's new ideas about neurons in an unpublished manuscript, "Project for a Scientific Psychology," written in late 1895, after he had begun to use psychoanalysis to treat patients and had uncovered the unconscious meaning of dreams. Even though Freud became fully immersed in psychoanalysis, his earlier experimental work had a lasting influence on his thought, and therefore on the evolution of psychoanalytic thought. Robert Holt, a psychologist interested in psychoanalysis, has put it this way:

> In many respects Freud seems to have undergone a profound re-orientation as he turned from being a neuroanatomical researcher to a clinical neurologist who experimented with psychotherapy, finally becoming the first psychoanalyst. We would be poor psychologists, however, if we imagined that there was not at least as much continuity as change in this development. Twenty years of passionate investment in the study of the nervous system were not easily tossed aside by Freud's decision to become a psychologist instead and to work with a purely abstract, hypothetical model.

Freud called the period he spent studying nerve cells in simple organisms like crayfish, eels, and primitive fish "the happiest hours of my student life." He left those basic research studies after he met and fell in love with Martha Bernays, whom he later married. In the nineteenth century, one needed an independent income in order to take on a career in research. In view of his poor financial position, Freud turned instead to the establishment of a medical practice that would earn him sufficient income to support a wife and family. Perhaps if a scientific career could have ensured a living wage then, as it does today, Freud would be known as a neuroanatomist and a co-founder of the neuron doctrine, instead of as the father of psychoanalysis.

5

THE NERVE CELL SPEAKS

Had I become a practicing psychoanalyst, I would have spent much of my life listening to patients talk about themselves—about their dreams and waking memories, their conflicts and their desires. This is the introspective method of "talk therapy" that Freud pioneered to arrive at deeper levels of self-understanding. By encouraging the free association of thoughts and memories, the psychoanalyst helps patients unpack the unconscious memories, traumas, and impulses that underlie their conscious thoughts and behavior.

In Grundfest's laboratory I soon appreciated that to understand how the brain functions, I would have to learn how to listen to neurons, to interpret the electrical signals that underlie all mental life. Electrical signaling represents the language of mind, the means whereby nerve cells, the building blocks of the brain, communicate with one another over great distances. Listening in on those conversations and recording neuronal activity was, so to speak, objective introspection.

GRUNDFEST WAS A LEADER IN THE BIOLOGY OF SIGNALING. From him I learned that thinking about the signaling function of nerve cells has proceeded in four distinct phases, reaching from the eighteenth century to a particularly clear and satisfying resolution in the work of Alan Hodgkin and Andrew Huxley two hundred years later. Throughout, the question of how nerve cells communicate has attracted some of the best brains in science.

The first phase dates to 1791, when Luigi Galvani, a biologist from Bologna, Italy, discovered electrical activity in animals. Galvani left a frog's leg hanging on a copper hook from his iron balcony and found that the interaction of the two dissimilar metals, copper and iron, would occasionally cause the leg to twitch, as if it were animated. Galvani could also cause a frog's leg to twitch by stimulating it with a pulse of electricity. After further study, he proposed that nerve cells and muscle cells are themselves capable of generating a flow of electrical current and that the twitch of muscles is caused by the electricity generated by muscle cells—not by spirits or "vital forces," as was commonly believed at the time.

Galvani's insight and his achievement in bringing nervous activity out of the realm of vital forces and into natural science was elaborated in the nineteenth century by Hermann von Helmholtz, one of the first scientists to bring the rigorous methods of physics to bear on a range of problems in brain science. Helmholtz found that the axons of nerve cells generate electricity not as a by-product of their activity, but as a means of producing messages that are carried along their whole length. These messages are then used to carry sensory information about the outside world into the spinal cord and the brain and to transmit commands for action from the brain and spinal cord to the muscles.

In the course of this work, Helmholtz made an extraordinary experimental measurement that changed thinking about electrical activity in animals. In 1859 he succeeded in capturing the speed at which these electrical messages are conducted and found to his amazement that electricity conducted along a living axon is fundamentally different from the flow of electricity in a copper wire. In a metal wire, an electrical signal is conducted at close to the speed of light (186,000 miles per hour). Despite its speed, however, the strength of the signal deteriorates badly over long distances because it is propagated passively. If an axon relied on passive propagation, a sig-

5-1 Edgar, Lord Adrian (1889–1977) developed methods of recording action potentials, the electrical signals nerve cells use for communication. (Reprinted from *Essentials of Neural Science and Behavior*, Kandel, Schwartz, and Jessell, McGraw-Hill, 1995.)

nal from a nerve ending in the skin of your big toe would die out before it reached your brain. Helmholtz found that the axons of nerve cells conduct electricity much more slowly than wires do, and they do so by means of a novel, wavelike action that propagates actively at various speeds up to approximately 90 feet per second! Later studies showed that the electrical signals in nerves, unlike signals in wires, do not decrease in strength as they propagate. Thus, nerves sacrifice speed of conduction for active propagation, which ensures that a signal that arises in your big toe arrives at your spinal cord undiminished in size.

Helmholtz's findings raised a set of questions that would occupy physiology for the next hundred years: What do these propagated signals, later called action potentials, look like and how do they encode information? How can biological tissue generate electrical signals? Specifically, what carries the current for the signals?

THE FORM OF THE SIGNAL AND ITS ROLE IN ENCODING INFORMATION were addressed in the second phase, which began in the 1920s with Edgar Douglas Adrian's work. Adrian (figure 5-1) developed methods of recording and amplifying the action potentials propagated along the axons of individual sensory neurons on the skin, thereby making the elementary utterances of nerve cells intelligible for the first time. In the process, he made several remarkable discoveries about the action potential and how it leads to what we perceive as a sensation.

To record action potentials, Adrian used a thin piece of metal wire. He placed one end of the wire on the outside surface of the axon of a sensory neuron on the skin and then ran the wire to both an ink writer (so he could look at the shape and pattern made by the action potentials) and a loudspeaker (so he could hear them). Every time Adrian touched the skin, one or more action potentials were generated. Each time an action potential was generated, he heard a brief bang! bang! bang! over the loudspeaker and saw a brief electrical pulse on the ink writer. The action potential in the sensory neuron lasted only about $^1/_{1000}$ of a second and had two components: a swift upstroke to a peak, followed by an almost equally rapid downstroke that returned it to the starting point (figure 5-2).

The ink writer and the loudspeaker both told Adrian the same remarkable story: all of the action potentials generated by a single nerve cell are pretty much the same. They are about the same shape and amplitude, regardless of the strength, duration, or location of the stimulus that elicits them. The action potential is thus a constant, all-or-none signal: once the threshold for generating the signal is reached, it is almost always the same, never smaller or larger. The current produced by the action potential is sufficient to excite adjacent regions of the axon, thus causing the action potential to be propagated without failure or flagging along the whole length of the axon at speeds of up to 100 feet per second, pretty much as Helmholtz had earlier found!

The discovery of the all-or-none characteristic of the action potential raised more questions in Adrian's mind: How does a sensory neuron report the intensity of a stimulus—whether a touch is light or heavy, whether a light is bright or dim? How does it signal the dura-

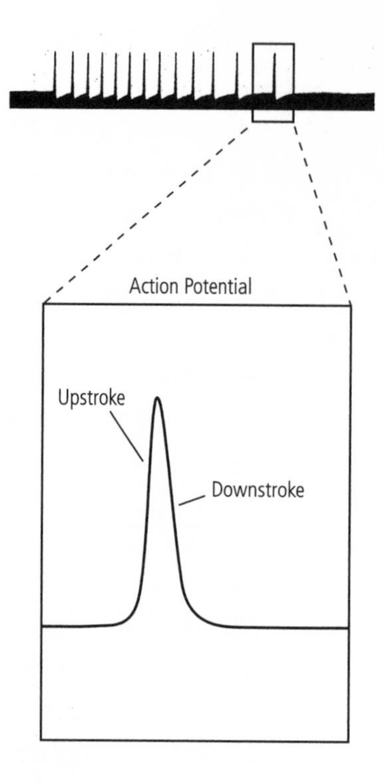

5-2 Edgar Adrian's recordings revealed the characteristics of the action potential. Recordings in single nerve cells showed that action potentials are all-or-none: once the threshold for generating an action potential is reached, the signal is always the same, both in amplitude and shape.

tion of the stimulus? More broadly, how do neurons differentiate one type of sensory information from another, such as touch from pain, light, smell, or sound? How do they differentiate sensory information for perception from motor information for action?

Adrian first addressed the question of intensity. In a landmark finding, he discovered that intensity results from the frequency with which action potentials are emitted. A mild stimulus, such as a gentle touch on the arm, will elicit just two or three action potentials per second, whereas a strong one, such as a pinch or bumping one's elbow, could fire a hundred action potentials per second. Similarly, the duration of a sensation is determined by the length of time over which the action potentials are generated.

Next, he explored how information is conveyed. Do neurons use different electrical codes to tell the brain that they are carrying information about different stimuli, such as pain or light or sound? Adrian found that they did not. There was very little difference among the action potentials produced by neurons in the various sensory systems.

Thus the nature and quality of a sensation—whether visual or tactile, for instance—does not depend upon differences in action potentials.

What, then, accounts for the differences in information carried by neurons? In a word, anatomy. In a clear confirmation of Cajal's principle of connection specificity, Adrian found that the nature of the information conveyed depends on the type of nerve fibers that are activated and the specific brain systems to which those nerve fibers are connected. Each class of sensation is transmitted along specific neural pathways, and the particular kind of information relayed by a neuron depends on the pathway of which it is a part. In a sensory pathway, information is transmitted from the first sensory neuron—a receptor that responds to an environmental stimulus such as touch, pain, or light—to specific and specialized neurons in the spinal cord or in the brain. Thus visual information is different from auditory information because it activates different pathways.

In 1928 Adrian summarized his work in his characteristically vivid style: "all impulses are very much alike, whether the message is destined to arouse the sensation of light, of touch, or of pain; if they are crowded together, the sensation is intense; if they are separated by any interval, the sensation is correspondingly feeble."

Finally, Adrian found that signals sent from motor neurons in the brain to the muscles are virtually identical to signals conveyed by sensory neurons from the skin to the brain: "the motor fibers transmit discharges which are almost an exact counterpart of those in the sensory fibers. The impulses . . . obey the same all-or-nothing principle." Thus, a rapid train of action potentials down a particular neural pathway causes a movement of our hands rather than a perception of colored lights because that pathway is connected to our fingertips, not to our retinas.

Adrian, like Sherrington, extended Cajal's neuron doctrine, which was based on anatomical observations, into the realm of function. But unlike Golgi and Cajal, who were locked in a bitter rivalry, Sherrington and Adrian were friends who lent each other support. For their discoveries regarding the function of neurons they shared the Nobel Prize in Physiology or Medicine in 1932. On hearing that he would share the prize with Sherrington, Adrian, who was a generation younger, wrote him:

> I won't repeat what you must be almost tired of hearing—how much we prize your work and yourself—but I must let you know what acute pleasure it gives me to be associated with you like this. I would not have dreamt of it, and in cold blood I would not have wished it, for your honor should be undivided, but as it is I cannot help rejoicing at my good fortune.

Adrian had listened in on the bang! bang! bang! of neuronal signaling and discovered that the frequency of these electrical impulses represents the intensity of a sensory stimulus, but several questions remained. What lies beneath the nervous system's remarkable ability to conduct electricity in this all-or-none manner? How are electrical signals turned on and off, and what mechanism is responsible for their rapid propagation along an axon?

THE THIRD PHASE IN THE HISTORY OF SIGNALING CONCERNS THE mechanisms underlying the action potential and begins with the membrane hypothesis, first proposed in 1902 by Julius Bernstein, a student of Helmholtz and one of the most creative and accomplished electrophysiologists of the nineteenth century. Bernstein wanted to know: What mechanisms give rise to these all-or-none impulses? What carries the charge for the action potential?

Bernstein understood that the axon is surrounded by the cell surface membrane and that even in the resting state, in the absence of any neural activity, there exists a steady potential, or difference in voltage, across this membrane. He knew that the difference, now called the resting membrane potential, was of great importance for nerve cells because all signaling is based on changes in this, resting potential. He determined that the difference across the membrane is about 70 millivolts, with the inside of the cell having a greater negative charge than the outside.

What accounts for this difference in voltage? Bernstein reasoned that something must carry electrical charge across the cell membrane. He knew that every cell in the body is bathed in the extracellular fluid. This fluid does not contain free electrons to carry current, as metal conductors do; instead it is rich in ions—electrically charged atoms such as sodium, potassium, and chloride. Moreover, the cytoplasm inside each cell also contains high concentrations of ions. These ions could carry current, Bernstein reasoned. He had the further insight that an imbalance in the concentration of ions inside and outside the cell could give rise to current across the membrane.

Bernstein knew from earlier studies that the extracellular fluid is salty: it contains a high concentration of sodium ions, which are positively charged, balanced by an equally high concentration of chloride ions, which are negatively charged. In contrast, the cytoplasm of the cell contains a high concentration of proteins, which are negatively charged, balanced by potassium ions, which are positively charged. Thus the positive and negative charges of ions on either side of the cell membrane are balanced, but different ions are involved.

For electrical charge to flow through the nerve cell membrane, the membrane must be permeable to some ions in the extracellular fluid or the cytoplasm. But which ions? After experimenting with various possibilities, Bernstein arrived at the bold conclusion that in the resting state, the cell membrane presents a barrier to all ions except one—potassium. The cell membrane, he argued, contains special openings, now known as ion channels; these channels allow potassium ions, and only potassium ions, to flow along a concentration gradient from the inside of the cell, where they are present in high concentrations to the outside, where they are present in low concentrations. Since potassium is a positively charged ion, its movement out of the cell leaves the inside surface of the membrane with a slight excess of negative charge resulting from the proteins inside the cell.

Even as potassium moves out of the cell, however, it is drawn back toward the inside of the cell by the net negative charge it leaves behind. Thus the outside surface of the cell membrane becomes lined with positive charges from the potassium ions that have diffused out of the cell, and the inside of the membrane becomes lined with negative charges from the proteins attempting to draw potassium ions back into the cell. This balance of ions maintains the stable membrane potential of –70 millivolts (figure 5-3).

These fundamental discoveries about how nerve cells maintain their resting membrane potential led Bernstein to ask: What happens

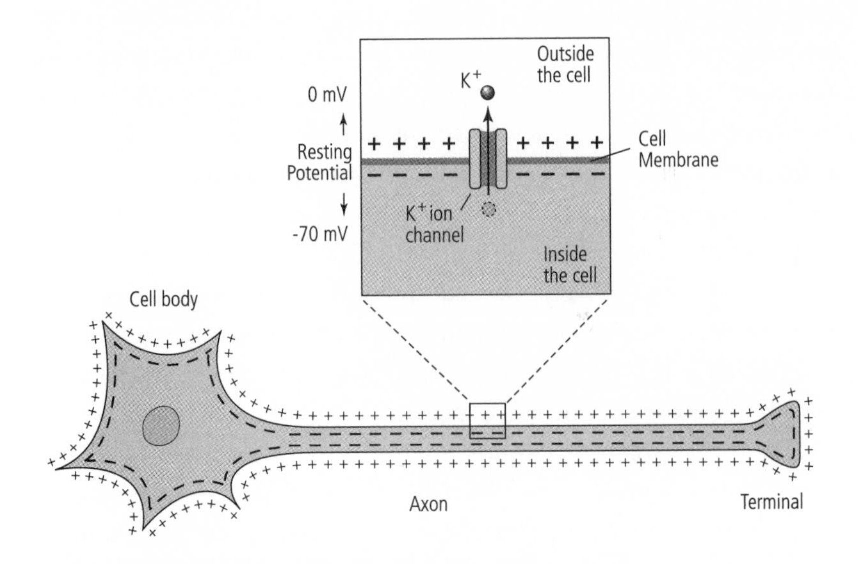

5-3 Bernstein's discovery of the resting membrane potential.
Julius Bernstein deduced that there is a difference in voltage between the inside and the outside of a nerve cell, even in its resting state. He proposed that the nerve cell membrane must have a special channel through which positively charged potassium ions (K^+) can leak out of the cell and that this loss of positive charge leaves the inside surface of the cell membrane negatively charged and creates the resting membrane potential.

when a neuron is stimulated sufficiently to generate an action potential? He applied electrical current from a battery-operated stimulator to a nerve cell axon to generate an action potential and inferred that the selective permeability of the cell membrane breaks down very briefly during the action potential, allowing all ions to enter and leave the cell freely and reducing the resting membrane potential to zero. According to this reasoning, by moving the cell membrane from its resting potential of –70 millivolts to 0 millivolts, the action potential should be 70 millivolts in amplitude.

The membrane hypothesis formulated by Bernstein proved powerful, in part because it was based on well-established principles of ions moving in solution and in part because it was so elegant. The resting potential and the action potential did not require elaborate biochemical reactions but simply utilized the energy stored in the concentration gradients of the ions. In a larger sense, Bernstein's formulation joined those of Galvani and Helmholtz in providing compelling evidence that the laws of physics and chemistry can explain even some aspects of how mind functions—the signaling of the nervous system and therefore the control of behavior. There was no need or room for "vital forces" or other phenomena that could not be explained in terms of physics and chemistry.

THE FOURTH PHASE WAS DOMINATED BY THE IONIC HYPOTHESIS and by the thinking of Alan Hodgkin, Adrian's most brilliant student, and Andrew Huxley, Hodgkin's own talented student and colleague (figure 5-4). Hodgkin and Huxley's working relationship was collaborative and synergistic. Hodgkin had piercing biological and historical insights into how nerve cells function. A fine experimentalist and a superb theoretician, he was always looking for the larger meaning beyond the immediate finding. Huxley was a technically gifted mathematical whiz. He devised new ways of recording and visualizing the

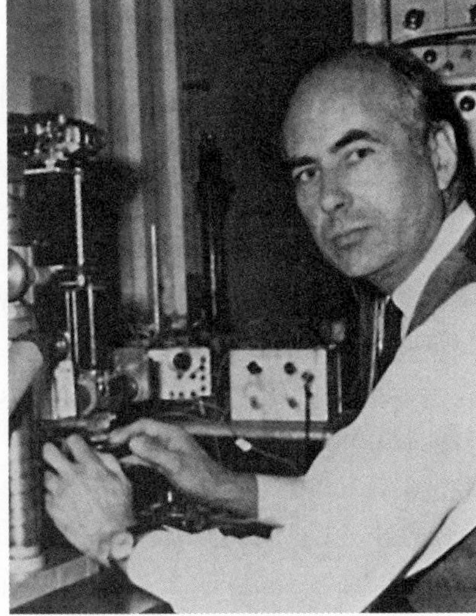

5-4 Alan Hodgkin (1914–1998) and Andrew Huxley (b. 1917) carried out a series of classic studies in the giant axon of squid nerve cells. Besides confirming Bernstein's notion that the resting membrane potential is caused by the movement of potassium ions out of the cell, they discovered that the action potential is caused by the movement of sodium ions into the cell. (Courtesy of Jonathan Hodgkin and A. Huxley.)

activity of single cells, and he developed mathematical models to describe the data he and Hodgkin obtained. They were perfect collaborators, more than the sum of their parts.

Hodgkin's enormous gifts were evident early in his career, and by the time he started his collaboration with Huxley in 1939, he had already made a major contribution to neuronal signaling. He obtained his Ph.D. in 1936 from Cambridge University in England and wrote his dissertation on the "Nature of Conduction in Nerve." There he showed, in elegant quantitative detail, that the current generated by an action potential is large enough to jump over an anesthetized segment of the axon and prompt the unanesthetized portion beyond it to generate an action potential. These experiments provided the final insights into how action potentials, once initiated, can propagate without failure or flagging. They do so, Hodgkin demonstrated, because the current generated by the action potential is substantially greater than the current required to excite a neighboring region.

The research described in Hodgkin's dissertation was so important and so beautifully executed that it immediately brought him, at the age of twenty-two, to the attention of the international scientific community. A. V. Hill, a Nobel laureate and one of England's leading physiologists, sat on Hodgkin's dissertation committee and was so impressed that he sent the dissertation to Herbert Gasser, president of the Rockefeller Institute. In his covering letter, Hill referred to Hodgkin as "very remarkable. . . . It is almost an unheard-of thing for an experimental scientist at Trinity College in Cambridge to get a fellowship in his fourth year, but this youngster has done it."

Gasser found young Hodgkin's dissertation to be a "beautiful job of experimentation" and invited him to spend 1937 as a visiting scientist at Rockefeller. During that year Hodgkin befriended Grundfest, who worked in the laboratory next door. Hodgkin also visited a number of other laboratories in the United States and in so doing learned about the squid's giant axon, which he would subsequently use to great advantage. Finally, he met the woman he would ultimately marry, the daughter of a Rockefeller Institute professor. Not a small set of accomplishments for just one year!

Hodgkin and Huxley's first great insight came in 1939, when they traveled to the marine biological station at Plymouth, England, to study how the action potential is generated in the giant axon of the squid. The British neuroanatomist J. Z. Young had recently discovered that the squid, one of the fastest swimmers in the sea, has a giant axon that is a full millimeter ($^1/_{25}$ of an inch) in diameter, making it about a thousand times wider than most axons in the human body. It is about the width of a piece of thin spaghetti and can be seen with the naked eye. Young, a comparative biologist, understood that animals evolve specializations that enable them to survive more effectively in their environments, and he realized that the squid's specialized axon, which powers its rapid escape from predators, could prove a godsend for biologists.

Hodgkin and Huxley sensed immediately that the giant axon of the squid might be just what they needed to pursue the neural scientist's dream of recording the action potential from inside the cell as well as from outside it, thereby revealing how the action potential is generated. Because that axon is so large, they could thread one electrode into the cytoplasm of the cell while keeping another on the outside. Their recordings confirmed Bernstein's inference that the resting membrane potential is about –70 millivolts and that it depends on the movement of potassium ions through ion channels. But when they stimulated the axon electrically to produce an action potential, as Bernstein had done, they discovered to their amazement that it was 110 millivolts in amplitude, not the 70 millivolts Bernstein had predicted. The action potential had increased the electrical potential at the cell membrane from –70 millivolts at rest to +40 millivolts at its peak. This astonishing discrepancy had a profound implication: Bernstein's hypothesis that the action potential represents a generalized breakdown of the cell membrane's permeability to all ions had to be incorrect. Rather, the membrane must still be acting selectively during the action potential, allowing some ions but not others to move across it.

This was an extraordinary insight. Since action potentials are the key signals for conveying information about sensations, thoughts, emotions, and memory from one region of the brain to another, the question of how the action potential is generated became, in 1939, the dominant question in all of brain science. Hodgkin and Huxley thought deeply about it, but before they could test any of their ideas, World War II intervened and both were called into military service.

It was not until 1945 that the two men could return to their research on the action potential. Working briefly with Bernard Katz of University College, London (while Huxley was preparing to get married), Hodgkin discovered that the upstroke—the rise and eventual height of the action potential—depends on the amount of sodium in the extracellular fluid. The downstroke—the decline of the action potential—is affected by the concentration of potassium. This finding suggested to them that some ion channels in the cell are selectively permeable to sodium and are open only during the upstroke of an action potential, whereas other ion channels are open only during the downstroke.

To test this idea more directly, Hodgkin, Huxley, and Katz applied a voltage clamp, a newly developed technique for measuring ion currents across the cell membrane, to the squid's giant axon. They again confirmed Bernstein's finding that the resting potential is created by the unequal distribution of potassium ions on either side of the cell membrane. Moreover, they confirmed their earlier finding that when the cell membrane is sufficiently stimulated, sodium ions move into the cell for about $^{1}/_{1000}$ of a second, changing the internal voltage from –70 millivolts to +40 millivolts and producing the rise of the action potential. The increased sodium inflow is followed almost immediately by a dramatic increase in potassium outflow, which produces the decline of the action potential and returns the voltage inside the cell to its initial value.

How does the cell membrane regulate the change in sodium and potassium ion permeabilities? Hodgkin and Huxley postulated the existence of a previously unimagined class of ion channels, channels with hinged "doors," or "gates," that open and close. They proposed that as an action potential propagates along an axon, the gates of the sodium and then the potassium channels open and close in rapid succession. Hodgkin and Huxley also realized that, because the opening and closing of the gates is very fast, the gating must be regulated by the voltage difference across the cell membrane. They therefore referred to these sodium and potassium channels as *voltage-gated chan-*

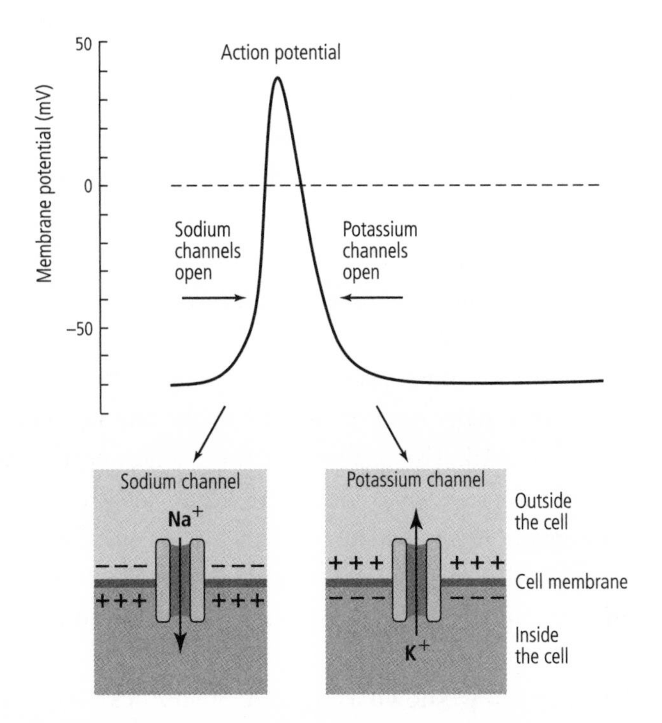

5-5 The Hodgkin-Huxley model of the intracellularly recorded action potential. The influx of positively charged sodium ions (Na^+) changes the cell's internal voltage and produces the upstroke of the action potential. Almost immediately, potassium channels open and potassium ions (K^+) flow out of the cell, producing the downstroke and returning the cell to its original voltage.

nels. In contrast, they called the potassium channels that Bernstein had discovered and that are responsible for the resting membrane potential *non-gated potassium channels*, because they have no gates and are unaffected by the voltage across the cell membrane.

When the neuron is at rest, the voltage-gated channels are shut. When a stimulus reduces the cell's resting membrane potential sufficiently, say from –70 millivolts to –55 millivolts, the voltage-gated sodium channels open and sodium ions rush into the cell, creating a brief but dramatic surge in positive charge that moves the membrane potential from –70 millivolts to +40 millivolts. In response to the same change in membrane potential, the sodium channels close after a split second's delay and the voltage-gated potassium channels open briefly, increasing the outflow of positive potassium ions and rapidly returning the cell to its resting state of –70 millivolts (figure 5-5).

Each action potential ultimately leaves the cell with more sodium on the inside and more potassium on the outside than is optimal. Hodgkin found that this imbalance is remedied by a protein that transports excess sodium ions out of the cell and potassium ions back into the cell. Eventually, the original concentration gradients of sodium and potassium are reestablished.

Once an action potential has been generated in one region of the axon, the current it generates excites the neighboring region to generate an action potential. The resulting chain reaction ensures that the action potential is propagated along the entire length of the axon, from the site where it was initiated to its terminals near another neuron (or muscle cell). In this way a signal for a visual experience, a movement, a thought, or a memory is sent from one end of the neuron to the other.

For their work, now known as the ionic hypothesis, Hodgkin and Huxley shared the Nobel Prize in Physiology or Medicine in 1963. Hodgkin later said the prize should have gone to the squid, whose giant axon made their experiments possible. But this modesty overlooks the extraordinary insights that the two men provided—insights that gave the scientific community, including new converts like me, confidence that we could understand signaling in the brain at a deeper level.

WHEN MOLECULAR BIOLOGY WAS APPLIED TO BRAIN SCIENCE, it revealed that the voltage-gated sodium and potassium channels are actually proteins. These proteins span the width of the cell membrane and contain a fluid-filled pathway, the ion pore, through which ions pass. Ion channels are present in every cell of the body, not just neurons, and they all use essentially the same mechanism Bernstein proposed to generate the resting membrane potential.

Much as the neuron doctrine had done earlier, the ionic hypothesis strengthened the link between the cell biology of the brain and other areas of cell biology. It offered the final proof that nerve cells can be understood in terms of the physical principles common to all cells. Most important, the ionic hypothesis set the stage for exploring the mechanisms of neuronal signaling on the molecular level. The generality and predictive power of the ionic hypothesis unified the cellular study of the nervous system: it did for the cell biology of neurons what the structure of DNA did for the rest of biology.

In 2003, fifty-one years after the ionic hypothesis was formulated, Roderick MacKinnon at Rockefeller University received the Nobel Prize in Chemistry for obtaining the first three-dimensional picture of the atoms that form the protein of two ion channels—a non-gated potassium channel and a voltage-gated potassium channel. Several features revealed by MacKinnon's highly original structural analysis of these two proteins had been predicted with amazing prescience by Hodgkin and Huxley.

SINCE THE MOVEMENT OF IONS THROUGH CHANNELS IN THE cell membrane is critical to the functioning of neurons, and the functioning of neurons is critical to mental functioning, it is not surprising that mutations in the genes that code for ion channel proteins should lead to disease. In 1990 it became possible to pinpoint the molecular defects responsible for human genetic diseases with relative ease. Shortly thereafter, several ion channel defects that underlie neurological disorders of muscle and of the brain were identified in rapid succession.

These disorders are now referred to as channelopathies, or disorders of ion channel function. For example, a disorder called familial idiopathic epilepsy, an inherited epilepsy of newborns, has been found to be associated with mutations in genes coding for a potassium channel. Recent progress in exploring channelopathies and the development of specific treatments for them can be attributed directly to the large body of basic scientific knowledge about ion channel function already on hand, thanks to Hodgkin and Huxley.

6

CONVERSATION BETWEEN NERVE CELLS

I arrived in Harry Grundfest's laboratory in 1955 in the wake of a major controversy about how neurons communicate with one another. Hodgkin and Huxley's epochal work had solved the long-standing mystery of how electrical signals are generated within neurons, but how does signaling occur *between* neurons? For one neuron to "talk" to the next neuron in line, it would have to send a signal across the synapse, across the gap between the cells. What kind of signal could that be?

Until they were proven wrong in the early 1950s, Grundfest and other leading neurophysiologists firmly believed that this small signal across the gap between two cells was *electrical*, the result of the flow into the postsynaptic neuron of the electrical current produced by an action potential in the presynaptic neuron. But starting in the late 1920s, evidence began to accumulate that the signal between certain nerve cells might be *chemical* in nature. This evidence came from studies of nerve cells in the autonomic, or involuntary, nervous system. The autonomic nervous system is considered part of the peripheral nervous system because the bodies of its nerve cells lie in clusters, called peripheral autonomic ganglia, which are located just outside the spinal cord and brain stem. The autonomic nervous system controls vital involuntary actions such as breathing, heart rate, blood pressure, and digestion.

The new evidence gave rise to the chemical theory of synaptic transmission and led to a controversy humorously referred to as "soup versus spark," with the "sparkers," such as Grundfest, believing that synaptic communication is electrical and the "soupers" maintaining that it is chemical.

THE CHEMICAL THEORY OF SYNAPTIC TRANSMISSION AROSE from studies by Henry Dale and Otto Loewi. In the 1920s and early 1930s, they investigated the signals sent by the autonomic nervous system to the heart and certain glands. Working independently, they discovered that when an action potential in a neuron of the autonomic nervous system reaches the terminals of the axon, it causes a chemical to be released into the synaptic cleft. That chemical, which we now call a neurotransmitter, moves across the synaptic cleft to the target cell, where it is recognized and captured by specialized receptors on the outer surface of the target cell membrane.

Loewi, a German-born physiologist living in Austria, examined the two nerves, or bundles of axons, that control heart rate: the vagus nerve, which slows it down, and the accelerans nerve, which speeds it up. In a crucial experiment on a frog, he stimulated the vagus nerve, causing it to fire action potentials that led to a slowing of the frog's heart rate. He quickly collected the fluid around the frog's heart during and just after stimulating the vagus nerve and injected that fluid into the heart of a second frog. Remarkably, the second frog's heart rate also slowed down! No action potential had been fired to slow the second frog's heart rate; instead, some substance that was released by the first frog's vagus nerve transmitted the heart-slowing signal.

Loewi and the British pharmacologist Dale went on to show that the substance released by the vagus nerve is the simple chemical acetylcholine. Acetylcholine acts as a neurotransmitter, slowing the heart by binding to a specialized receptor. The substance released by the accelerans nerve to speed up heart rate is related to adrenaline, another simple chemical. For providing the first evidence that the signals sent across synapses from one neuron to another in the autonomic nervous system are carried by specific chemical transmitters, Loewi and Dale shared the Nobel Prize in Physiology or Medicine in 1936.

Two years after winning the Nobel Prize, Loewi personally experienced the disdain that the Austrian Nazis had for science and scholarship. The day after Hitler drove into Austria to the cheers of millions of my fellow citizens, Loewi was thrown in jail because he was a Jew. A scientist who had been professor of pharmacology at the University of Graz for twenty-nine years, Loewi was released two months later on the condition that he transfer his share of the Nobel Prize, still in a bank in Sweden, to a Nazi-controlled bank in Austria and leave the country immediately. He did so, moving to the New York University Medical School, where years later I had the privilege of hearing him lecture about his discovery of chemical signaling in the heart.

Loewi and Dale's pioneering work on the autonomic nervous system convinced many neural scientists with a pharmacological bent that the cells in the central nervous system probably also use neurotransmitters to communicate across the synaptic cleft. However, some electrophysiologists, John Eccles and Harry Grundfest among them, remained skeptical. They acknowledged the importance of chemical transmission in the autonomic nervous system, but they were convinced that signaling between cells in the brain and spinal cord was simply too quick to be chemical in nature. They, therefore continued to favor the theory of electrical transmission in the central nervous system. Eccles hypothesized that the current produced by an action potential in the presynaptic neuron crosses the synaptic cleft and enters the postsynaptic cell, where it is amplified, leading to the firing of action potentials.

AS METHODS OF RECORDING ELECTRICAL SIGNALS CONTINUED to improve, a small electrical signal was discovered at the synapse between motor neurons and skeletal muscle, proving that the action potential in the presynaptic neuron does not directly initiate the action potential in the muscle cell. Rather, the presynaptic action potential prompts a much smaller, distinctive signal, known as the synaptic potential, in the muscle cell. Synaptic potentials proved to be different from action potentials in two ways: they are much slower and their amplitude can vary. Thus on a loudspeaker like Adrian's, a synaptic potential would sound like a soft, slow, protracted hissing rather than the sharp bang! bang! bang! of an action potential, and its volume would vary. The discovery of the synaptic potential proved that nerve cells use two different kinds of electrical signals. They use the action potential for long-range signaling, to carry information from one region of the nerve cell to another, and they use the synaptic potential for local signaling, to convey information across the synapse.

Eccles immediately recognized that synaptic potentials are responsible for Sherrington's "integrative action of the nervous system." At any given moment, a cell in any neural pathway is bombarded by many synaptic signals, both excitatory and inhibitory, but it has only two options: to fire or not to fire an action potential. Indeed, the fundamental task of a nerve cell is integration: it sums up the excitatory and inhibitory synaptic potentials it receives from presynaptic neurons and generates an action potential only when the total of excitatory signals exceeds the total of inhibitory signals by a certain critical minimum. Eccles saw that it is the capability of nerve cells to integrate all the excitatory and inhibitory synaptic potentials from the nerve cells that converge on it that assures the singleness of action in behavior that Sherrington described.

By the mid-1940s both sides of the debate agreed that a synaptic potential occurs in all postsynaptic cells and that it is the critical link between the action potential in the presynaptic neuron and the action potential in the postsynaptic cell. Yet that discovery just focused the argument more sharply: Was the synaptic potential in the central nervous system initiated electrically or chemically?

Dale and his colleague William Feldberg, another émigré from Germany, provided a key breakthrough when they discovered that acetylcholine, used in the autonomic nervous system to slow the heart, is also released by motor neurons in the spinal cord to excite skeletal muscles. This finding piqued the interest of Bernard Katz in exploring whether acetylcholine was responsible for the synaptic potential in skeletal muscle.

A prize-winning medical student at the University of Leipzig, Katz had fled Hitler's Germany in 1935 because he was Jewish. He went to

England and joined A. V. Hill's laboratory at University College, London. Katz arrived at the English port of Harwich that February without a passport, which was, he remembered, "a terrifying experience." Three months after his arrival, Katz attended a meeting in Cambridge where he had a ringside seat at a soup-versus-spark dustup. "To my great astonishment," he would later write, "I witnessed what seemed almost a stand-up fight between J. C. Eccles and H. H. Dale, with the chairman, [Lord] Adrian, acting as a most uncomfortable and reluctant referee." John Eccles, the leader of the sparkers, had presented a paper vigorously disputing a central claim of Henry Dale, the leader of the soupers, and his colleagues: that acetylcholine acts as a transmitter of signals at synapses in the nervous system. "I had some difficulty in following the argument, as I was not fully acquainted with the terminology," Katz recalled. "The word *transmitter* conveyed to me something to do with radio communications, and as this did not make sense, the matter was a bit confusing."

Indeed, Katz's confusion aside, one of the problems with chemical transmission was that no one knew how an electrical signal in the presynaptic terminal could cause the release of a chemical transmitter and how that chemical signal could then be converted into an electrical signal in the postsynaptic neuron. Over the next two decades Katz joined the effort to address these two questions and to extend the work of Dale and Loewi from the autonomic nervous system to the central nervous system.

However, as with Hodgkin and Huxley, the threat of war interrupted Katz's work. In August 1939, one month before World War II broke out, Katz, feeling uncomfortable as a German alien in London, accepted an invitation from John Eccles to join him in Sydney, Australia.

As it happened, Stephen Kuffler, another scientist who left Europe to escape the Nazis and who also greatly influenced my thinking, also ended up in Sydney and joined Eccles's laboratory (figure 6-1). A Hungarian-born, Viennese-trained physician-turned-physiologist, Kuffler left Vienna in 1938 because, in addition to having one Jewish grandfather, he was a socialist. Kuffler, a junior tennis champion in Austria, later joked that the real reason Eccles had asked him to join the laboratory was that he needed a skilled tennis partner.

6-1 Three pioneers of synaptic transmission worked together in Australia during World War II and then went on to make major contributions individually. Stephen Kuffler (left, 1918–1980) characterized the properties of the dendrites of crayfish, John Eccles (middle, 1903–1997) discovered synaptic inhibition in the spinal cord, and Bernard Katz (right, 1911–2002) uncovered the mechanisms of synaptic excitation and of chemical transmission. (Courtesy of Damien Kuffler.)

Although Eccles and Katz were vastly more experienced scientists, Kuffler amazed them with his surgical skills. He could dissect out individual muscle fibers to study the synaptic input from one motor axon to one muscle fiber, a real tour de force.

KATZ, KUFFLER, AND ECCLES SPENT THE WAR YEARS TOGETHER arguing about chemical versus electrical transmission between nerve cells and muscle. Eccles attempted to reconcile the evidence of chemical transmission, which he insisted must be a slow process, with the rapidity of nerve-muscle signaling. He hypothesized that the synaptic potential has two components: an initial, rapid component mediated by an electrical signal, and a prolonged, residual action mediated by a chemical transmitter such as acetylcholine. Katz and Kuffler became newly minted soupers when they discovered evidence that the chemical acetylcholine is responsible even for the initial component of the synaptic potential in muscle. In 1944, as World War II was nearing its end, Katz returned to England and Kuffler immigrated to the United States. In 1945 Eccles accepted a major professorship and moved to the University of Dunedin, in New Zealand, to set up a new laboratory.

As experiments cast more and more doubt on the electrical theory of synaptic transmission, Eccles, a large, athletic, and normally energetic and enthusiastic man, became dispirited. After we became friends in the late 1960s, he recalled how, in this despondent state, he underwent a great intellectual transformation for which he was forever grateful. That transformation occurred in the faculty club of the university, where Eccles went regularly for a break after the day's work. On one such occasion in 1946, he met Karl Popper, the Viennese philosopher of science who had immigrated to New Zealand in 1937 anticipating that Hitler might annex Austria. In the course of their conversation, Eccles told Popper about the chemical-electrical transmission controversy and how he appeared to be on the losing side of a long and, for him, fundamental argument.

Popper was fascinated. He assured Eccles that there was no reason for despair. On the contrary, he urged him to be jubilant. No one was challenging Eccles's research findings—the challenge was to his theory, his interpretation of the research findings. Eccles was doing science at its best. It is only when the facts become clear and competing interpretations of them can be brought into sharp focus that opposing hypotheses can clash. And only when sharply focused ideas clash can one of them be found wrong. Being on the wrong side of an interpretation was unimportant, Popper argued. The greatest strength of the scientific method is its ability to disprove a hypothesis. Science proceeds by endless and ever refining cycles of conjecture and refutation.

One scientist proposes a new idea about nature and then other scientists work to find observations that support or refute this idea.

Eccles had every reason to be pleased, Popper argued. He urged Eccles to go back to the laboratory and refine his ideas and his experimental attack on electrical transmission even further so that he could, if necessary, actually disprove the idea of electrical transmission himself. Eccles later wrote about this encounter:

> I learned from Popper what for me is the essence of scientific investigation—how to be speculative and imaginative in the creation of hypotheses, and then to challenge them with the utmost rigor, both by utilizing all existing knowledge and by mounting the most searching experimental attacks. In fact I learned from him even to rejoice in the refutation of a cherished hypothesis, because that, too, is a scientific achievement and because much has been learned by the refutation.
>
> Through my association with Popper I experienced a great liberation in escaping from the rigid conventions that are generally held with respect to scientific research. . . . When one is liberated from these restrictive dogmas, scientific investigation becomes an exciting adventure opening up new visions; and this attitude has, I think, been reflected in my own scientific life since that time.

ECCLES DID NOT HAVE TO WAIT LONG FOR HIS HYPOTHESIS TO be proved false. When Katz returned to University College, London, he provided direct evidence that the acetylcholine released by the motor neuron gives rise to and fully accounts for all phases of the synaptic potential. Acetylcholine does this by diffusing rapidly across the synaptic cleft and binding quickly to receptors on the muscle cell. Later, the acetylcholine receptor was shown to be a protein with two major components: an acetylcholine-binding component and an ion channel. When acetylcholine is recognized by and bound to the receptor, it causes the ion channel to open.

Katz went on to show that the novel ion channels gated by a chemical transmitter differ from voltage-gated sodium and potassium channels in two ways: they respond only to specific chemical transmitters, and they allow *both* sodium and potassium ions to flow through. The simultaneous passage of sodium and potassium ions changes the muscle cell's resting membrane potential from –70 millivolts to nearly zero. Moreover, even though the synaptic potential is produced by a chemical, it is rapid, as Dale had predicted. When sufficiently large, it produces an action potential that causes the muscle fiber to contract (figure 6-2).

Together, the work of Hodgkin, Huxley, and Katz showed that there are two fundamentally different types of ion channels. Voltage-gated channels generate action potentials that carry information *within* neurons, while chemical transmitter–gated channels transmit information *between* neurons (or between neurons and muscle cells) by generating synaptic potentials in postsynaptic cells. Thus Katz discovered that by producing the synaptic potential, transmitter-gated ion channels in effect translate chemical signals from motor neurons into electrical signals in muscle cells.

Just as there are diseases of voltage-gated ion channels, so there are diseases of transmitter-gated channels. For instance, myasthenia gravis, a serious autoimmune disease that occurs primarily in men, produces antibodies that destroy the acetylcholine receptors in muscle cells and thus weaken muscle action. Muscular weakness can become so severe that patients cannot keep their eyes open.

SYNAPTIC TRANSMISSION IN THE SPINAL CORD AND BRAIN IS decidedly more complex than signaling between motor neurons and muscle. Eccles had spent the years 1925 to 1935 working directly with Sherrington on the spinal cord. He returned to those studies full-time in 1945, and by 1951 had obtained intracellular recordings from motor neurons. Eccles confirmed Sherrington's finding that motor neurons receive both excitatory and inhibitory signals and that these signals are produced by distinctive neurotransmitters acting on distinctive receptors. In the motor neuron, excitatory neurotransmitters released by the presynaptic neurons lower the resting membrane potential of the postsynaptic cell from –70 millivolts to –55 millivolts, the threshold for firing an action potential, while inhibitory neurotransmitters increase the membrane potential from –70 millivolts to –75 millivolts, making it much more difficult for the cell to fire an action potential.

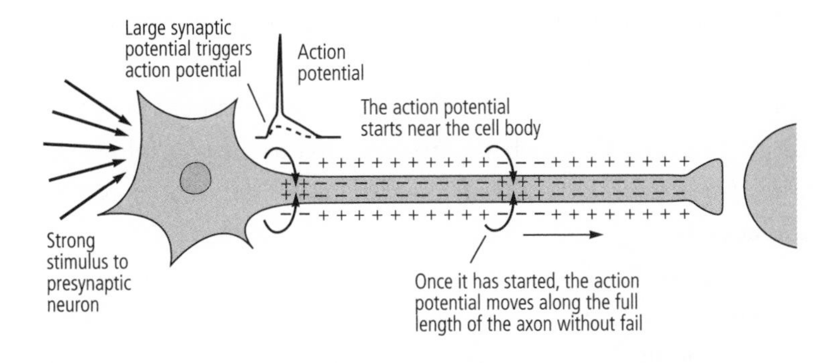

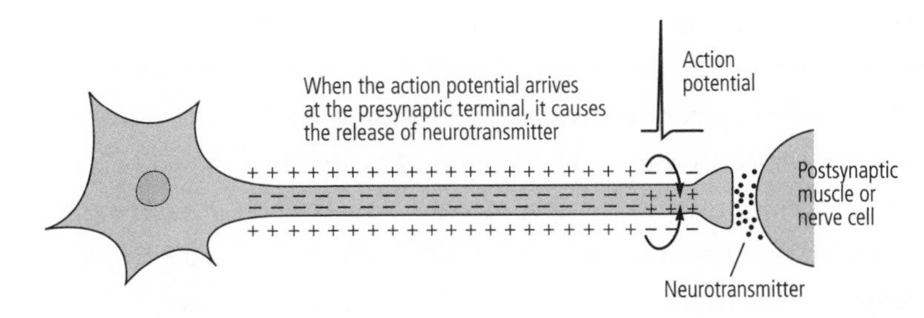

6-2 The propagated action potential.

We now know that the major excitatory neurotransmitter in the brain is the amino acid glutamate, while the main inhibitory transmitter is the amino acid GABA (gamma-aminobutyric acid). A variety of tranquilizing drugs—benzodiazepines, barbiturates, alcohol, and general anesthetics—bind to GABA receptors and produce a calming effect on behavior by enhancing the receptors' inhibitory function.

Eccles thus confirmed Katz's finding that excitatory synaptic transmission is chemically mediated, and he showed that inhibitory synaptic transmission is also chemically mediated. When he described these findings in later years, Eccles wrote, "I had been encouraged by Karl Popper to make my hypothesis as precise as possible, so that it would call for experimental attack and falsification. It turned out that it was I who was to succeed in this falsification." Eccles celebrated his discoveries by abandoning the electrical hypothesis he had so vigorously championed and converting wholeheartedly to the chemical hypothesis, arguing with equal enthusiasm and vigor for its universality.

It was at this point, in October 1954, that Paul Fatt, one of Katz's outstanding collaborators, wrote a masterly review of synaptic transmission. Fatt took a farsighted view, pointing out that it was premature to conclude that all synaptic transmission is chemical. He concluded, "Although there is every indication that chemical transmission occurs across those junctions . . . which are most familiar to the physiologist, *it is probable that electrical transmission occurs at certain other junctions* [emphasis added]."

Three years later, Fatt's prediction was convincingly demonstrated by Edwin Furshpan and David Potter, two postdoctoral fellows in Katz's laboratory who found an actual case of electrical transmission between two cells of the nervous system in the crayfish. Thus, as sometimes happens in scientific controversies, both sides of the argument had merit. We now know that most synapses, including those under scrutiny at the time of the controversy, are chemical in nature. But some neurons form electrical synapses with other nerve cells. At such synapses there are small bridges between the two cells that allow electrical current to pass from one cell to the other, very much as Golgi had predicted.

The existence of two forms of synaptic transmission raised questions in my mind that would reemerge in my thinking later. Why do chemical synapses predominate in the brain? Do chemical and electrical transmission have different roles in behavior?

IN THE FINAL PHASE OF AN EXTRAORDINARY CAREER, KATZ turned his attention from the synaptic potential in the target cell to the release of neurotransmitters from the signaling cell. He wanted to know how an electrical event in the presynaptic terminal, the action potential, leads to the release of a chemical transmitter. Here, he made two more remarkable discoveries. First, as an action potential propagates along the axon into the presynaptic terminal, it leads to the opening of voltage-gated channels that admit calcium ions. The influx of calcium ions into the presynaptic terminals sets off a series of molecular steps that lead to the release of the neurotransmitter. Thus, in the signaling cell, voltage-gated calcium channels opened by the action potential start the process of translating an electrical signal into a chemical signal, just as in the receiving cell, transmitter-gated channels translate chemical signals back into electrical signals.

Second, Katz discovered that transmitters such as acetylcholine are

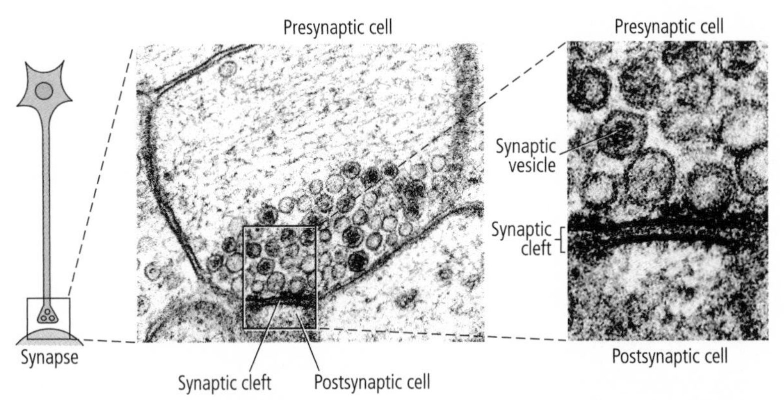

6-3 How signals travel from cell to cell. The first images of a synapse showed that the presynaptic terminal contains synaptic vesicles, which were later found to enclose about 5,000 molecules of neurotransmitter. These vesicles cluster near the membrane of the presynaptic terminal, where they prepare to release the transmitter into the space between the two cells, the synaptic cleft. After crossing the synaptic cleft, the neurotransmitters bind to receptors on dendrites of the postsynaptic cell. (Reprinted from *Cell*, vol. 10, 1993, page 2, Jessell and Kandel. Used with permission from Elsevier. Center image courtesy of C. Bailey and M. Chen.)

not released from the axon terminal as single molecules but in small, discrete packets containing about five thousand molecules each. Katz called these packets *quanta* and postulated that each one is packaged in a membrane-bound sac that he called the synaptic vesicle. In 1955, images of the synapse taken by Sanford Palay and George Palade with an electron microscope confirmed Katz's prediction, showing that the presynaptic terminal is packed with vesicles that were later shown to contain neurotransmitters (figure 6-3).

To test this idea further, Katz made a brilliant strategic decision. He shifted from studying the nerve-muscle synapse of the frog to the giant synapse of the squid. Using this advantageous system, Katz was able to infer what calcium ions do when they flow into the presynaptic terminal: they cause the synaptic vesicles to fuse with the surface membrane of the presynaptic terminal and open a pore in the membrane through which the vesicles release their transmitter into the synaptic cleft (figure 6-4).

THE REALIZATION THAT THE WORKINGS OF THE BRAIN—THE ability not only to perceive, but to think, learn, and store informa-

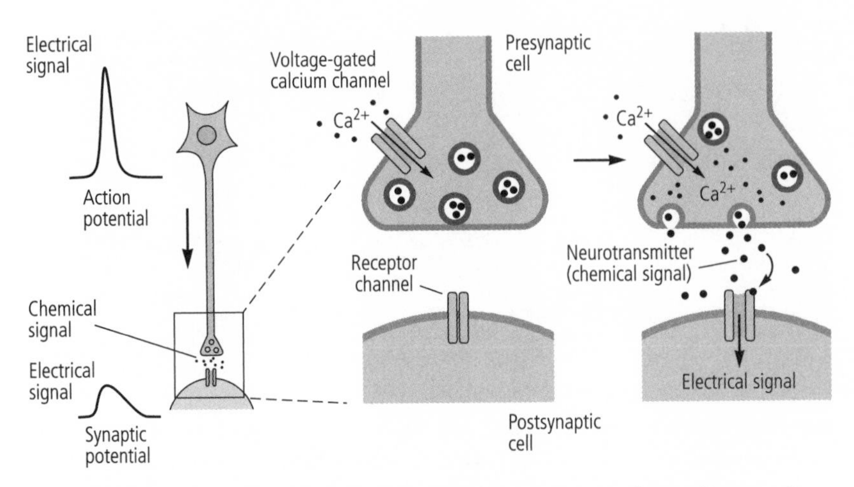

6-4 From electrical to chemical signals and back again. Bernard Katz discovered that when an action potential enters a presynaptic terminal, it causes calcium channels to open, letting calcium ions flow into the cell. This leads to the release of neurotransmitters into the synaptic cleft. The neurotransmitter binds to receptors on the surface of the postsynaptic cell, and the chemical signals are reconverted into electrical signals.

tion—may occur through chemical as well as electrical signals expanded the appeal of brain science from anatomists and electrophysiologists to biochemists. In addition, since biochemistry is a universal language of biology, synaptic transmission piqued the interest of the biological science community as a whole, not to mention students of behavior and mind, like me.

How fortunate for brain science throughout the world that England, Australia, New Zealand, and the United States opened their doors to the remarkable scholars of the synapse cast out by Austria and Germany, including Loewi, Feldberg, Kuffler, and Katz. I am reminded of a story told about Sigmund Freud when he arrived in England and was shown the beautiful house on the outskirts of London that he was to live in. On seeing the tranquility and civility that his forced emigration had brought him to, he was moved to whisper with typical Viennese irony, "Heil Hitler!"

7

SIMPLE AND COMPLEX NEURONAL SYSTEMS

Soon after I arrived at Columbia in 1955, Grundfest suggested that I work alongside Dominick Purpura, a young physician whom he had encouraged to change careers from neurosurgery to basic research on the brain (figure 7-1). When I met Dom, he had just made the decision to focus his research on the cerebral cortex, the most highly developed region of the brain. Dom was interested in mind-altering drugs, and the first experiments I helped him with concerned the role of the psychedelic agent LSD (lysergic acid diethylamide) in producing visual hallucinations.

LSD was discovered in the 1940s. By the mid-1950s it had become extremely well-known because of its widespread recreational use. Aldous Huxley had publicized its mind-altering properties in his book *The Doors of Perception*, in which he described how LSD enhanced his own awareness of visual experiences, giving rise to powerful, brightly colored images and a greater sense of clarity. The ability of LSD and related psychedelic drugs to alter perception, thought, and feeling in ways that are not ordinarily experienced except in dreams and exalted religious states makes them markedly different from other classes of drugs. People taking LSD often have the sense that their mind has expanded and split in two: one part is organized, experiencing the

7-1 Dominick Purpura (b. 1927) trained as a neurosurgeon but switched to full-time research and became a major contributor to the physiology of the cortex. I worked with him in 1955–56 during my first stay in the Grundfest laboratory. He later became an academic leader at Stanford University and then at the Albert Einstein School of Medicine. (From Eric Kandel's personal collection.)

enhanced perceptual effects; the other part is passive, observing the events as a neutral outsider. Attention is typically turned inward, and the clear distinction between self and non-self is lost, giving the user of LSD a mystical sense of being part of the cosmos. In many people the perceptual distortions take the form of visual hallucinations; in some people LSD can even cause a psychotic reaction resembling schizophrenia. Because of these remarkable properties, Dom wanted to know how LSD worked.

A year earlier D. W. Woolley and E. N. Shaw, two pharmacologists at the Rockefeller Institute, had found that LSD binds to the same receptor as serotonin, a substance that had recently been discovered in the brain and was thought to be a neurotransmitter. For their studies they used a preparation favored by experimental pharmacologists, the smooth muscle of the rat uterus, which they found would undergo spontaneous contractions in response to serotonin. LSD counteracted this effect of serotonin, and it did so by displacing serotonin from its receptor. This led Woolley and Shaw to suggest that LSD might counteract serotonin in the brain. They further suggested that since LSD can cause psychotic reactions, it might do so by preventing the normal action of serotonin in the brain. If that were so, they argued, serotonin might well be required for our sanity—for normal mental functioning.

Although Dom had no problem with the idea of using smooth uterine muscle to test ideas about chemicals in the brain, he thought a more relevant test about brain functioning in mental health and illness would be to look at the brain directly to see how psychedelic drugs act. Specifically, he wanted to know whether LSD affects synaptic activity in the visual cortex, the area of the cortex concerned with visual perception, where presumably the dramatic visual distortions and hallucinations occur. He asked me to help him explore the action of serotonin on a neural pathway in cats that ends in the visual cortex.

We anesthetized the animals, opened their skulls to expose the brain, and placed electrodes on the surface of the visual cortex. We found that in the visual cortex, serotonin and LSD did not act in opposition to each other, as they did in the smooth muscle of the uterus. Not only did both have the same action, inhibiting synaptic signaling, but each enhanced the other's inhibitory activity. Thus our studies, and subsequent studies from other laboratories, seemed to disprove Woolley and Shaw's notion that the disorienting visual effects of LSD were due to the drug's blocking the action of serotonin in the visual system. (We now know that serotonin acts on as many as eighteen different types of receptors throughout the brain and that LSD seems to produce its hallucinatory action by stimulating one of these receptors, located in the frontal lobe of the brain.)

This was quite a nice result. In the course of these studies I learned from Dom how to set up experiments with cats and how to operate electrical recording and stimulating equipment. To my surprise, I found my first laboratory experiences to be absorbing, quite unlike the rather dry science I had been taught in college and medical school classrooms. In the laboratory, science is a means for formulating interesting questions about nature, discussing whether those questions are important and well formulated, and then designing a series of experiments to explore possible answers to a particular question.

The questions Grundfest and Purpura were asking were not immediately related to the ego, superego, or id, but they made me realize that neural science was beginning to be able to test ideas about aspects of major mental illnesses, such as the perceptual distortions and hallucinations of schizophrenia.

More important, I found discussions with Grundfest and Purpura fascinating—they were penetrating and sometimes marvelously gossipy about other scientists' work, their careers, their sex lives. Dom was extremely bright, technically strong, and highly entertaining (I later called him the Woody Allen of neurobiology). I began to realize that what makes science so distinctive, particularly in an American laboratory, is not just the experiments themselves, but also the social context, the sense of equality between student and teacher, and the open, ongoing, and brutally frank exchange of ideas and criticism. Grundfest and Purpura admired each other and were involved together in the design of the experiment, but Grundfest would criticize Dom's data as if he were a rival from another laboratory. Grundfest was at least as demanding about the experiments from his and Dom's laboratory as he was about other people's experiments.

In addition to learning about the important new ideas emerging from biological studies of the brain, I learned methodology and strategy from Grundfest and Purpura, and later from Stanley Crain, a young colleague of Grundfest's. In a larger sense, much as the painful memories of my youth in Vienna in 1938 were to obsess me in later years, these early positive research experiences and the ideas to which I was exposed when I was twenty-five years old had a major impact on my thinking and my life's work.

The findings regarding serotonin and LSD encouraged Dom to carry his analysis to the edge of what was technically possible in the mammalian cortex. We had used flashes of light to activate the visual cortex. Those stimuli activated a pathway that ended on the dendrites of neurons in the visual cortex. Very little was known about dendrites. In particular, it was not known whether they could generate action potentials like those in the axon. Based on their studies, Purpura and Grundfest proposed that dendrites have limited electrical properties: they can produce synaptic potentials, but they cannot generate action potentials.

In reaching this conclusion Grundfest and Purpura were tentative, however, because they were uncertain that the experimental methods they were using were adequate to the task of studying dendrites. To detect changes in synaptic transmission produced by LSD, Grundfest and Purpura ideally needed to obtain intracellular recordings from the dendrites of the neurons in the visual cortex one dendrite at a time. This required using small glass electrodes of the sort used by Katz in single muscle fibers and by Eccles in single motor neurons. After some discussion they concluded that intracellular recordings were unlikely to succeed, because the neurons in the visual cortex were much smaller than the cells studied by Katz and Eccles. The slender dendrites, which are only one-twentieth the size of the cell body, seemed impossibly difficult recording targets.

IT WAS IN THE CONTEXT OF THESE DISCUSSIONS THAT I ONCE again encountered Stephen Kuffler. One evening Grundfest threw into my lap an issue of the *Journal of General Physiology* that contained three papers by Kuffler based on his work with single nerve cells and their dendrites in the crayfish. I found the idea of a contemporary neurophysiologist working on crayfish simply remarkable: one of Freud's first scientific papers, published in 1882, when he was only twenty-six, was on the nerve cells of the crayfish! It was in the course of this study that Freud almost discovered, independently of Cajal, that the nerve cell body and all of its processes are a single unit, the signaling unit of the brain.

I read Kuffler's papers as best I could. Even though I did not understand them fully, one thing popped out immediately: Kuffler was doing what Purpura and Grundfest aspired to do but could not achieve in the mammalian brain. He was studying the dendrites of a single, isolated nerve cell. Here, without any other nerve cells present, Kuffler could actually see the individual branches of the dendrites and could record the conseqences of electrical changes in them.

Kuffler's papers drove home the point that selecting an anatomically simple system is crucial to the success of an experiment and that invertebrate animals are a rich source of simple systems. The papers also reminded me that the choice of an experimental system is one of the most important decisions a biologist makes, a lesson I had learned earlier from Hodgkin and Huxley's work on the squid's giant axon and Katz's work on the squid's giant synapse.

These insights had a great impact on me, and I was eager to test the new research strategies directly for myself. I had no specific idea in mind, but I was beginning to think like a biologist. I appreciated that all animals have some form of mental life that reflects the architecture of their nervous system, and I knew I wanted to study nervous system function at the cellular level. All I knew at this point is that someday I might want to test an idea with an invertebrate animal.

AFTER GRADUATING FROM MEDICAL SCHOOL IN 1956, I SPENT A year as a medical intern at Montefiore Hospital in New York City. In the spring of 1957, during a brief elective period in my internship, I returned to Grundfest's lab and spent six weeks with Stanley Crain, a master of simple systems. I sought out Crain because he was a cell biologist who had searched for appropriate experimental systems to solve important problems. He was one of the first to study the properties of single, isolated nerve cells removed from the brain and grown in tissue culture apart from all other cells. It doesn't get much simpler than that!

Knowing of my growing curiosity about invertebrate animals, particularly about the crayfish, Grundfest suggested that I set up an electrophysical recording system with Crain's help. I could use the system to replicate Hodgkin and Huxley's experiment by recording from the large axon of the crayfish, which controls the animal's tail and thus its escape from predators. This crayfish axon is smaller than that of the squid but nonetheless very large.

Crain showed me how to manufacture glass microelectrodes for insertion into individual axons and how to obtain and interpret electrical recordings from them. It was in the course of those experiments—which were almost laboratory exercises, since I was not exploring new ground scientifically or conceptually—that I first began to feel the excitement of working on my own. I connected the output from the amplifier I was using to record the electrical signal to a loudspeaker, as Adrian had done thirty years earlier. Whenever I penetrated a cell, I, too, could hear the crack of an action potential. I am not fond of the sound of gunshots, but I found the bang! bang! bang! of action potentials intoxicating. The idea that I had successfully impaled an axon and was actually listening in on the brain of the crayfish as it conveyed messages seemed marvelously intimate. I was becoming a true psychoanalyst: I was listening to the deep, hidden thoughts of my crayfish!

The beautifully straightforward results I obtained from early experiments on the simple nervous system of the crayfish—measurements of the resting membrane and action potentials, confirmation that the action potential is all-or-none and that it does not simply nullify the resting membrane potential but overshoots it—made a profound impression on me and confirmed the importance of selecting just the right animal for my studies. My results were completely unoriginal, but to me they were wonderful.

Based on my two brief periods in his laboratory, Grundfest offered to nominate me for a research position at the National Institute of Mental Health (NIMH), the psychiatric component of NIH, as an alternative to being drafted into the armed forces. During the years following the Korean War, physicians were drafted to provide medical care for members of the armed services and their families. The Public Health Service, then part of the Coast Guard, was an alternative form of active duty for those who were deemed eligible, and NIH was one of the installations that belonged to the Public Health Service. With Grundfest's recommendation, I was accepted by Wade Marshall, chief of the Laboratory of Neurophysiology at NIMH, and was slated to arrive in July 1957.

IN THE LATE 1930s WADE MARSHALL WAS PROBABLY THE MOST promising and accomplished young scientist working on the brain in the United States (figure 7-2). In a now classic series of studies he asked: How are the touch receptors of the body surface—the hands, face, chest, back—represented in the brain of cats and monkeys? Marshall and his colleagues discovered that the internal representation of touch is organized spatially: neighboring areas on the body surface are preserved in the brain.

By the time Marshall began his research, a great deal was already known about the anatomy of the cerebral cortex. The cortex is a convoluted structure that covers the two symmetrical hemispheres of the forebrain and is divided into four parts, or lobes (frontal, parietal, temporal, and occipital) (figure 7-3). Unfolded, the human cerebral cortex is about the size of a large cloth dinner napkin, only somewhat thicker. It

7-2 Wade Marshall (1907–1972) was the first scientist to map the detailed sensory representation of touch and vision in the cerebral cortex. He moved to the NIH in 1947 and became head of the Laboratory of Neurophysiology at NIMH in 1950, where I worked for him from 1957 to 1960. (Courtesy of Louise Marshall.)

contains about 100 billion neurons, each with about a thousand synapses, making a total of about 1 quadrillion synaptic connections.

Marshall began his studies of touch sensation as a graduate student at the University of Chicago in 1936. He discovered that moving the hairs on a cat's leg or touching its skin produces an electrical response in specific groups of neurons in the somatosensory cortex, a region of the parietal lobe that governs the sense of touch. These studies showed only that the sense of touch is represented in the brain, but Marshall immediately realized that he could advance his analysis much further. He wanted to know whether neighboring areas of the skin are represented in neighboring areas of the somatosensory cortex or scattered at random across it.

For guidance in answering that question, Marshall sought postdoctoral training under Philip Bard, chairman of the department of physi-

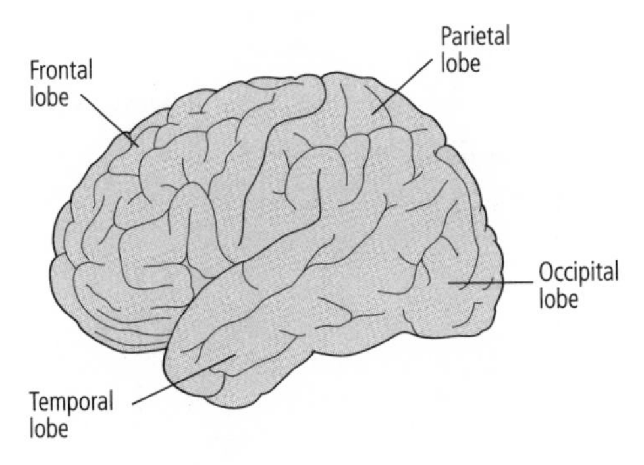

7-3 The four lobes of the cerebral cortex. The frontal lobe is part of the neural circuit governing social judgments, planning and organization of activities, aspects of language, control of movement, and a form of short-term memory called working memory. The parietal lobe receives sensory information about touch, pressure, and space around the body and helps integrate that information into coherent perceptions. The occipital lobe is involved with vision. The temporal lobe is involved with auditory processing and aspects of language and memory.

ology at the Johns Hopkins Medical School and a major figure in American biology. Marshall joined Bard in carrying out studies on monkeys in which they discovered that the entire body surface is represented in the somatosensory cortex in the form of a point-for-point neural map. Parts of the body surface that are adjacent to one another, such as the fingers, are represented near each other in the somatosensory cortex. A few years later a remarkably gifted Canadian neurosurgeon, Wilder Penfield, extended the study from monkeys to people, revealing that the parts of the body surface most sensitive to touch are represented by the largest areas of the somatosensory cortex (figure 7-4).

Next, Marshall found that the light receptors in the retina of the eye are also represented in an orderly way in the primary visual cortex, a region of the occipital lobe. Finally, Marshall showed that the temporal lobe has a sensory map for sound frequencies, with different pitches represented systematically in the brain.

These studies revolutionized our understanding of how sensory

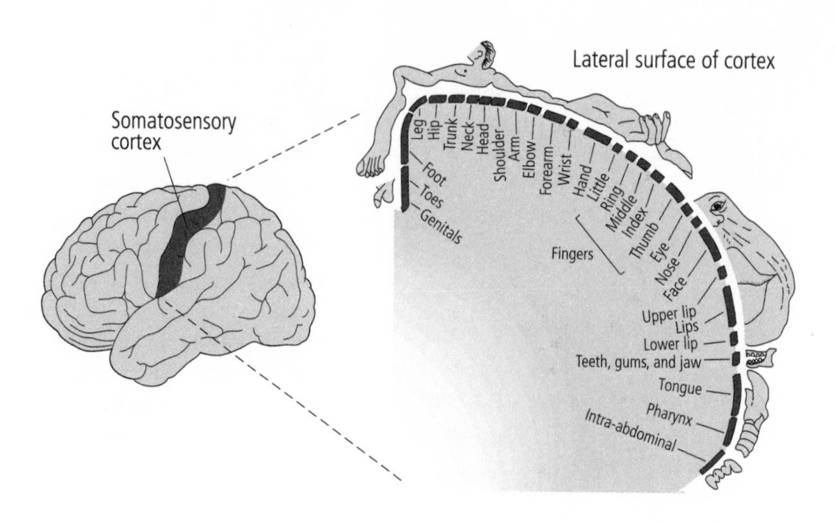

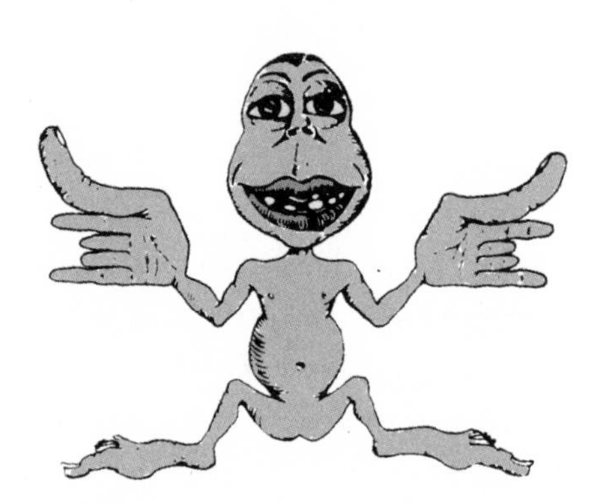

7-4 A sensory map of the body, as represented in the brain. The somatosensory cortex—a strip in the parietal lobe of the cerebral cortex—receives sensations of touch. Each part of the body is represented separately. Fingers, mouth, and other particularly sensitive areas take up most space. Wilder Penfield called this cross-sectional map a sensory homunculus. A composite representation of the sensory homunculus as a figurine drawing (below) is based on the cross-sectional map. It shows a human with large hands, fingers, and mouth. (From *Mechanics of the Mind*, Colin Blakemore. © Cambridge University Press, 1977.)

information is organized and represented in the brain. Marshall showed that, even though the different sensory systems carry different types of information and end up in different regions of the cerebral cortex, they share a common logic in their organization: all sensory information is organized topographically in the brain in the form of precise maps of the body's sensory receptors, such as, the retina of the eye, the basilar membrane in the ear, or the skin on the body surface.

These sensory maps are most easily understood by the representation of touch in the somatosensory cortex. Touch begins with receptors in the skin that translate the energy of a stimulus—for example, the energy transmitted by a pinch—into electrical signals in sensory neurons. The signals then travel along precise pathways to the brain, passing through several processing, or relay, stages in the brain stem and thalamus before terminating in the somatosensory cortex. At each stage the signals traveling from adjacent points on the skin are carried by nerve fibers that run alongside each other. In this way stimulation of two adjacent fingers, for instance, activates adjacent populations of nerve cells in the brain.

Knowledge of the brain's sensory maps and an understanding of how they are organized topographically is exceedingly helpful in treating patients. Because these maps are incredibly precise, clinical neurology has long been an accurate diagnostic discipline, even though it has relied, until the recent development of brain imaging, only on the simplest, most primitive tools—a wad of cotton to test for touch, a safety pin to test for pain, a tuning fork to test for vibration, and a hammer to test reflex actions. Disturbances in the sensory and motor systems can be located with remarkable accuracy because of the one-to-one relationship between sites on the body and areas of the brain.

A dramatic example of this relationship is the Jacksonian sensory march, which characterizes a certain type of epileptic seizure first described by British neurologist John Hughlings Jackson in 1878. In such a seizure, numbness and sensations such as burning or prickling begin in one place and spread throughout the body. For example, numbness might begin at the fingertips and spread over the next minute to the hand, up the arm, across the shoulder, into the back, and down the leg on the same side. This sequence of sensations is explained by the arrangement of the sensory map of the body: the seizure, which is a wave of abnormal electrical activity in the brain, starts in the lateral area of the somatosensory cortex, where the hand is represented, and propagates across the cortex toward the midline, where the leg is represented.

Marshall's marvelous scientific achievements came at a price, however. The experiments were physically demanding, often lasting more than twenty-four hours at a time. Frequently deprived of sleep, he became exhausted. In addition, there was tension with Bard. In 1942 Marshall collapsed with an acute psychotic paranoid episode after having actually threatened Bard physically. This episode of mental illness required that Marshall be hospitalized for eighteen months.

When Marshall returned to neuroscience in the late 1940s, he moved to a completely new set of problems: spreading cortical depression, an experimentally induced, reversible silencing of electrical activity in the cerebral cortex. By the time I arrived at NIH, he had passed the peak of his brilliant career. He still enjoyed doing occasional experiments, but he had lost his scientific drive and the clarity of his vision, and he focused much of his energy and interest on administrative matters, which he did well.

Although eccentric, moody, and somewhat suspicious in unpredictable ways, Marshall was a generous lab chief who was extremely supportive of the young people for whom he was responsible. I learned a great deal from him about the modesty and rigor that belong in a scientific laboratory. He had high standards of scientific conduct and a fine sense of humor about himself, which was reflected in the wonderful aphorisms he would call up at appropriate moments. One of his favorites, used whenever one of his findings was challenged, was, "We were confused and they were confused, but we were more accustomed to being confused." On other occasions he would mutter, "Things will go along like this for a while and then they'll get worse!"

In addition to humility, I learned from Marshall that with great personal strength and the passage of time, one might substantially recover from a severe mental illness (therapeutic drugs were not yet available). I also learned how much a person who had recovered from such a devastating disorder could accomplish. Many young people who went on to have superb careers in science and I myself owe our start and much of our later success to the personal and professional example of Wade Marshall. Despite my obvious inexperience, he did not insist that I work only on problems that interested him. Rather, he allowed me to think about what I wanted to do—which was to study how learning and memory are achieved in the cells of the brain. Science gives one a structured opportunity to try out ideas—and, if one is not afraid of falling on one's face, to try out ideas that are raw, important, and bold. Marshall gave me the freedom to try to think creatively.

Grundfest, Purpura, Crain, Marshall, and later Steve Kuffler influenced me greatly. They transformed my life. They and Mr. Campagna, who paved the way for me to go to Harvard, illustrate the importance of student–teacher relationships in one's intellectual development. Equally, they underscore the role of chance influences and of the generosity of spirit that encourages young people to thrive. Young people, for their part, must strive to have an open mind and seek out places where they will be surrounded by first-rate intellects.

8

DIFFERENT MEMORIES, DIFFERENT BRAIN REGIONS

By the time I arrived at Wade Marshall's laboratory, I had progressed from the naïve notion of trying to find the ego, id, and superego in the brain to the slightly less vague idea that finding the biological basis of memory might be an effective approach to understanding higher mental processes. It was clear to me that learning and memory are central to psychoanalysis and psychotherapy. After all, aspects of many psychological problems are learned, and psychoanalysis is based on the principle that what is learned can be unlearned. In a large sense, learning and memory are central to our very identity. They make us who we are.

At that time, however, the biology of learning and memory was in a muddle. It was dominated by the thinking of Karl Lashley, a professor of psychology at Harvard who had convinced many scientists that there were no specific areas within the cerebral cortex that were specialized for memory.

Shortly after I arrived at NIMH, two research scientists changed all that. In a newly published research paper, Brenda Milner, a psychologist at the Montreal Neurological Institute of McGill University, and William Scoville, a neurosurgeon in Hartford, Connecticut, reported that they had tracked memory to specific regions in the brain. This news had a powerful impact on me and on many others, for it meant that perhaps there was at last a decisive end to a very old controversy about the human mind.

UNTIL THE MIDDLE OF THE TWENTIETH CENTURY, THE SEARCH for where memory resides in the brain was shaped by two competing views of how the brain—and especially the cerebral cortex—works. One held that the cerebral cortex is composed of discrete regions with specific functions—one representing language, another vision, and so on. The other view was that mental capacities are products of the combined activity of the entire cerebral cortex.

The first person to champion the idea that particular mental abilities are located in specific regions of the cortex was Franz Joseph Gall, a German physician and neuroanatomist who taught at the University of Vienna from 1781 to 1802. Gall made two enduring conceptual contributions to the science of mind. First, he argued that *all* mental processes are biological, and so arise from the brain. Second, he proposed that the cerebral cortex has many distinct regions that govern specific mental functions.

Gall's theory that all mental processes are biological put him at odds with dualism, the dominant theory of his time. Promulgated in 1632 by René Descartes, a mathematician and the father of modern philosophy, this theory holds that human beings have a dual nature: the body, which is material, and the soul, which resides outside the body and is immaterial and indestructible. This dual nature reflects two types of substances. The *res externa*—the physical substance that fills the body, including the brain—runs through the nerves and imbues the muscles with animal spirits. The *res cogitans*—the nonphysical stuff of thinking—is uniquely human. It gives rise to rational thought and consciousness, and it reflects in its nonphysical character the spiritual nature of the soul. Reflex actions and many other physical behaviors are carried out by the brain, while mental processes are carried out by the soul. Descartes believed these two agents interacted through the pineal gland, a small structure located deep in the middle of the brain.

The Roman Catholic Church, feeling its authority threatened by new discoveries in anatomy, embraced dualism because it separated

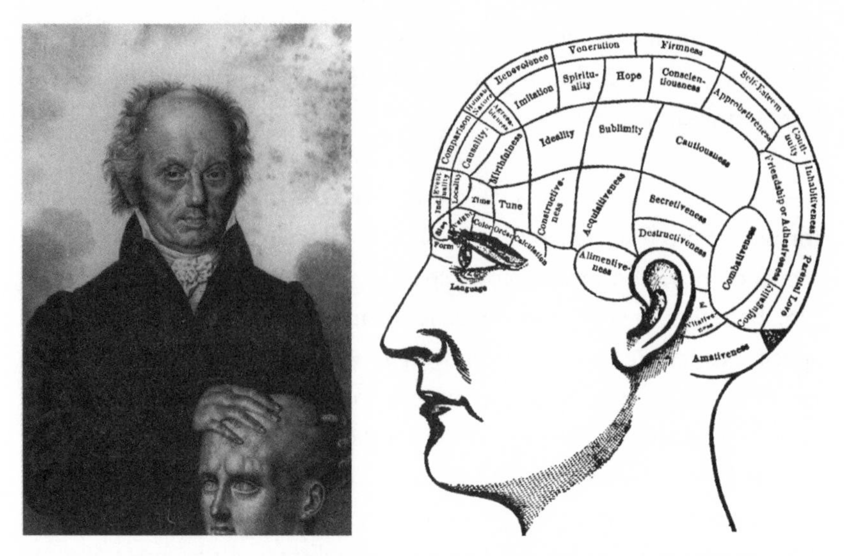

8-1 Phrenology. Franz Joseph Gall (1758–1828) assigned different mental functions to specific regions of the brain, based on his observations. Gall later developed phrenology, a system that related personality to bumps on the skull. (Gall image courtesy of Anthony A. Walsh.)

the realms of science and religion. Gall's radical argument for a materialist view of mind appealed to the scientific community because it ended the concept of the nonbiological soul, but it threatened the powerful conservative elements of society. Indeed, Emperor Francis I forbade Gall from lecturing in public and expelled him from Austria.

Gall also speculated on which areas of the cortex do different things. Academic psychology at the time had settled on twenty-seven mental faculties. Gall assigned these faculties to twenty-seven different regions of the cortex, each of which he called a "mental organ." (Additional regions were added later by Gall and others.) These mental faculties—such as factual memory, cautiousness, secretiveness, hope, belief in God, sublimity, parental love, and romantic love—were both abstract and complex, yet Gall insisted that each was controlled by a single, distinct region of the brain. This theory of localized function opened a debate that persisted through the next century.

Although correct in principle, Gall's theory was flawed in its details. First, most of the "faculties" considered discrete mental functions in

Gall's time are far too complex to arise from single regions of the cerebral cortex. Second, Gall's method of assigning functions to specific areas of the brain was misguided.

Gall distrusted studies of the behavior of people who had lost parts of their brain, so he ignored clinical findings. Instead, he developed an approach based on studies of the skull. He believed that each area of the cerebral cortex grew with usage and that this growth caused the overlying skull to protrude (figure 8-1).

Gall developed his idea in stages, beginning when he was young. In school, he had formed the impression that his most intelligent classmates had prominent foreheads and eyes. In contrast, a very romantic and enchanting widow he encountered had a prominent back of the head. Thus, Gall came to believe that great intelligence creates greater mass in the front of the brain, whereas great romantic passion produces greater mass in the back. In each case the overlying skull was enlarged by growth of the brain. Gall believed that by examining the bumps and ridges on the skulls of people well endowed with specific faculties, he could identify the centers of those faculties.

He systematized his thinking further when, as a young physician, he was put in charge of an insane asylum in Vienna. There he examined the skulls of criminals and found a bump above the ear that was remarkably similar to one found in carnivorous animals; Gall associated the bump with a part of the brain he believed was responsible for sadistic and destructive behavior. This approach to identifying the sites of mental faculties led to phrenology, a discipline that correlated personality and character with the shape of the skull.

By the late 1820s Gall's ideas and the discipline of phrenology had become extremely popular, even among the public. Pierre Flourens, a French experimental neurologist, decided to put them to the test. Using various animals for his experiments, Flourens removed, one by one, the areas of the cerebral cortex that Gall had associated with specific mental functions, but he failed to find any of the behavioral deficits Gall had predicted. In fact, Flourens was unable to associate any deficits in behavior with specific regions of the cortex. It was only the size of the area removed, not its location or the complexity of behavior involved, that mattered.

Flourens therefore concluded that all regions of the cerebral hemispheres were equally important. The cortex is equipotential, he argued, meaning that every region is able to perform any of the brain's functions. Thus, an injury to a particular region of the cerebral cortex would not affect one capacity more than another. "All perceptions, all volitions occupy the same seat in these [cerebral] organs; the faculty of perceiving, of conceiving, of willing merely constitutes thereby a faculty which is essentially one," Flourens wrote.

Flourens's views spread rapidly. Their swift acceptance was certainly due in part to the credibility of his experimental work, but it also represented a religious and political reaction against Gall's materialistic view of the brain. If the materialistic view was correct, there was no need to postulate a soul as a necessary mediator of human cognitive functions.

THE DEBATE BETWEEN THE FOLLOWERS OF GALL AND THE followers of Flourens colored thinking about the brain over the next several decades. It was not resolved until the second half of the nineteenth century, when the question attracted the attention of two neurologists, Pierre-Paul Broca in Paris and Carl Wernicke in Breslau, Germany. In the course of their studies of patients with specific language deficits, or aphasias, Broca and Wernicke made several important discoveries. Taken together, these discoveries form one of the most exciting chapters in the study of human behavior—namely, the first insight into the biological basis of a complex cognitive ability, language.

Rather than exploring the normal brain to test Gall's ideas, as Flourens had done, Broca and Wernicke studied disease states—what physicians at that time referred to as nature's experiments. They succeeded in associating specific disorders of language with damage to particular areas of the cerebral cortex, thus providing convincing evidence that at least some higher mental functions arise there.

The cerebral cortex has two important characteristics. First, although its two hemispheres appear to be mirror images of each other, they differ in both structure and function. Second, each hemisphere is concerned primarily with sensing and moving the opposite side of the body. Thus, sensory information that arrives at the spinal

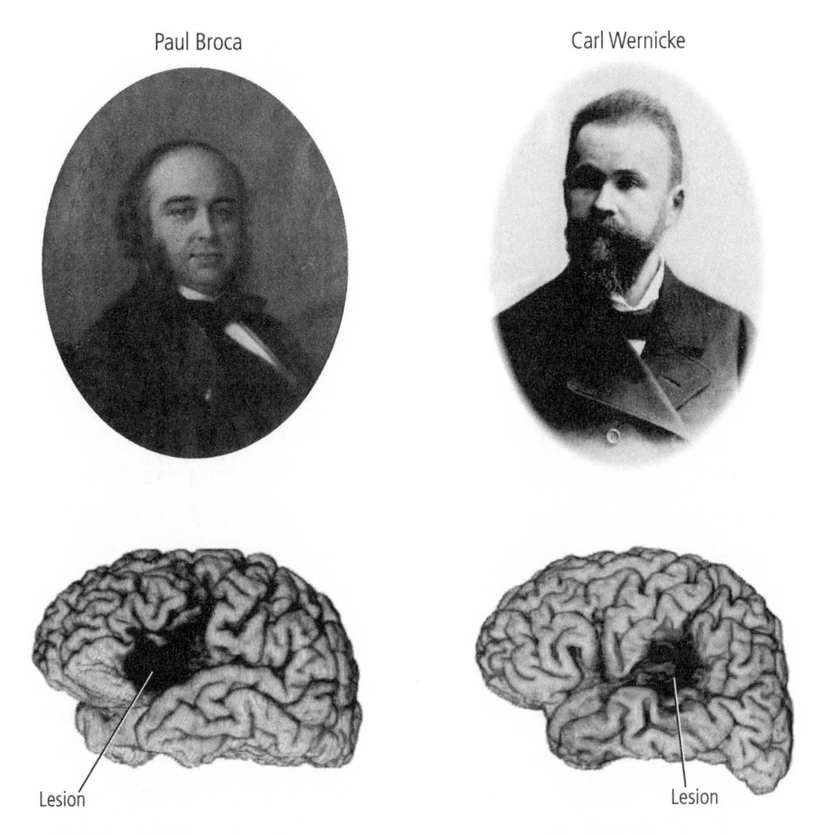

8-2 Two pioneers in the study of brain function in language. (Portraits reprinted from *Essentials of Neural Science and Behavior*, Kandel, Schwartz, and Jessell, McGraw-Hill, 1995. Brain images courtesy of Hanna Damasio.)

cord from the left side of the body—from the left hand, say—crosses over to the right side of the nervous system on its way to the cerebral cortex. Similarly, motor areas in the right hemisphere control movements of the left half of the body.

Broca (figure 8-2), who was also a surgeon and anthropologist, founded what we now call neuropsychology, a science that examines alterations in mental processes produced by brain damage. In 1861 he described a fifty-one-year-old Parisian shoemaker named Leborgne who had suffered a stroke twenty-one years earlier. As a result of his stroke,

Leborgne had lost the ability to speak fluently, although he indicated with facial expressions and actions that he understood the spoken language quite well. Leborgne had none of the conventional motor deficits that would affect speech. He had no difficulty moving his tongue, mouth, or vocal cords. In fact, he could utter isolated words, whistle, and sing a melody without difficulty, but he could not speak grammatically or create complete sentences. Moreover, his difficulty was not limited to the spoken language; Leborgne could not express his ideas in writing either.

Leborgne died one week after Broca first examined him. At the postmortem examination Broca discovered a damaged area, or lesion, in a region of the frontal lobe now called Broca's area (figure 8-2). He went on to study, after their deaths, the brains of eight other patients who had been unable to speak. Each had a similar lesion in the frontal lobe of the left cerebral hemisphere. Broca's findings provided the first empirical evidence that a well-defined mental capacity could be assigned to a specific region of the cortex. Since all the patients' lesions were in the left hemisphere, Broca established that the two hemispheres, though apparently symmetrical, have different roles. This discovery led him to announce, in 1864, one of the most famous principles of brain function: "*Nous parlons avec l'hémisphère gauche*!" (We speak with the left hemisphere!)

Broca's discovery stimulated a search for the location of other behavioral functions in the cortex. Nine years later two German physiologists, Gustav Theodor Fritsch and Eduard Hitzig, galvanized the scientific community when they showed that dogs would move their limbs in predictable ways when a specific region of the cerebral cortex was stimulated electrically. Moreover, Fritsch and Hitzig identified the small areas of the cortex that controlled the individual muscle groups responsible for the movements.

In 1879 Carl Wernicke (figure 8-2) described a second type of aphasia. This disorder is not an impairment in the production of speech, but a disruption in the comprehension of spoken or written language. Moreover, while people with Wernicke's aphasia can speak, what they say is completely incoherent to anyone else. Like Broca's aphasia, this aphasia is caused by a lesion in the left side of the brain, but in this case the damage is in the back of the brain, in a region now called Wernicke's area (figure 8-2).

Based on his own and Broca's work, Wernicke put forward a theory of how the cortex is wired for language. That theory, although simpler than our current understanding of language, is nevertheless consistent with how we view the brain today. As a first principle, Wernicke proposed that any complex behavior is the product not of a single region but of several specialized, interconnected areas of the brain. For language, these are Wernicke's area (comprehension) and Broca's area (expression). The two areas, as Wernicke knew, are connected by a neural pathway (figure 8-3). He also understood that large, interconnected networks of specialized regions, such as those governing language, enable people to experience mental activity as seamless.

The idea that different regions of the brain are specialized for different purposes is central to modern brain science, and Wernicke's model of a network of interconnected, specialized regions is a dominant theme in the study of the brain. One reason this conclusion eluded investigators

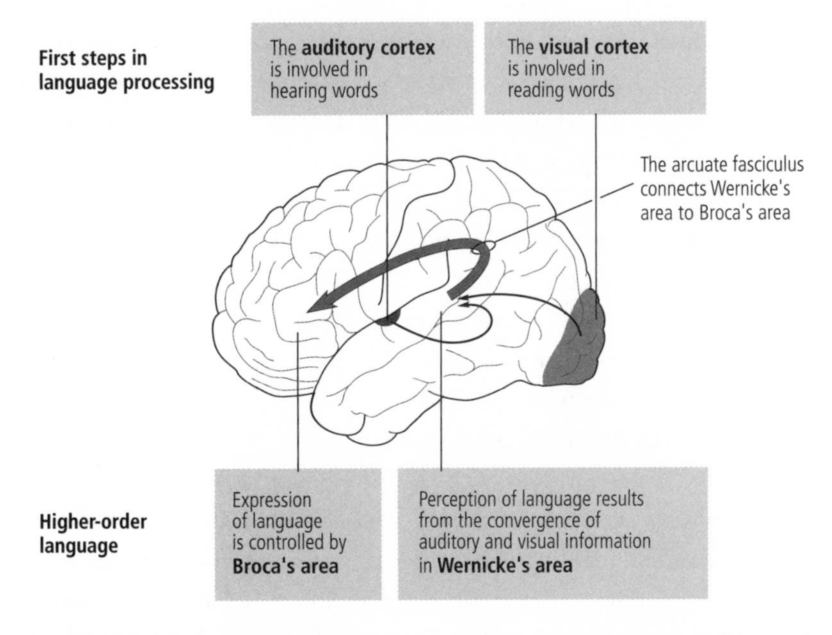

8-3 Complex behavior, such as language, involves several interconnected areas of the brain.

for so many years can be found in another organizational principle of the nervous system: brain circuitry has a built-in redundancy. Many sensory, motor, and cognitive functions are served by more than one neural pathway—the same information is processed simultaneously and in parallel in different regions of the brain. When one region or pathway is damaged, others may be able to compensate, at least partially, for the loss. When compensation occurs and no behavioral deficits are obvious, researchers have difficulty linking a damaged site in the brain to a behavior.

Once it was known that language is produced and understood in specific regions of the brain, regions governing each of the senses were identified, providing a foundation for Wade Marshall's later discoveries of sensory maps for touch, vision, and hearing. It was only a matter of time before these searches turned to memory. Indeed, the fundamental question of whether memory is a unique neural process or is associated with motor and sensory processes remained open.

INITIAL ATTEMPTS TO PINPOINT A REGION OF THE BRAIN responsible for memory, or even to delineate memory as a unique mental process, failed. In a series of famous experiments in the 1920s, Karl Lashley trained rats to run through a simple maze. He then removed different areas of the cerebral cortex and retested the rats twenty days later to see how much training they had retained. Based on these experiments, Lashley formulated the law of mass action, which holds that the severity of memory impairment is correlated with the size of the cortical area removed, not with its specific location. Thus Lashley wrote, echoing Flourens a century earlier, "It is certain that the maze habit, when formed, is not localized in any single area of the cerebrum [the cerebral cortex] and that its performance is somehow conditioned by the quantity of tissue which is intact."

Many years later Lashley's results were reinterpreted by Wilder Penfield and Brenda Milner at the Montreal Neurological Institute. As more scientists experimented with rats, it became clear that mazes were not suitable for studying the location of memory function. Maze learning is an activity that involves many different sensory and motor capabilities. When an animal is deprived of one kind of sensory cue (for example, touch), it can still recognize a place reasonably well using other senses (such as vision or smell). In addition, Lashley focused his efforts on the cerebral cortex, the outer layer of the brain; he did not explore the structures that lie deeper in the brain. Subsequent research has shown that many forms of memory require one or more of these deeper regions.

The first suggestion that some aspects of human memory can be stored in specific regions of the brain arose in 1948 from Penfield's neurosurgical work (figure 8-4). Penfield had trained in physiology with Charles Sherrington while on a Rhodes scholarship. He began to use surgery to treat focal epilepsy, a disorder that produces seizures in limited regions of the cortex. He developed a technique, still used today, of removing epileptic tissue while avoiding or minimizing damage to areas involved in the patient's mental processes.

Because the brain contains no pain receptors, surgery can be carried out with a local anesthetic. Thus Penfield's patients remained fully conscious during surgery and were able to report their experiences. (In describing this to Sherrington, who had spent his career working on cats and monkeys, Penfield could not resist adding, "Imagine having an experimental preparation that can talk back to you.") Penfield applied

8-4 Wilder Penfield (1891–1976) exposed the surface of the brain in conscious patients during surgery for epilepsy. He then stimulated different parts of the cortex and through the patients' responses identified the temporal lobe as a potential site for memory storage. (Courtesy of the Penfield Archive and the Montreal Neurological Institute.)

weak electrical stimulation to various areas of his patients' cerebral cortex during surgery and determined the effects of this stimulation on their ability to speak and comprehend language. Through his patients' responses, he could pinpoint Broca's and Wernicke's areas and try to avoid damage to them when removing epileptic tissue.

Over the years, Penfield explored much of the surface of the cerebral cortex in more than a thousand persons. On occasion, in response to electrical stimulation, a patient would describe complex perceptions or experiences: "It sounded like a voice saying words, but it was so faint that I couldn't get it." Or, "I am seeing a picture of a dog and cat . . . the dog is chasing the cat." Such responses were rare (occurring in only about 8 percent of cases) and were invariably elicited only from the temporal lobes of the brain, never from other areas. The responses suggested to Penfield that the experiences elicited by electrical stimulation of the temporal lobes are snippets of memory, of the stream of experience in a person's life.

Lawrence Kubie, the psychoanalyst whom I knew through Ernst Kris, traveled to Montreal and used a tape recorder to monitor Penfield's patients' utterances. Kubie became convinced that the temporal lobe stored a particular type of unconscious information called the preconscious unconscious. I read an important paper by Kubie when I was in medical school and heard him lecture several times while I was working in Grundfest's lab, and I was influenced by his enthusiasm for the temporal lobe.

In time, Penfield's view that the temporal lobes store memory was called into question. First, all of his patients had abnormal brains because of their epilepsy; moreover, in almost half of the cases the mental experience evoked by stimulation was identical to the hallucinatory mental experiences that often accompanied seizures. These findings convinced most brain scientists that Penfield was eliciting seizurelike phenomena with his electrical stimulation—specifically, that he might be eliciting the auras (hallucinatory experiences) characteristic of the early phase of an epileptic attack. Second, the reports of mental experiences included elements of fantasy as well as improbable or impossible situations; they were more like dreams than memories. Finally, removing the brain tissue under the stimulating electrode did not erase the patient's memory.

Nevertheless, a number of neurosurgeons were inspired by Penfield's work, among them William Scoville, who obtained direct evidence that the temporal lobes are critical to human memory. In the paper that I had read on arriving at NIH, Scoville and Brenda Milner reported the extraordinary story of a patient known to science only by his initials, H.M.

AT THE AGE OF NINE, H.M. WAS KNOCKED DOWN BY SOMEONE riding a bicycle. He sustained a head injury that led eventually to epilepsy. Over the years, his seizures worsened, until he was having as many as ten blackouts and one major seizure a week. By age twenty-seven, he was severely incapacitated.

Because H.M.'s epilepsy was thought to have originated within the temporal lobe (specifically, the medial temporal lobe), Scoville decided, as a last resort, to remove the inner surface of that lobe on both sides of the brain, as well as the hippocampus, which lies deep within the temporal lobe. The surgery succeeded in relieving H.M.'s seizures, but it left him with a devastating memory loss from which he never recovered. After his operation in 1953, H.M. remained the same intelligent, kind, and amusing man he had always been, but he was unable to convert any new memories into permanent memory.

In a series of studies, Milner (figure 8-5) documented in exquisite detail the memory ability H.M. had lost, the memory ability he retained, and the areas of the brain responsible for each. She found that what H.M. retained was remarkably specific. To begin with, he had perfectly good short-term memory, lasting for minutes. He could readily remember a multidigit number or a visual image for a short period after learning it, and he could carry on a normal conversation, provided it did not last too long or move among too many topics. This short-term memory function was later called working memory and shown to involve an area known as the prefrontal cortex, which had not been removed from H.M. Second, H.M. had perfectly good long-term memory for events that had occurred before his surgery. He

8-5 Brenda Milner (b. 1918), whose studies of H.M. opened up the modern study of memory storage by localizing memory to a particular site in the brain. Milner identified the roles of the hippocampus and the medial temporal lobe in explicit memory and provided the first evidence of implicit memory storage. (Reprinted from *Essentials of Neural Science and Behavior*, Kandel, Schwartz, and Jessell, McGraw-Hill, 1995.)

could remember the English language, his IQ was good, and he recalled vividly many events from his childhood.

What H.M. lacked, and lacked to the most profound degree, was the ability to convert new short-term memory into new long-term memory. Without this ability he forgot events shortly after they happened. He could retain new information as long as his attention was not diverted from it, but a minute or two after his attention was directed to something else, he could not remember the previous subject or anything he thought about it. Less than an hour after eating he could not remember anything he had eaten or even the fact that he had had a meal. Brenda Milner studied H.M. monthly for almost thirty years, and each time she entered the room and greeted him he failed to recognize her. He did not recognize himself in recent photographs or in the mirror because he remembered himself only as he was prior to surgery. He had no memory of his changed appearance: his identity has been frozen for over fifty years, from the time of his surgery until today. Milner was to say of H.M., "He couldn't acquire the slightest new piece of knowledge. He lives today chained to the past, in a sort of childlike world. You can say his personal history stopped with the operation."

From her systematic studies of H.M., Milner extracted three important principles of the biological basis of complex memory. First, memory is a distinct mental function, clearly separate from other perceptual, motor, and cognitive abilities. Second, short-term memory and long-term memory can be stored separately. Loss of medial temporal lobe structures, particularly loss of the hippocampus, destroys the ability to convert new short-term memory to new long-term memory. Third, Milner showed that at least one type of memory can be traced to specific places in the brain. Loss of brain substance in the medial temporal lobe and the hippocampus profoundly disrupts the ability to lay down new long-term memories, whereas losses in certain other regions of the brain do not affect memory.

Milner thus disproved Lashley's theory of mass action. It is only in the hippocampus that the various strands of sensory information necessary for forming long-term memory come together. Lashley never went below the surface of the cortex in his experiments. Moreover, Milner's finding that H.M. had good long-term memory for events that happened prior to the surgery showed clearly that the medial temporal lobe and the hippocampus are not the permanent storage sites of memory that has been in long-term storage for some time.

We now have reason to believe that long-term memory is stored in the cerebral cortex. Moreover, it is stored in the same area of the cerebral cortex that originally processed the information—that is, memories of visual images are stored in various areas of the visual cortex, and memories of tactile experiences are stored in the somatosensory cortex (figure 8-6). That explains why Lashley, who used complex tasks based on several different sensory modalities, could not erase his rats' memories fully by removing selected sections of the cortex.

For many years Milner thought H.M.'s memory defect was complete, that he was unable to convert any short-term memory to long-term memory. But in 1962 she demonstrated another principle of the biological basis of memory—the existence of more than one kind of memory. Specifically, Milner found that in addition to conscious memory, which requires the hippocampus, there is an unconscious memory that resides outside the hippocampus and the medial temporal lobe.

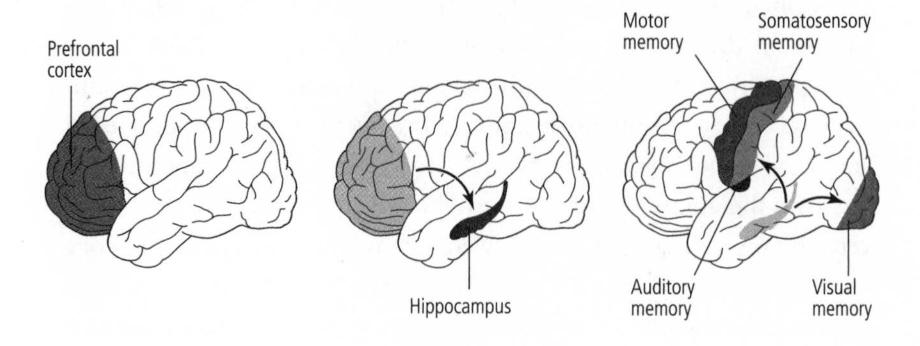

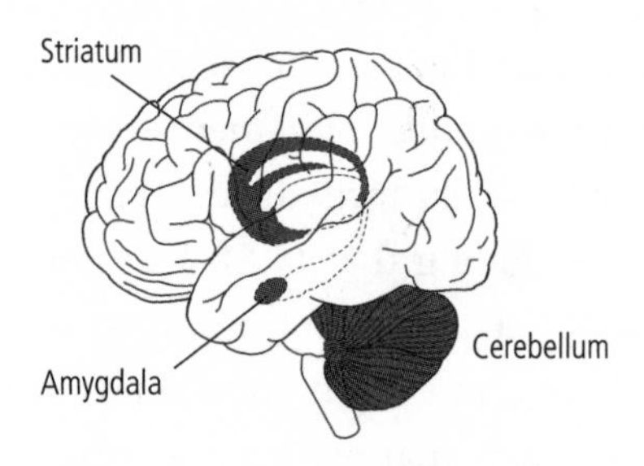

8-6 Explicit and implicit memories are processed and stored in different regions in the brain. In the short term, explicit memory for people, objects, places, facts, and events is stored in the prefrontal cortex. These memories are converted to long-term memories in the hippocampus and then stored in the parts of the cortex that correspond to the senses involved—that is, in the same areas that originally processed the information. Implicit memories of skills, habits, and conditioning are stored in the cerebellum, striatum, and amygdala.

(This distinction had been proposed on behavioral grounds in the 1950s by Jerome Bruner at Harvard, one of the fathers of cognitive psychology.)

Milner thus demonstrated this distinction by showing that the two forms of memory require different anatomical systems (figure 8-6). She found that H.M. could learn and remember some things over the long term—that is, he had a kind of long-term memory that does not depend on the medial temporal lobe or the hippocampus. He learned to trace the outline of a star in a mirror and his skill at tracing improved from day to day, just as it would in a person without brain damage (figure 8-7). Yet, even though his performance improved at the beginning of each day's test, H.M. could never remember having performed the task on an earlier day.

The ability to learn a drawing skill proved to be just one of a number of abilities that were intact in H.M. Moreover, this and other learning abilities described by Milner proved to be remarkably general and applied equally well to other people with damage to the hippocampus and the medial temporal lobe. Thus Milner's work revealed that we process and store information about the world in two fundamentally different ways (figure 8-6). It also revealed once again that, as with Broca and Wernicke, one learns a great deal from the careful study of clinical cases.

Larry Squire, a neuropsychologist working at the University of California, San Diego, expanded on Milner's finding. He did parallel experiments on memory storage in humans and in animals. These studies and those of Daniel Schacter, now at Harvard University, described the biology of two major classes of memory.

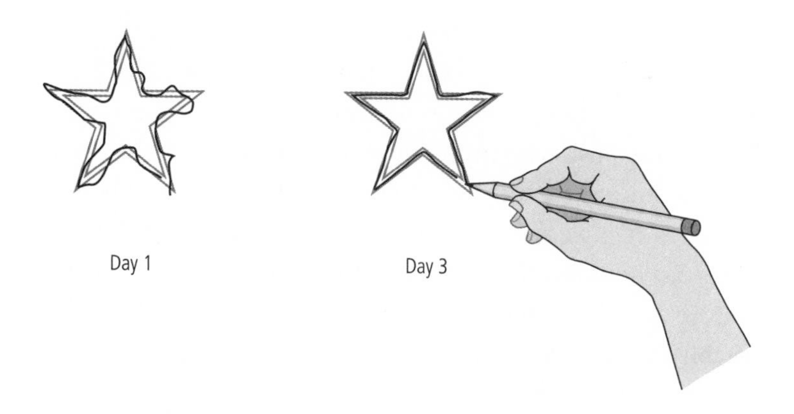

8-7 Despite his obvious loss of memory, H.M. could learn and retain new skills. During his first attempt on Day 1 (left), H.M. made many errors tracing a star he could see only in a mirror. During his first attempt on Day 3 (right), H.M. retained what he had learned through practice—even though he had no recollection of the task.

What we usually think of as conscious memory we now call, following Squire and Schacter, explicit (or declarative) memory. It is the conscious recall of people, places, objects, facts, and events—the memory that H.M. lacked. Unconscious memory we now call implicit (or procedural) memory. It underlies habituation, sensitization, and classical conditioning, as well as perceptual and motor skills such as riding a bicycle or serving a tennis ball. This is the memory H.M. retained.

Implicit memory is not a single memory system but a collection of processes involving several different brain systems that lie deep within the cerebral cortex (figure 8-6). For example, the association of feelings (such as fear or happiness) with events involves a structure called the amygdala. The formation of new motor (and perhaps cognitive) habits requires the striatum, while learning new motor skills or coordinated activities depends on the cerebellum. In the simplest animals, including invertebrates, implicit memory for habituation, sensitization, and classical conditioning can be stored within the reflex pathways themselves.

Implicit memory often has an automatic quality. It is recalled directly through performance, without any conscious effort or even awareness that we are drawing on memory. Although experiences change perceptual and motor abilities, those experiences are virtually inaccessible to conscious recollection. For example, once you learn to ride a bicycle, you simply do it. You do not consciously direct your body: "Now push with my left foot, now my right. . . ." If we paid that much attention to each movement, we would probably fall off the bicycle. When we speak, we do not consider where in the sentence to place the noun or the verb. We do it automatically, unconsciously. This is the type of reflexive learning studied by the behaviorists, notably Pavlov, Thorndike, and Skinner.

Many learning experiences recruit both explicit and implicit memory. Indeed, constant repetition can transform explicit memory into implicit memory. Learning to ride a bicycle initially involves conscious attention to one's body and the bicycle, but eventually riding becomes an automatic, unconscious motor activity.

Philosophers and psychologists had already anticipated the distinction between explicit and implicit memory. Hermann Helmholtz, the first person to measure the speed of conduction of the action potential, also worked on visual perception. In 1885 he pointed out that a great deal of mental processing for visual perception and for action occurs on an unconscious level. In 1890, in his classic work *The Principles of Psychology*, William James expanded on this idea, writing separate chapters on habit (unconscious, mechanical, reflexive action) and memory (conscious awareness of the past). In 1949 the British philosopher Gilbert Ryle distinguished between knowing *how* (the knowledge of skills), and knowing *what* (knowledge of facts and events). Indeed, a central premise of Freudian psychoanalytic theory, enunciated in 1900 in *The Interpretation of Dreams*, is an extension of Helmholtz's idea that experiences are recorded and recalled not only as conscious memories, but also as unconscious memories. Unconscious memories are ordinarily inaccessible to consciousness, but they nevertheless exert powerful effects on behavior.

While Freud's ideas were interesting and influential, many scientists were not convinced of their truth in the absence of experimental inquiry into how the brain actually stores information. Milner's star-tracing experiment with H.M. was the first time a scientist had uncovered the biological basis of a psychoanalytic hypothesis. By showing that a person who has no hippocampus (and therefore no ability to store conscious memories) can nonetheless remember an action, she validated Freud's theory that the majority of our actions are unconscious.

WHENEVER I RETURN TO BRENDA MILNER'S PAPERS ON H.M., I am impressed yet again by how much these studies clarified our thinking about memory. Pierre Flourens in the nineteenth century and Karl Lashley well into the twentieth century thought of the cerebral cortex as a bowl of porridge, in which all regions were similar in how they worked. For them memory was not a discrete mental process that could be studied in isolation. But when other scientists began to track not only cognitive processes but also various memory processes to distinct regions of the brain, the theory of mass action was dismissed once and for all.

Thus by 1957, having read Milner's initial paper and having some insight into where memory is stored in the brain, the question of how memory is stored in the brain had become the next meaningful scientific question for me. As I settled into Wade Marshall's laboratory, I began to think that this question would be an ideal challenge. Moreover, I thought that the answer would best be pursued by examining the cells involved in the storage of specific explicit memories. I would plant my flag midway between my interests in clinical psychoanalysis and the basic biology of nerve cells and proceed to investigate the territory of explicit memory "one cell at a time."

9

SEARCHING FOR AN IDEAL SYSTEM TO STUDY MEMORY

Prior to Brenda Milner's discoveries, many behaviorists and some cognitive psychologists had followed the lead of Freud and Skinner and abandoned biology as a useful guide to the study of learning and memory. They had done so not because they were dualists, like Descartes, but because they thought that biology was unlikely to play a significant role in studies of learning in the near future. Indeed, Lashley's influential work made it seem that the biology of learning was essentially incomprehensible. In 1950, toward the end of his career, he wrote, "I sometimes feel, in reviewing the evidence on the localization of the memory trace, that the necessary conclusion is that *learning is just not possible* [italics added]."

Milner's work changed all that. Her discoveries that certain regions of the brain are necessary for some forms of memory provided the first evidence of *where* different memories are processed and stored. But the question of *how* memory is stored remained unanswered, and it fascinated me. Although I had only the most rudimentary preparation for researching how memory is stored in the nervous system, I was eager to give it a try—and the environment at NIH encouraged a certain boldness. All around me research on various problems in the spinal cord, first outlined by Sherrington, was being conducted at the cellular level. Ultimately, cellular studies of memory had to answer a number of key questions: What changes occur in the brain when we learn? Do different types of learning involve different changes? What are the biochemical mechanisms of memory storage? These questions were swirling in my mind, but such questions do not translate easily into useful experiments.

I wanted to begin where Milner had left off. I wanted to tackle the most complex and interesting aspect of memory—the formation of long-term memory for people, places, and things that she found lacking in H.M. I therefore wanted to focus on the hippocampus, which Milner had shown was essential for the formation of new long-term memories. But my ideas about how to tackle the biology of memory in the hippocampus were not only vague, they were naïve.

As a first step, I asked a simple question: Do nerve cells that participate in memory storage have easily recognizable distinguishing features? Were the nerve cells of the hippocampus—cells that are presumably critical for memory storage—physiologically different from motor neurons in the spinal cord, the only other well-studied neurons in the mammalian central nervous system? Possibly, I thought, the properties of hippocampal neurons would reveal something about how memory is recorded.

I was emboldened to try this technically demanding study because Karl Frank, who worked in the lab next door to me, and John Eccles in Australia were using microelectrodes to study individual motor neurons in the spinal cord of cats. Their electrodes were identical to the ones I had used to listen in on crayfish cells. Although Frank himself thought that studying the hippocampus was formidable and risky, he was not discouraging.

Marshall had only one laboratory and two postdoctoral fellows, Jack Brinley and me. Jack had obtained a medical degree from the University of Michigan and had started work on a Ph.D. in biophysics at Johns Hopkins University just before he came to NIH. His proposed thesis was on the movement of potassium ions through the membrane of neurons in the autonomic nervous system. Because Wade liked the cerebral cortex, Jack shifted his focus a bit and studied potassium movement through the cerebral cortex in response to spreading cortical depression, a seizure process that Marshall had been interested in for several years. This was a perfectly good problem, but not one that interested me. Jack felt the same way about the hippocampus. So we worked out a compromise: we would share the lab. He would use it half of the time and I would help him, and I would use it the other half and he would help me.

This arrangement was working well when all of a sudden Marshall thrust upon us a third person, a new postdoctoral fellow, Alden Spencer, who had just graduated from the University of Oregon Medical School. The idea that now the lab would be shared by three independent projects with each of us having even less time in the lab to work on our own studies produced apprehension in both Jack's heart and mine. We each worked feverishly to persuade Alden to join our particular project.

To my delight, it took little effort to convince Alden that we should work together on the hippocampus. Part of my success, which I did not realize until later, was due to the fact that Alden never for a moment considered working with Jack, whose project required using a radioactive form of potassium. Alden was a bit of a hypochondriac and was frightened to death of radioactivity.

MY RESEARCH TOOK AN EXTREMELY FORTUNATE TURN WITH Alden's arrival. Born in Portland, he was a liberal in the best Oregonian tradition of independent thought based on moral rather than narrow political considerations (figure 9-1). Alden's father, a perpetual student, at once a freethinker and a religious man, was a conscientious objector during World War I and was recruited into the noncombatant corps. After the war, he went to divinity school in British Columbia and served for a while as pastor of a small church. He then went back to school at Stanford University, where he studied mathematics and statistics and later worked as a statistician for the civil service in Oregon.

Alden completely changed my narrow view of life outside the East Coast. He was strongly independent, with an original turn of mind, a great interest in music and art, and an enthusiasm for life that made him exciting to be with. He had novel insights about most things he

9-1 Alden Spencer (1931–1977), with whom I had the privilege of collaborating at the NIMH from 1958 to 1960 and who later joined me at NYU Medical School and at Columbia University. Alden made major contributions to the understanding of the hippocampus, the modification of simple reflex responses by learning, and the perception of touch. (From Eric Kandel's personal collection.)

experienced: a lecture, a concert, a tennis match. His creativity was so abundant and came to him so readily that he was always extending himself to something new, immersing himself in yet another problem. Alden had a considerable musical talent as well, having played the clarinet in the Portland Symphony Orchestra. His wife, Diane, was a fine pianist. Moreover, Alden was extremely modest and expressed these deep creative interests in an utterly unpretentious way. Denise and I soon became good friends with them, and the four of us routinely attended the Library of Congress's weekly chamber music recitals featuring the renowned Budapest String Quartet.

Among Alden's many talents were surgical skills, a good knowledge of the anatomical organization of the brain, and insights into what questions were scientifically important. Although he had no experience with intracellular recording, he had done some excellent electrophysiological research on the brain, studying how the pathways between the thalamus and the cortex contribute to the various brain rhythms displayed on EEGs (electroencephalograms). Alden was great company. We talked science incessantly and reinforced each other's audacity. Once we decided it was important, we were not reluctant to tackle any difficult problem, such as trying to obtain recordings from individual cortical neurons in an intact brain.

Soon after we started we had our first successful experiment. I shall never forget it. I worked all morning and part of the afternoon to complete the surgery that exposed a cat's hippocampus. Late in the afternoon Alden took over and was advancing the recording electrode into the hippocampus. I was sitting in front of the oscilloscope, the instrument that displayed the electrical signals, and I also controlled the stimulators that could activate pathways into and out of the hippocampus. As I had done in Stanley Crain's laboratory, I connected the recording electrode to a loudspeaker so that any electrical signal we might obtain could be heard as well as seen. We were trying to record from pyramidal cells, the major class of neurons in the hippocampus. These cells receive and process the information coming into the hippocampus and send it on to the next relay point. We also set up a camera to photograph the display screen of the oscilloscope.

Suddenly we heard the loud bang! bang! bang! of action potentials, a sound I recognized immediately from my experiments on crayfish. Alden had penetrated a cell! We quickly realized it was a pyramidal cell because the axons of these neurons are bundled into a pathway (called the fornix) that leads out of the hippocampus, and I had positioned electrodes on that pathway. Every stimulus I applied elicited a beautiful, large action potential. This method of stimulating the outgoing axon and causing pyramidal cells to fire proved to be a powerful way of identifying those cells. We also managed to excite pyramidal cells by stimulating the pathway that carries information *into* the hippocampus. We thus obtained a remarkable amount of information in the roughly ten minutes during which we recorded signals from pyramidal cells. We ran the camera continuously to ensure that every moment of the recording, every synaptic potential and every action potential in the pyramidal cells, was captured on film.

Alden and I were euphoric—we had obtained the first intracellular signals ever recorded from the region of the brain that stores our fondest memories! We almost danced around the lab. The mere accomplishment of recording from these cells successfully for several minutes met our most optimistic expectations. In addition, our data looked fascinating and somewhat different from what Eccles and Frank had found in the motor neurons of the spinal cord.

This experiment and the ones that followed were physically exhausting, sometimes lasting twenty-four hours. It was a good thing we had both just finished a medical internship, where working twenty-four hours at a stretch was not uncommon. We carried out three experiments a week and used the intervening two days, often only partial days because Jack was using the laboratory, to analyze data, to discuss the results, and just to talk. Many experiments were unsuccessful, but we eventually developed simple technical improvements that allowed us to obtain high-quality recordings once or twice a week.

By applying the powerful methodologies of cell biology to the hippocampus, Alden and I easily picked some low-hanging intellectual fruit. To begin with, we found that unlike motor neurons, a certain class of hippocampal neurons fires spontaneously, even without receiving instructions from sensory or other neurons. More interesting, we found that action potentials in the pyramidal cells of the hippocampus originate at more than one site within the cell. In the motor neuron, action potentials are initiated only at the base of the axon, where it emerges from the cell body. We had good evidence to suggest that action potentials in pyramidal cells of the hippocampus can also begin in the dendrites, and that they can be initiated in response to stimulation of the perforant pathway, an important direct synaptic input to the pyramidal cells from a part of the cortex called the entorrhinal cortex.

This proved to be an important discovery. Up to that time most neural scientists, including Dominick Purpura and Harry Grundfest, thought that dendrites could not be excited and therefore could not generate action potentials. Willifred Rall, a major theorist and model builder at NIH, had developed a mathematical model showing how the dendrites of motor neurons function. This model was based on the fundamental assumption that the cell membrane of dendrites is passive: it does not contain voltage-gated sodium channels and therefore cannot support an action potential. The intracellular signals we recorded were the first evidence to the contrary, and our finding later proved to be a general principle of neuronal function.

Our technical success and these intriguing findings brought enthusiastic encouragement and unstinting praise from our senior colleagues at NIH. John Eccles, who was emerging as the leading cellular physiologist in the mammalian brain, stopped by to see us when he visited NIH and was generous in his comments. Eccles invited Alden and me to join him in Australia to continue with him our work on the hippocampus, an offer we refused after much hesitation. Wade Marshall asked me to give a seminar at NIMH to summarize Alden's and my efforts; I did, to a packed conference room, and it was warmly received. But even in our headiest moments, we realized that ours was a typical NIH story. Young, inexperienced people were given the opportunity to try things on their own, knowing that wherever they turned, experienced people were available to help.

Not everything was wine and roses, however. Soon after I arrived, another young scientist, Felix Strumwasser, went to work at a neighboring laboratory. Unlike the rest of the young research associates, who were medical doctors, Felix had a Ph.D. in neurophysiology from the University of California, Los Angeles. While most of us knew relatively little about brain science, Felix was extremely knowledgeable. We became good friends and had dinner at each other's homes. I learned a great deal from him. In fact, in my conversations with Felix, I sharpened my thinking about how to tackle neurobiological studies of learning. Felix also got me to think about the hypothalamus, a region of the brain concerned with emotional expression and hormone secretions. The hypothalamus was becoming important in clinical discussions of how to treat stress and mental depression.

I was therefore taken aback—and hurt—when the day after I gave the seminar on our work, Felix stopped talking to me. I could not understand what had happened. Only with time did I realize that science is filled not simply with a passion for ideas but also with the ambition and strivings of people at different stages of their careers. Many years later, Felix renewed our friendship and explained that he had been chagrined when two relatively inexperienced scientists—incompetents, in his eyes—had been able to produce interesting and important experimental results.

As the afterglow of our beginner's luck began to fade, Alden and I realized that as fascinating as our findings were, they were leading us in directions unrelated to memory. In fact, we found that the properties of hippocampal neurons were not sufficiently different from those of spinal motor neurons to account for the ability of the hippocampus to store memories. It took us a year to realize what should have been obvious from the start: the cellular mechanisms of learning and memory reside not in the special properties of the neuron itself, but in the connections it receives and makes with other cells in the neuronal circuit to which it belongs. As we learned through reading and discussions with each other to think more deeply about the biological mechanisms of learning and memory, we concluded that the role of the hippocampus in memory must arise in some other way, perhaps from the nature of the information it receives, the way its cells are interconnected, and from how that circuitry and the information it carries are affected by learning.

That change in our thinking led us to change our experimental approach. To understand how the neural circuitry of the hippocampus affects memory storage, we needed to know how sensory information reaches the hippocampus, what happens to it there, and where it goes after it leaves the hippocampus. This was a formidable challenge. Practically nothing was known then about how sensory stimuli reach the hippocampus or how the hippocampus sends information to other areas of the brain.

We therefore carried out a series of experiments to examine how various sensory stimuli—tactile, auditory, and visual—affect the firing pattern of the pyramidal neurons of the hippocampus. We saw only occasional, sluggish responses—nothing comparable to the brisk responses reported by other investigators in the neural pathways of the somatosensory, auditory, and visual cortices. In a final attempt to understand how the hippocampus might participate in memory storage, we explored the properties of the synapses that incoming axons of the perforant pathway form with the nerve cells in the hippocampus. We stimulated those axons repetitively at the rate of 10 impulses per second and observed an increase in synaptic strength that lasted about 10 to 15 seconds. We then stimulated them at the rate of 60 to 100 impulses per second and produced an epileptic seizure. These were all interesting findings, but they were not what we were looking for!

As we became more familiar with the hippocampus, we realized that finding out how its neural networks process learned information and how learning and memory storage change those networks was an extraordinarily difficult task that would take a very long time.

I was initially drawn to the hippocampus because of my interest in psychoanalysis, which tempted me to tackle the biology of memory in its most complex and intriguing form. But it became clear to me that the reductionist strategy used by Hodgkin, Katz, and Kuffler to study the action potential and synaptic transmission also applied to research on learning. To make any reasonable progress toward understanding how memory storage occurs, it would be desirable, at least initially, to study the simplest instance of memory storage and to study it in an animal with the simplest possible nervous system, so I could trace the flow of information from sensory input to motor output. I therefore searched for an experimental animal—perhaps an invertebrate such as a worm, a fly, or a snail—in which simple but modifiable behaviors were controlled by simple neuronal circuits made up of a small number of nerve cells.

But what animal? Here Alden and I parted intellectual company. He was committed to mammalian neurophysiology and wanted to stay with the mammalian brain. He felt that although invertebrates are instructive, the organization of the invertebrate brain is so fundamentally different from that of the vertebrate brain that he did not want to work on them. Moreover, the components of the vertebrate brain were already well described. Biological solutions valid for the rest of the animal kingdom would draw his interest and his admiration, but unless they were true for the vertebrate brain, the human brain, they would not draw his effort. Alden therefore turned to one of the simple subsystems of the spinal cord of the cat and examined spinal reflexes, which are modified through learning. Over the next five years Alden made important contributions in this area of research in collaboration with the psychologist Richard Thompson. However, even the relatively simple reflex circuits in the spinal cord proved too difficult for a detailed cellular analysis of learning, and by 1965 Alden had turned from the spinal cord and the study of learning to other areas of research.

EVEN THOUGH IT MEANT SWIMMING AGAINST THE TIDE OF current thinking, I yearned for a more radical, reductionist approach to the biology of learning and memory storage. I was convinced that the biological basis of learning should be studied first at the level of individual cells and, moreover, that the approach was most likely to succeed if it focused on the simplest behavior of a simple animal. Many years later, Sydney Brenner, a pioneer of molecular genetics who introduced the worm *Caenorhabditis elegans* to biology, was to write:

> What you need to do is find which is the *best* system to experimentally solve the problem, and as long as it [the problem] is general enough you will find the solution there.
>
> The choice of an experimental object remains one of the most important things to do in biology and is, I think, one of the great ways to do innovative work. . . . The diversity in the living world is so large, and since everything is connected in some way, let's find the *best* one.

In the 1950s and 1960s, however, most biologists shared Alden's reluctance to apply a strictly reductionist strategy to the study of behavior because they thought it would have no relevance for human behavior. People have mental abilities not found in simpler animals, and these biologists believed that the functional organization of the human brain must be quite different from that of simpler animals. Although this view holds some truth, I thought it overlooked the fact—amply demonstrated in the fieldwork of ethologists like Konrad Lorenz, Niko Tinbergen, and Karl von Frisch—that certain elementary forms of learning are common to all animals. It seemed likely to me that, in the course of evolution, humans had retained some of the cellular mechanisms of learning and memory storage found in simpler animals.

Not surprisingly, I was discouraged from pursuing this research strategy by a number of senior scientists in neurobiology, including Eccles. His concern reflected in part the hierarchy of acceptable research questions in neurobiology at that time. Although some scientists were studying behavior in invertebrates, that work was not considered important—indeed, it was largely ignored—by most people working on the mammalian brain. Of even greater concern to me was the skepticism of knowledgeable psychologists and psychoanalysts that anything interesting about higher-order mental processes like learning and memory could be found by focusing on individual nerve cells—particularly the cells of an invertebrate. I had made up my mind, however. The only remaining question was which invertebrate would best suit the cellular study of learning and memory.

Besides being a good place to do research, NIH was also a good place to learn about new developments in biology. During the course of any given year, most good scientists working on the brain visit the NIH campus. As a result, I was able to speak with many people and to attend seminars in which I learned about the experimental advantages of various invertebrate animals, such as crayfish, lobsters, honeybees, flies, land snails, and the nematode worm *Ascaris*.

I remembered vividly Kuffler's description of the advantages of the crayfish's sensory neuron for studying the properties of dendrites. But I ruled out crayfish: although they have a few very large axons, their nerve cell bodies are not very large. I wanted an animal with a simple reflex that could be modified by learning and that was controlled by a small number of large nerve cells whose pathway from input to output could be identified. In that way, I could relate changes in the reflex to changes in the cells.

After about six months of careful consideration, I settled on the giant marine snail *Aplysia* as a suitable animal for my studies. I had been greatly impressed with two lectures I had heard about the snail, one given by Angelique Arvanitaki-Chalazonitis, a senior, highly accomplished scientist who had discovered the usefulness of *Aplysia* for studying the signaling characteristics of nerve cells, and the other by Ladislav Tauc, a younger person who brought a new biophysical perspective to the study of how nerve cells function.

Aplysia was first mentioned by Pliny the Elder in his encyclopedic study *Historia Naturalis*, written during the first century A.D. It was mentioned again in the second century by Galen. These ancient scholars called it *lepus marinus*, or sea hare, because, when sitting still and contracted, it resembles a rabbit. When I began to examine *Aplysia* myself, I found, as others had before me, that it releases copious amounts of purple ink when disturbed. That ink was once erroneously thought to be the royal purple used to dye the stripe on the togas of the Roman emperors. (The royal purple dye is actually secreted by the clam *Murex*.) Because of *Aplysia*'s tendency to ink so profusely, some ancient naturalists also thought it was holy.

The American species of *Aplysia* that lives off the California coast (*A. californica*), and which I have spent most of my career studying, measures more than one foot in length and weighs several pounds (figure 9-2). It assumes the reddish brown coloration of the seaweed it feeds on. It is a large, proud, attractive, and obviously highly intelligent beast—just the sort of animal one would select for studies of learning!

What drew my attention to *Aplysia* was not its natural history or physical beauty but several other features outlined by Arvanitaki-Chalazonitis and by Tauc in their lectures on the European species (*A. depilans*). They both emphasized that the brain of *Aplysia* has a small number of cells, about 20,000, compared with about 100 billion in the

9-2 ***Aplysia californica*****: the giant marine snail.** (Courtesy of Thomas Teyke.)

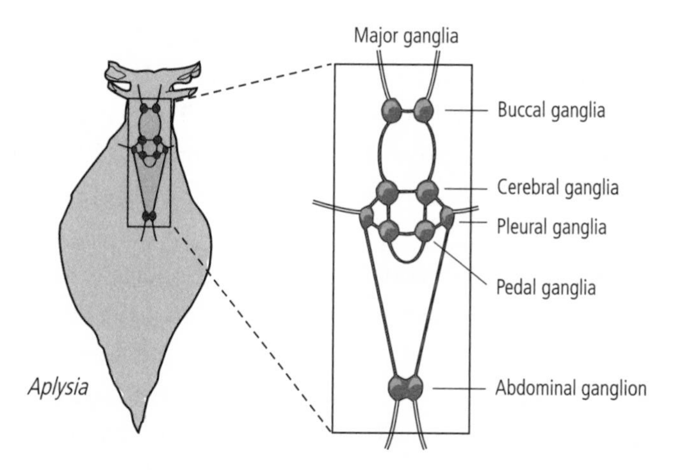

9-3 The brain of *Aplysia* is very simple. It has 20,000 neurons grouped into nine separate clusters, or ganglia. Since each ganglion has a small number of cells, researchers can isolate simple behaviors that are controlled by it. They can then study changes in particular cells as a behavior is altered by learning.

mammalian brain. Most of these cells are grouped into nine clusters, or ganglia (figure 9-3). Since individual ganglia were thought to control several simple reflex responses, it seemed to me that the number of cells committed to a single simple behavior was likely to be small. In addition, some of *Aplysia*'s cells are the largest in the animal kingdom, making it relatively easy to insert microelectrodes into them to record electrical activity. The pyramidal cells of the cat hippocampus, whose activity Alden and I had recorded, are among the largest nerve cells in the mammalian brain, yet they are only 20 micrometers in diameter ($^{1}/_{1250}$ of an inch) and can be seen only under a high-powered microscope. Some cells in *Aplysia*'s nervous system are fifty times that size and can be seen with the naked eye.

Arvanitaki-Chalazonitis had found that a few nerve cells in *Aplysia* are uniquely identifiable—that is, the same cells can easily be recognized by sight under the microscope in every single snail. In time I realized that the same thing is true of most other cells in its nervous system, heightening the prospect of mapping the entire neural circuitry controlling a behavior. As it turned out, the circuitry controlling *Aplysia*'s most elementary reflexes was quite simple. Moreover, I later found that stimulating a single neuron often produced a large synaptic potential in its target cells, a clear sign and measure of the strength of the synaptic connection between the two cells. These large synaptic potentials made it possible to map neural connections cell by cell and eventually enabled me to work out for the first time the precise wiring diagram of a behavior.

Many years later, Chip Quinn, one of the first scientists to conduct genetic studies of learning in the fruit fly, noted that the ideal experimental animal for biological studies of learning must have "no more than three genes, be able to play the cello or at least recite classical Greek, and learn these tasks with a nervous system containing only ten large, differently colored and therefore easily recognizable neurons." I have often thought that *Aplysia* meets these criteria to a surprising degree.

At the time I decided to work on *Aplysia*, I had never dissected the snail or recorded the electrical activity of its neurons. Moreover, no one in the United States was working on *Aplysia*. The only two people in the world who were studying it in 1959 were Tauc and Arvanitaki-Chalazonitis. Both of them were in France, Tauc in Paris and Arvanitaki-Chalazonitis in Marseilles. Denise, ever the Parisian chauvinist, thought Paris the better choice. Living in Marseilles, she said, would be like living in Albany instead of New York City. So the decision was for Tauc. Before leaving NIH in May 1960, I visited Tauc and we arranged that I would join him in September 1962, as soon as I had completed my residency in psychiatry at Harvard Medical School.

AS I LEFT NIH IN JUNE 1960, I FELT A DEEP SADNESS, SOMEWHAT similar to that which I experienced when I graduated from Erasmus Hall High School. I had come as a novice, and I left as a limited but nonetheless working scientist. At NIH I had walked the walk. I found that I liked it and that I was quite successful at what I tried. But I was genuinely surprised by my success. For a long time I felt that it was due to pure chance, my good fortune, my enjoyable and productive collaboration with Alden, the generous psychological support of Wade Marshall, and the youth-oriented scientific culture of NIH. I had had a number of ideas that proved to be useful, but I thought they were beginner's luck. I was very much afraid that I would run out of ideas and would not last in science.

This insecurity about my ability to come up with new ideas was not helped by the fact that John Eccles and several other senior scientists whom I respected thought I was making a grave error in switching from a highly promising start in the mammalian hippocampus to a new beginning with an invertebrate whose behavior had not been well studied. But three factors drove me onward. First, there was the Kuffler-Grundfest principle of biological research: for every biological problem there is a suitable organism in which to study it. Second, I was now a cell biologist. I wanted to think about how cells functioned during learning, and I wanted to spend my time reading, thinking, and discussing ideas with others. I did not want to spend hours simply setting up an experiment time and again as Alden and I had done on the hippocampus, only to find an occasional good cell to study. I liked the idea of big cells and, despite the risks involved, I was convinced that *Aplysia* was the right system and that I had the tools to study behavior in this snail effectively.

Finally, I had learned something in marrying Denise. I had been reluctant and fearful of marriage, even to Denise, whom I loved much more than any other woman I had ever thought of marrying. But Denise was confident that our marriage would work, so I took a leap of faith and went ahead. I learned from that experience that there are many situations in which one cannot decide on the basis of cold facts alone—because facts are often insufficient. One ultimately has to trust one's unconscious, one's instincts, one's creative urge. I did this again in choosing *Aplysia*.

10

NEURAL ANALOGS OF LEARNING

After visiting Ladislav Tauc briefly in Paris, Denise and I went to Vienna in May 1960 so I could show her the city where I was born. This was the first time I had returned since I left in April 1939. We walked around the beautiful Ringstrasse, the major boulevard where many of the most important public buildings of the city—the opera house, the university, the parliament—are located. We enjoyed our visit to the Kunsthistorisches Museum, a richly baroque building with its beautiful marble staircase and a superb art collection first put together by the Hapsburg royal family. One of the high points of this great museum is a room containing Pieter Bruegel the Elder's paintings of the seasons of the year. We visited the Oberes Belvedere and enjoyed the world's best collection of the Austrian expressionists—Klimt, Kokoschka, and Schiele, the three modern painters whose images are indelibly impressed in the minds of most Viennese art lovers of my generation.

Most important, we went to the apartment in Severingasse 8 where my family and I had lived. We found it occupied by a young woman and her husband. She allowed us to enter and look around. Despite the fact that legally the apartment still belonged to my family, since we had never sold it, I felt awkward imposing on this nice person. We stayed only a short time, but it was long enough for me to be surprised at how tiny the apartment was. I remembered the space as being rather small—the living room and dining room in which I had steered my shiny blue remote-controlled car on my ninth birthday all those years ago—but I was astonished at how small it was in reality, a common trick of memory distortion. We then walked to Schulgasse to visit my elementary school, only to find that it had been replaced by a government agency. The walk, which I remembered from my school days as being a bit of a hike, took us all of five minutes. It was a similarly short walk to Kutschkergasse, where my father's store had been.

Denise and I were standing across the street from the store and I was pointing it out to her when an older man I did not know came up and said, "You must be Hermann Kandel's son!"

I was dumbfounded. I asked him how he could possibly have inferred that, since my father had never returned to Vienna and I had left it as a child. He identified himself as living three buildings away and then said simply, "You look so much like him." Neither he nor I had the courage to discuss the intervening years—and in retrospect I am sorry not to have done so.

I was quite moved by the visit. Denise was interested, but she later told me that were it not for my deep and continuing fascination with Vienna, she would have found the city boring compared with Paris. Her comment reminded me of an evening early in our friendship, when Denise first invited me to her mother's house for dinner. Joining us that evening was Denise's imposing Aunt Sonia, a large, intellectually powerful, and slightly arrogant woman who worked for the United Nations and who had been secretary of the Socialist party in France prior to World War II.

As we sat down for a drink before dinner, she turned to me inquisitorially and asked in her strong French accent, "Where do you come from?"

"Vienna," I replied.

Without changing her overall condescending expression, she forced a small smile and said, "That's nice. We used to call that little Paris."

Many years later my friend Richard Axel, who introduced me to molecular biology, was preparing for his first trip to Vienna. Before I could prime him on its virtues, one of his other friends passed on to Axel his pronouncement about Vienna: "It's the Philadelphia of Europe."

It is clear to me that neither of these people really understood Vienna—its lost grandeur, its enduring beauty, or its present-day complacency and latent anti-Semitism.

UPON RETURNING FROM VIENNA, I BEGAN MY RESIDENCY training in psychiatry at the Massachusetts Mental Health Center of Harvard Medical School. I had actually committed to starting it a year earlier, but because work on the hippocampus was moving along so well, I had written to Jack Ewalt, director of the mental health center and professor of psychiatry at Harvard Medical School, asking if it were possible to have a one-year extension. He replied immediately that I should stay as long as necessary. That third year at NIH proved crucial, not only for my collaborative work with Alden, but also for my maturation as a scientist.

With this beginning in mind and a subsequent exchange of cordial letters, I visited Ewalt upon my arrival. I asked him if it might be possible to have some space and modest resources to set up a laboratory. Suddenly, the atmosphere was transformed. It was as if I were having a conversation with a completely different person. He looked at me and then pointed to the pile of résumés of the twenty-two other residents who were about to begin their training and bellowed, "Who do you think you are? What makes you think that you are better than any one of these?"

I was completely taken aback by the content of his remarks and even more by the tone. In all my years as an undergraduate at Harvard and a medical student at NYU, none of my professors had ever talked to me like that. I assured him that while I had no illusions about my clinical skills compared with those of my peers, I did have three years of research experience that I did not want to lie dormant. Ewalt told me to go to the wards and take care of patients.

I left his office confused and depressed, and I briefly entertained the idea of switching to the Boston Veterans Administration Hospital residency training program. Jerry Lettvin, a neurobiologist and friend to whom I described the conversation with Ewalt, urged me to take the position at the Veterans Administration, stating, "Working at the Massachusetts Mental Health Center is like swimming in a whirlpool. It's impossible to change things or to make progress." Nevertheless, because of the excellent reputation of its residency program, I decided to swallow my pride and stay.

It proved a wise decision. A few days later, I went across the street to the medical school and discussed my situation with Elwood Henneman, a professor of physiology. He offered me space in his laboratory. Within several weeks, Ewalt approached me and said that he gathered from his colleagues at the medical school, referring to Henneman and Stephen Kuffler, that I was a good person to invest in. "What do you need?" he said, "How can I help you?" He then made available all the resources necessary to continue my research in Henneman's laboratory throughout the two years of residency training.

The residency training turned out to be at once stimulating and a bit disappointing. My fellow residents were a gifted group who remained my friends over the years. Many of them went on to have major careers in academic psychiatry. The group included Judy Livant Rappaport, who became a leading researcher in the mental disorders of children; Paul Wender, who pioneered the modern era of genetic studies of schizophrenia; Joseph Schildkraut, who developed the first biological model of depression; George Valliant, who helped outline some of the factors predisposing people to physical and mental illness; Alan Hobson and Ernst Hartmann, important contributors to the study of sleep; and Tony Kris (Anna's brother), a leading psychoanalyst who wrote an influential book on the nature of transference.

The clinical supervision was outstanding, if somewhat narrow in scope. In the first year we treated patients who were sufficiently ill to require hospitalization, some of whom were suffering from schizophrenia. We saw only a limited number of patients, and we had the rare opportunity to work with those very ill patients intensively in psychotherapy, seeing them for one-hour sessions two or even three times a week. Although we did not really improve their mental functioning, we learned a great deal about schizophrenia and depressive illnesses by simply listening to them. Elvin Semrad, the head of clinical services, and most of our supervisors were heavily oriented toward psychoanalytic theory and practice. Few of them thought in biological terms, few were familiar with psychopharmacology, and most discouraged us from reading the psychiatric or even the psychoanalytic literature because they thought we should learn from our patients and not from our books. "Listen to the patient, not the literature" was the prevailing pedagogical motto.

To a degree they had a point. Our patients taught us a great deal about the clinical and dynamic aspects of severe mental illness. We learned, above all, to listen very carefully and intelligently to what the patients told us about themselves and their lives. Most important, we learned to respect patients as individuals with distinctive assets and distinctive problems.

But we learned next to nothing about the fundamentals of diagnosis or the biological underpinnings of psychiatric disorders. We received only a rudimentary introduction to the use of drugs in the treatment of mental illness. In fact, we often were discouraged from using drugs in treatment because Semrad and our supervisors feared it would interfere with psychotherapy.

In response to this weakness in the program, the other residents and I organized a discussion group on descriptive psychiatry that met monthly at the house Kris and Hartmann shared. We took turns presenting original essays prepared for the occasion. I presented a paper on a group of acute mental disorders called the amentias, which follow head trauma and chemical intoxication. In some of these disorders, such as acute alcoholic hallucinosis, patients suffer from a psychosis that resembles schizophrenia but is completely reversible as the alcohol wears off. My point was that a psychotic reaction is not unique to schizophrenia but can be an end point for a number of disorders.

Prior to our arrival, the mental health center had almost never invited outside speakers to address the residents or the faculty. This was a reflection of the vaunted self-confidence of Harvard and Boston at large, which is best represented by the canard of the Boston matron who, when asked about her travels, responded, "Why should I travel? I'm already here."

Kris, Schildkraut, and I initiated academic grand rounds, conferences that brought together all of the researchers and physicians of the hospital as well as important people from other institutions. While at NIH, I had been spellbound by a lecture in which Seymour Kety, the former intramural director at the National Institute of Mental Health and the person who had recruited Wade Marshall, reviewed the contributions of genes to schizophrenia. I thought we might kick off our lecture series with that topic. But in 1961 I could not find a single psychiatrist in all of Boston who knew anything about genetics and mental illness. Somehow, I learned that Ernst Mayr, the great evolutionary biologist at Harvard, had been a friend of the late Franz Kallman, a pioneer in the genetics of schizophrenia. Mayr generously agreed to come and give us two splendid lectures on the genetics of mental illness.

I had entered medical school convinced that psychoanalysis had a promising future. Now, with my NIH experience behind me, I found myself questioning my decision to become a psychoanalyst. I also missed being in the laboratory. I yearned for new data and was eager to have findings to discuss with other scientists. But most of all I questioned the usefulness of psychoanalysis in the treatment of schizophrenia, an area that even Freud was not optimistic about.

In those days, residents did not work very hard: from 8:30 A.M. to 5:00 P.M., with only a rare shift on evenings or weekends. As a result, I was able to follow up on an idea first suggested to me by Felix Strumwasser, namely, that I study hypothalamic neuroendocrine cells. These are atypical and fairly rare cells in the brain. They look like neurons, but instead of signaling other cells directly through synaptic connections, they release hormones into the bloodstream. Neuroendocrine cells were particularly interesting to me because some research hinted that the neuroendocrine cells of the hypothalamus are disturbed in major depressive illnesses. I had learned that goldfish have very large neuroendocrine cells, and during my spare time I carried out a somewhat original series of experiments showing that those cells generate action potentials and receive synaptic signals from other nerve cells, just as ordinary neurons do. Denise helped me set up the tank for the goldfish, and she made me a nice net for catching them from a dishrag and a wire hanger.

My studies provided direct evidence that neuroendocrine cells that release hormones are in fact both fully functioning endocrine cells and fully functioning nerve cells. They have all the complex signaling capability of nerve cells. The studies were well received because they showed something new. More important for me, I had done them completely on my own in a back room in Henneman's laboratory, working at odd hours when other people were not usually there. After completing these studies, I began to feel more assured about my competence. But moving from the hippocampus to a project on neuroendocrine cells was not wildly original for me. I applied much of the same thinking here as I had at NIMH. How long would this limited burst of creativity last? I wondered, as I continued to worry that I would soon run out of ideas.

That was the least of my worries, however. Shortly after our son, Paul, was born, in March 1961, Denise and I had a serious crisis, by far the most serious of our life together. I thought we had an unusually harmonious relationship. She had strongly supported me as I was struggling to define my career, and she was working as a postdoctoral fellow at the Massachusetts Mental Health Center in a program designed to train research sociologists in issues related to mental health. We saw each other in passing during the day as well as at night.

But one Sunday afternoon she showed up as I was working in the lab and simply exploded on me. Carrying Paul in her arms, she screamed, "You can't go on like this! You are only thinking of yourself and your work! You are just ignoring the two of us!"

I was startled and deeply hurt. I was so transfixed by my science, both enjoying it and also worrying about it when experiments failed, as they often did, that it never dawned on me that I was neglecting or in any way devaluing Denise and Paul or withdrawing my love from them. I was upset and angry about being confronted so harshly and so suddenly. I sulked, pouted, and took days to recover. Only gradually did I come to realize what my actions must have seemed like from Denise's point of view. In response, I decided to spend more time at home with her and Paul.

On this and on many subsequent occasions, Denise succeeded in turning my attention from what could easily have become—and occasionally was—a full-time preoccupation with science to more immediate involvements with our children. Both for Paul and for our daughter, Minouche, born in 1965, I was a concerned and involved parent but hardly an ideal one. I missed at least half of Paul's Little League baseball games, including one game in which he stepped up to bat with the bases loaded and hit a bases-clearing double. This feat was heard around the world in our house, and to this day I regret having missed it.

As I was approaching my seventy-fifth birthday in 2004, we celebrated it three months early so we could be joined at our summer home on Cape Cod by our children, their spouses, and our four grandchildren: Minouche and her husband, Rick Sheinfeld, and their children, Izzy, five, and Maya, three; and Paul and his wife, Emily, and their two daughters, Allison, twelve, and Libby, eight. Minouche, who graduated from Yale and from Harvard Law, practices public interest law in San Francisco, focusing on women's issues and women's rights. Rick is a lawyer for the city and specializes in hospital and health care issues. Paul studied economics at Haverford as an undergraduate and then went on to the Columbia Business School. He manages a set of funds for Dreyfus. Emily graduated from Bryn Mawr and the Parsons School of Design and runs her own interior design firm.

At my birthday dinner, I raised a toast to our children, their spouses, and my four grandchildren. I said that I was proud to see what principled and interesting people our children had turned out to be and how thoughtful they are as parents of their own children, given the fact that I was only a B+ father. Minouche, who enjoys tweaking me, yelled out, "Grade inflation!"

Minouche put my parenting into perspective on another occasion. "You were great, Pops, in giving me the sense that I could do anything intellectually. You read to me often when I was little and you always took a deep interest in what I thought and in my work at Horace Mann, in college, in law school, and even now. But not once, as I remember my childhood, did you ever take me on routine visits to the doctor!"

It was, and still is, understandably difficult for my children to understand—much less excuse—that I find doing science endlessly fascinating and that my involvement in it can expand almost infinitely. It has required conscious effort on my part and help from Denise and from my psychoanalysis to be more realistic and to structure my time so as to make room for the responsibilities and pleasures of my life with Minouche and Paul and with their children.

SPENDING MORE TIME AT HOME WITH DENISE AND PAUL ALSO gave me more time to think about how to approach the study of learning in *Aplysia*. Alden Spencer and I had found few differences in the basic properties of neurons that participate in memory storage and those that do not. Those findings supported the idea that memory relies not on the properties of the nerve cell per se but on the nature of the connections between neurons and how they process the sensory information they receive. This led me to think that in a circuit that mediates behavior, memory may result from changes in synaptic strength brought about by certain patterns of sensory stimulation.

The basic idea that some type of change in synapses might be important for learning had been proposed by Cajal in 1894:

> Mental exercise facilitates a greater development of the protoplasmic apparatus and of the nervous collaterals in the part of the brain in use. In this way, pre-existing connections between groups of cells could be reinforced by multiplication of the terminal branches. . . . But the pre-existing connections could also be reinforced by the formation of new collaterals and . . . expansions.

A modern form of this hypothesis was put forward in 1948 by the Polish neuropsychologist Jerzy Kornorski, a student of Pavlov. He argued that a sensory stimulus leads to two types of changes in the nervous system. The first, which he called excitability, follows the generation of one or more action potentials in a neuronal pathway in response to a sensory stimulus. The firing of action potentials briefly raises the threshold for generating additional action potentials in those neurons, a well-known phenomenon called the refractory period. The second, more interesting change, which Kornorski called plasticity, or plastic change, leads, he wrote, to "permanent functional transformations . . . in particular systems of neurons as a result of appropriate stimuli or their combination."

The idea that certain systems of neurons were highly adaptable and plastic and could therefore be changed permanently—perhaps because of a change in the strength of their synapses—was now very appealing to me. This raised the question in my mind: How do these changes come about? John Eccles had been very enthusiastic about the possibility that synapses change in response to excessive use, but when he tested the idea, he found that they changed for only a brief period of time. "Unfortunately," he wrote, "it has not been possible to demonstrate experimentally that excess use produces prolonged changes in synaptic efficacy." To be relevant for learning, I thought, synapses would have to change for long periods of time—in extreme cases, as long as the lifetime of the animal. It now dawned on me that perhaps Pavlov was so successful in producing learning because the simple patterns of sensory stimulation he used elicited certain natural patterns of activation that were particularly suited for producing long-term changes in synaptic transmission. This idea really caught my fancy. But how to test it? How was I going to elicit this optimal pattern of activity?

With further reflection I decided to try to simulate in the nerve cells of *Aplysia* the patterns of sensory stimulation that Pavlov had used in his learning experiments. Even if initiated by artificial means, such patterns of activity might reveal some of the long-term plastic changes of which synapses are capable.

AS I BEGAN TO THINK SERIOUSLY ABOUT THESE IDEAS, I REALIZED that I would need to reformulate Cajal's theory that learning modifies the strength of the synaptic connections between neurons. Cajal thought of learning as a single process. Because I was familiar with the behaviorist work of Pavlov and the later cognitive psychological studies of Brenda Milner, I realized that there are many different forms of learning produced by different patterns and combinations of stimuli and that these give rise to two very different forms of memory storage.

I therefore extended Cajal's idea in the following manner. I presumed that different forms of learning give rise to different patterns of neural activity and that each of these patterns of activity changes the strength of synaptic connections in a particular way. When such changes persist, the result is memory storage.

Restating Cajal's theory in these terms enabled me to consider ways of translating Pavlov's behavioral protocols into biological protocols. After all, habituation, sensitization, and classical conditioning—the three learning protocols described by Pavlov—are essentially a series of instructions on how a sensory stimulus should be presented, alone or in combination with another sensory stimulus, to produce learning. My biological studies would be designed to determine whether different patterns of stimuli, modeled on Pavlov's forms of learning, would give rise to different forms of synaptic plasticity.

In habituation, for example, an animal that is repeatedly presented with a weak or neutral sensory stimulus learns to recognize the stimulus as unimportant and ignores it. When a stimulus is strong, as in sensitization, the animal recognizes the stimulus as dangerous and learns to enhance its defensive reflexes in preparation for withdrawal and escape; even an innocuous stimulus presented shortly thereafter will elicit an enhanced defensive response. When a neural stimulus is paired with a potentially dangerous stimulus, as in classical conditioning, the animal learns to respond to the neutral stimulus as if it were a danger signal.

I thought that I should be able to elicit, in the neural pathways of *Aplysia*, patterns of activity similar to those evoked in animals undergoing training in these three learning tasks. I could then determine how synaptic connections are changed by patterns of stimuli that simulate different forms of learning. I called this approach neural analogs of learning.

I was guided to this idea by an experiment that was reported just as I was contemplating how to begin my experiments on *Aplysia*. In 1961, Robert Doty at the University of Michigan in Ann Arbor made a remarkable discovery about classical conditioning. He applied a weak electrical stimulus to a part of the dog's brain governing vision and found that it produced electrical activity in neurons of the visual cortex, but no movement. Another electrical stimulus applied to the motor cortex caused the dog's paw to move. After a number of trials in which the stimuli were paired, the weak stimulus alone elicited movement of the paw. Doty had clearly shown that classical conditioning in the brain does not require motivation: it simply requires the pairing of two stimuli.

This was a big step toward a reductionist approach to learning, but the neural analogs of learning that I wanted to develop required two further steps. First, instead of conducting experiments in whole animals, I would remove the nervous system and work on a single ganglion, a single cluster of about two thousand nerve cells. Second, I would select a single nerve cell—a target cell—in that ganglion to serve as a model of any synaptic changes that might occur as a result of learning. I would then apply different patterns of electrical pulses modeled on the different forms of learning to a particular bundle of axons extending from sensory neurons on *Aplysia's* body surface to the target cell.

To simulate habituation, I would apply repeated, weak electrical pulses to this neural pathway. To simulate sensitization, I would stimulate a second neural pathway very strongly, one or more times, and see how it affected the target cell's response to weak stimulation of the first pathway. Finally, to simulate classical conditioning, I would pair the strong stimulus to the second pathway with the weak stimulus to the first pathway in such a way that the strong stimulus would always follow and be associated with the weak stimulus. In this way, I could determine whether the three patterns of stimulation altered the synaptic connections with the target cell and, if so, how. Different changes in synaptic strength in response to the three different patterns of electrical stimulation would represent analogs—biological models—of the synaptic changes in *Aplysia's* nervous system brought about by training for the three different forms of learning.

I wanted these neural analogs to answer one key question: How are synapses changed by different patterns of carefully controlled electrical stimuli that mimic the sensory stimuli in the three major learning experiments? How, for example, are synapses modified when, as in classical conditioning, a weak stimulus to one pathway immediately precedes, and therefore predicts, a strong stimulus to another pathway?

To answer this question, I applied in January 1962 for an NIH postdoctoral fellowship that would enable me to work in Tauc's laboratory. My specific aim was

> to study the cellular mechanisms of electrophysiological conditioning and of synaptic usage in a simple nerve network. . . . This exploratory study will attempt to develop methods of conditioning a simple preparation and of analyzing some of the neural elements in this process. . . . The long-range goal is to "trap" a conditioned response in the smallest possible neural population in order to permit a multiple microelectrode investigation of the activity of the participating cells.

I ended my application with these words:

> It is an explicit hypothesis of this research that the potentiality for elementary forms of conditioned plastic change is an inherent and fundamental property of all central nervous collectivity, whether simple or complex.

I was testing the idea that the cellular mechanisms underlying learning and memory are likely to have been conserved through evolution and therefore to be found in simple animals, even when using artificial modes of stimulation.

The German composer Richard Strauss commented that he often wrote his best music after an argument with his wife. This has not generally proven to be the case for me. But the argument with Denise about spending more time with her and Paul did cause me to pause and think. As a consequence I learned from this argument the obvious lesson that hard thinking, especially if it leads to even one useful idea, is much more valuable than simply running more experiments. I was later reminded of a comment made about Jim Watson by Max Perutz, the Viennese-born British structural biologist: "Jim never made the mistake of confusing hard work with hard thinking."

In September 1962, with an NIH stipend assuring us a grand annual salary of $10,000, Denise, Paul, and I set off for a fourteen-month stint in Paris.

THREE

The century that is ending has been preoccupied with nucleic acids and proteins. The next one will concentrate on memory and desire. Will it be able to answer the questions they pose?

—François Jacob, *Of Flies, Mice and Men* (1998)

11

STRENGTHENING SYNAPTIC CONNECTIONS

Being in Paris was wonderful, and I grew accustomed to spending time walking around the city every weekend with Denise and with Paul, which made the experience of being in France rewarding for all of us. In addition, I was pleased to be doing science full-time again. Ladislav Tauc and I complemented each other's interests and areas of competence, making him an excellent person with whom to work. Besides being completely at home with *Aplysia*, Tauc had trained in physics and biophysics, areas that are fundamental to cellular physiology. I lacked a strong background in either area and learned a good deal about them from him.

Born in Czechoslovakia, Tauc (figure 11-1) had obtained his Ph.D. by studying the electrical properties of large plant cells, which have a resting potential and an action potential similar to those of nerve cells. He brought this interest to bear on *Aplysia*, studying the largest cell in the abdominal ganglion, a cell I later called R2, and describing the site within this neuron where the action potential is generated. Since his focus was on the biophysical properties of nerve cells, he had not studied neuronal circuits or animal behavior and had given little thought to learning and memory, the issues that dominated my thinking about the mammalian brain.

11-1 Ladislav Tauc (1925–1999) was a pioneer in the study of *Aplysia*. I worked with him for a fourteen-month period in Paris and in Arcachon, France in 1962–63. (Reprinted from *Cellular Basis of Behavior*, E. R. Kandel, W. H. Freeman and Company, 1976.

Like many good postdoctoral experiences, mine did more than simply enable me to benefit from a senior scientist's considerable background and experience. It also allowed me to contribute intellectually by bringing my own knowledge and experience to bear on our common work. Originally, Tauc was a bit skeptical about trying to study learning on the cellular level in *Aplysia*. But in time, he became enthusiastic about my plan of studying analogs of learning in single cells in the abdominal ganglion.

As I had planned when thinking about this research, I dissected out the abdominal ganglion, with its two thousand nerve cells, and mounted it in a small chamber bathed with aerated seawater. I placed microelectrodes inside one cell, usually cell R2, and then recorded that cell's responses to various sequences of stimuli applied to the neural pathways that converged on it. I used three patterns of stimulation, based on Pavlov's work in dogs, to develop three analogs of learning: habituation, sensitization, and classical conditioning. In classical conditioning, an animal learns to respond to a neutral stimulus in the same way it would respond to an effective, threatening or negative stimulus. That is, it forms an association between the neutral stimulus and the negative stimulus. In habituation and sensitization, an animal learns to respond to one type of stimulus without associating it with any other stimulus. The experiments proved even more effective than I had anticipated.

Through habituation, the simplest form of learning, an animal learns to recognize a stimulus that is harmless. When an animal perceives a sudden noise, it initially responds with several defensive changes in its autonomic nervous system, including dilation of the pupils and increased heart and respiratory rates (figure 11-2). If the noise is repeated several times, the animal learns that the stimulus can safely be ignored. The animal's pupils no longer dilate and its heart rate no longer increases when the stimulus is presented. If the stimulus is removed for a period of time and then presented again, the animal will respond to it again.

Habituation enables people to work effectively in an otherwise noisy environment. We become accustomed to the sound of the clock in the study and to our own heartbeat, stomach movements, and other bodily sensations. These sensations then enter our awareness rarely and only under special circumstances. In this sense, habituation is learning to recognize recurrent stimuli that can safely be ignored.

Habituation also eliminates inappropriate or exaggerated defensive responses. This is illustrated in the following fable (with apologies to Aesop):

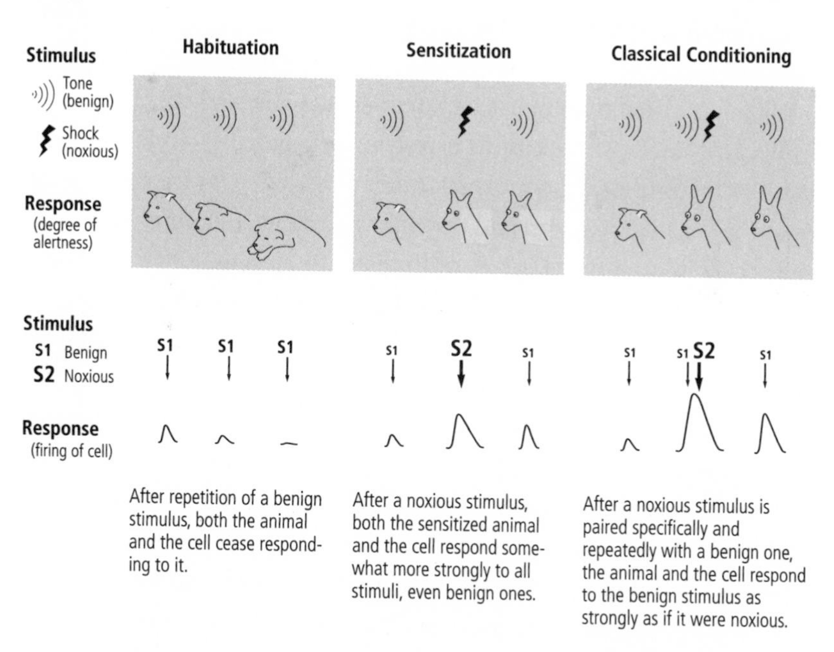

11-2 Three types of implicit learning. Habituation, sensitization and classical conditioning can be studied both in animals (above) and in individual nerve cells (below).

> A fox that had never yet seen a turtle was so frightened when he encountered one in the forest for the first time that he nearly died. On his meeting with the turtle for the second time, he was still much alarmed but not to the same extent as at first. On seeing the turtle for the third time, he was so increased in boldness that he went up to him and commenced a familiar conversation with him.

The elimination of responses that fail to serve a useful purpose focuses an animal's behavior. Immature animals often show escape responses to a variety of nonthreatening stimuli. Once they become habituated to such stimuli, they can focus on stimuli that are novel or associated with pleasure or danger. Habituation is therefore important in organizing perception.

Habituation is not restricted to escape responses: the frequency of sexual responses can also be decreased through habituation. Given free access to a receptive female, a male rat will copulate six or seven times over a period of one or two hours; afterward, he appears sexually exhausted and becomes inactive for thirty minutes or longer. This is sexual habituation, not fatigue. An apparently exhausted male will promptly resume mating if a new female is made available.

Because of its simplicity as a test for recognizing familiar objects, habituation is one of the most effective means of studying the development of visual perception and memory in infants. Infants characteristically respond to a novel image with dilated pupils and increased heart and respiratory rates. If shown an image repeatedly, however, they will stop responding to it. Thus an infant who has been repeatedly shown a circle will ignore it. But if the infant is then shown a square, its pupils will again dilate, and its heart and respiratory rates will increase, indicating that it can distinguish between the two images.

I modeled habituation by applying a weak electrical stimulus to a bundle of axons leading to cell R2 and then repeating that stimulus ten times. I found that the synaptic potential produced by the cell in response to the stimulus decreased progressively with repetition. By the tenth stimulus, the response was only about one-twentieth as strong as it had been initially, just as an animal's behavioral response abates when a neutral stimulus is presented repeatedly (figure 11-2). I called this process homosynaptic depression: depression because the synaptic response was decreased, and homosynaptic because the depression occurred in the same neural pathway that was stimulated (*homo* means "the same" in Greek). After withholding the stimulus for ten or fifteen minutes, I applied it again and saw that the cell's response returned to almost its initial strength. I called this process recovery from homosynaptic depression.

Sensitization is the mirror image of habituation. Instead of teaching an animal to ignore a stimulus, sensitization is a form of learned fear: it teaches the animal to attend and respond more vigorously to almost any stimulus after having been subjected to a threatening stimulus. Thus, immediately after a shock has been delivered to an animal's foot, for instance, the animal will exhibit exaggerated withdrawal and escape responses to a bell, a tone, or a soft touch.

Like habituation, sensitization is common in people. After hearing a gun go off, a person will show an exaggerated response and will jump when he hears a tone or senses a touch on the shoulder. Konrad Lorenz elaborates on the survival value of this learned form of arousal for even simple animals: "An earthworm that has just avoided being eaten by a blackbird . . . is indeed well advised to respond with a considerably lowered threshold to similar stimuli because it is almost certain that the bird will still be nearby for the next few seconds."

In modeling sensitization I applied a weak stimulus to the same neural pathway leading to cell R2 that I had used in my earlier experiments on habituation. I stimulated it once or twice to elicit a synaptic potential that would serve as a baseline measure of the cell's responsiveness. I then applied a series of five stronger stimuli (designed to simulate disagreeable or noxious stimuli) to a different pathway leading to cell R2. After I had presented the stronger stimuli, the cell's synaptic response to stimulation of the first pathway was greatly enhanced, indicating that synaptic connections in that pathway had been strengthened. The enhanced response lasted up to thirty minutes. I called this process heterosynaptic facilitation: facilitation because synaptic strength was enhanced, and heterosynaptic because the enhanced response to stimulation of axons in the first pathway was brought about by strongly stimulating a different pathway (*hetero* means "different" in Greek) (figure 11-2). The enhanced response to the first pathway depended only on the greater strength of the stimulus to the different pathway and not on any pairing of weak and strong stimuli. Thus it resembled behavioral sensitization, a nonassociative form of learning.

Finally, I tried to simulate aversive classical conditioning. This form of classical conditioning teaches an animal to associate an unpleasant stimulus, such as an electric shock, with a stimulus that ordinarily elicits no response. The neutral stimulus must always precede the aversive stimulus and in this way will come to predict it. For example, Pavlov used a shock to a dog's paw as an aversive stimulus. The shock caused the animal to raise and withdraw its leg, a fear response. Pavlov found that after several trials in which he paired the shock with the ringing of a bell—first sounding the bell and then administering the shock—the dog would withdraw its leg whenever the bell sounded, even if no shock followed. Thus, aversive classical conditioning is an associative form of learned fear (figure 11-2).

Aversive classical conditioning resembles sensitization in that activity in one sensory pathway enhances activity in another, but it differs in two ways. First, in classical conditioning an association is formed between a pair of stimuli that occur in rapid sequence. Second, classical conditioning enhances an animal's defensive responses to the neutral stimulus only, not to environmental stimuli in general, as sensitization does.

Therefore, in my experiments on aversive classical conditioning in *Aplysia*, I repeatedly paired a weak stimulus to one neural pathway with a strong stimulus to another pathway. The weak stimulus came first and acted as a warning of the strong stimulus. The pairing of the two stimuli greatly enhanced the cell's response to the weak stimulus; moreover, that enhanced response was far greater than the cell's enhanced response to the weak stimulus in the sensitization experiments (figure 11-2). The added boost was critically dependent on the timing of the weak stimulus, which had unfailingly to precede and predict the strong stimulus.

These experiments confirmed what I had suspected—namely, that a pattern of stimulation designed to mimic the patterns used to induce learning in behavioral studies can change the effectiveness of a neuron's communication with other nerve cells. The experiments clearly showed that synaptic strength is not fixed—it can be altered in different ways by different patterns of activity. Specifically, the neural analogs of sensitization and aversive classical conditioning strengthened a synaptic connection, whereas the analog of habituation weakened the connection.

Thus, Tauc and I had discovered two important principles. First, the strength of the synaptic communication between nerve cells can be changed for many minutes by applying different patterns of stimulation derived from specific training protocols for learned behavior in animals. Second, and even more remarkable, the same synapse can be strengthened or weakened by different patterns of stimulation. These findings encouraged Tauc and me to write in our paper in the *Journal of Physiology*:

> The fact that the connections between nerve cells can be strengthened for over half-an-hour with an experimental protocol designed to simulate a behavioral conditioning paradigm also suggests that the concomitant changes in the synaptic strength may underlie certain simple forms of information storage in the intact animal.

What most impressed us was how readily the strength of synapses could be altered by different patterns of stimuli. This suggested that synaptic plasticity is built into the very nature of the chemical synapse, its molecular architecture. In the broadest terms it suggested that the flow of information in the various neural circuits of the brain could be modified by learning. We did not know whether synaptic plasticity was an element of actual learning in the intact, behaving animal, but our results suggested that the possibility was very much worth pursuing.

Aplysia was proving to be more than a wonderfully informative experimental system; it was an extremely enjoyable one to work with as well. What had started out as an infatuation based on the hope of finding a suitable animal was turning into a serious commitment. Moreover, because *Aplysia*'s cells are large (cell R2, in particular, is gigantic—1 millimeter in diameter and visible to the naked eye), the technical demands of the experiments were less severe than those on the hippocampus.

The experiments were also more leisurely. Because placing a tiny electrode into such a gigantic cell causes essentially no damage, one can effortlessly record from cell R2 for five to ten hours. I could go to lunch and come back to find the cell still in perfect health, waiting for me to pick up the experiment where I had left off. This compared very favorably to the many nights Alden and I had had to work to obtain an occasional recording of ten to thirty minutes from the pyramidal cells of the hippocampus. A typical experiment in *Aplysia* could be completed in six to eight hours: as a result, the experiments became great fun.

In this mood, after a season of working on *Aplysia*, I was reminded of a story Bernard Katz had told me about the great physiologist A. V. Hill, his mentor at University College, London. On Hill's first visit to the United States, in 1924, shortly after having won the Nobel Prize at age thirty-six for his work on the mechanism of muscular contraction, he gave a talk on the subject at a scientific meeting. At the end of the talk, an elderly gentleman rose and asked him about the practical use of his research.

Hill pondered for a moment as to whether he should enumerate the many instances in which great benefits for mankind have arisen from experiments undertaken purely to satisfy intellectual curiosity. Rather than take this path, however, he simply turned to the man and said with a smile, "To tell the truth, sir, we don't do it because it's useful; we do it because it's amusing."

On a personal level, these studies were pivotal to my confidence as an independent scientist. When I first arrived and talked about learning and analogs of learning, the other postdoctoral fellows' eyes simply glazed over. In 1962, talking to most cellular neurobiologists about learning was a little bit like talking to the moon. By the time I left, however, the tenor of discussions in the laboratory had changed.

I also felt that I was developing a style of doing science. Even though I still felt myself inadequately trained in some areas, I proved to be quite bold in approaching scientific problems. I did experiments that I thought were interesting and important. Without quite knowing it, I had found my voice, much as a writer must feel after having written a number of satisfactory stories. With that finding came self-assurance, a sense that I could make a go of it in science. After my fellowship with Tauc, I never again had the fear that I would run out of ideas. I had many moments of disappointment, despondency, and exhaustion, but I always found that by reading the literature and showing up at my lab looking at the data as they emerged day by day and discussing them with my students and postdoctoral fellows, I would gain a notion of what to do next. We would then discuss these ideas over and over again. When I tackled the next problem, I would immerse myself in reading about it.

As I had done when selecting *Aplysia* to study, I learned to trust my instincts, to unconsciously follow my nose. Maturation as a scientist involves many components, but a key one for me was the development of taste, much as it is in the enjoyment of art, music, food, or wine. One needs to learn what problems are important. I sensed myself developing taste, distinguishing what was interesting from what was not—and among the things that were interesting, I also learned what was doable.

BEYOND THE PLEASURE OF THE SCIENCE, OUR FOURTEEN MONTHS' stay in France was a transforming experience for Denise and me. Because we so enjoyed Paris and because *Aplysia* was so easy to work with, I did not work on weekends for the first time in years and was home for dinner every night at seven. We used our leisure time to see Paris and its environs. We began to visit art galleries and museums on a regular basis, and we purchased, after much financial agonizing, our first works of art. One was a wonderful self-portrait in oil by Claude Weisbusch, an Alsatian artist who had recently won an award as the young painter of the year and who used rapid, nervous strokes reminiscent of Kokoschka. We also bought a tender Mother and Child, an oil by Akira Tanaka. Our largest investment was a beautiful etching by Picasso of the artist and his models, number 82 of the Vollard Suite, published in 1934. In this marvelous etching, each of the four women is drawn in a different style. Denise thought she could recognize three of these women as having been important to Picasso at different points in his early life: Olga Koklova, Sarah Murphy, and Marie-Thérèse Walter. We still greatly enjoy looking at these three beautiful works.

The French species of *Aplysia* that Ladislav Tauc worked with came from the Atlantic Ocean. The system for supplying the snails was not very reliable, so it was difficult to obtain them in Paris. We therefore spent close to the entire autumn of 1962 and 1963 in Arcachon, a beautiful little resort not far from Bordeaux. I conducted most of my experiments on *Aplysia* in Arcachon and then analyzed the data in Paris, where I also carried out some experiments on the land snail.

As if several months in Arcachon were not vacation enough, Tauc, the members of his laboratory, and all of France considered vacationing during the month of August sacred. We joined in that belief. We rented a house on the Mediterranean in the Italian town of Marina di Pietra Santa, about an hour and a half from Florence, and we visited the city three or four times a week. On other holidays, we would travel near and far. We went to Versailles, just outside of Paris, and to Cahors, in the south of France, to visit the convent where Denise was hidden during the war.

In Cahors, we spoke to a nun who remembered Denise and showed us pictures of her dormitory room, with ten cots arranged neatly on each side, and a photograph of Denise with the other girls in her class. The nun pointed out that one of the other girls was also Jewish but that neither Denise nor this girl knew of the other's identity. To protect them, none of the students was informed that there were Jews among them. Each of the Jewish girls was taken aside by the Mother Superior and shown a private escape route, a passage through a tunnel to be used if the Gestapo came searching for Jewish students.

About twenty miles from Cahors, in a very small village of two hundred inhabitants, we visited the baker Alfred Aymard and his wife, Louise, who had sheltered Denise's brother. That was certainly one of the most remarkable days in our year in France. Aymard, a Communist, took in Denise's brother not because he necessarily liked Jews, but because he hated the Nazis. Within a few months, he came to adore Jean-Claude and had a difficult time parting with him at the end of the war. The Bystryns sensed this difficulty and in the years after the war spent part of each summer vacationing with Aymard and his wife.

When we arrived for our visit, Aymard insisted that we stay overnight. He had recently suffered a stroke, which slowed his speech and left him partially paralyzed on the left side, but he was nonetheless jovial and extremely generous. He cleared out his and his wife's bedroom and ran an electric extension cord into the bedroom so we could have better light. Despite my repeated insistence that he stay in their room, Aymard and his wife insisted that we, the guests, should have the best room while the two of them slept in the kitchen. During dinner, we tried to repay his kindness with one story after another about Jean-Claude, whom Aymard still missed seventeen years later.

In another trip Denise and I shall not readily forget, we stayed for the night in Carcassonne, a medieval walled city in the south of France. We arrived in the late evening and had a difficult time finding a room. Finally, we located one in a small hotel. The room, however, had only a single, rather large bed. We put Paul in the center, changed into our sleeping clothes, and climbed into bed on either side of him. Accustomed to sleeping alone, Paul instantly rebelled and started to scream in protest. We tried repeatedly to calm him down; when this failed, we climbed out of the bed and lay down on either side of it, surrendering it to him. Denise and I at first appreciated the quiet we had won by lying on the floor. But after ten minutes of discomfort, we realized that we could not easily fall asleep there. So we turned from being progressive parents to being disciplinarians. We climbed back into bed and resolutely refused to leave. Within minutes, all was calm and the three of us slept through the night.

Living in France also enabled me to see my brother on a regular basis. When we had arrived in New York from Vienna in 1939, Lewis was fourteen and had been an academic star throughout his school years. Despite his academic ambitions, he sensed that his major efforts should be to help support our family, since my father's income was small and the Depression had not yet ended. So rather than enrolling in an academic curriculum, he went to the New York High School for Specialty Trades and learned to be a printer, a trade he enjoyed because he liked books so much. Through high school and his first two years at Brooklyn College, Lewis worked part-time for a printer; this gave him some money for our family plus a bit extra to feed his addiction to Wagnerian opera, a habit he satisfied by purchasing standing-room tickets. When he was nineteen, he was drafted into the U.S. Army and sent to Europe, where he fought and was wounded by shrapnel in the Battle of the Bulge, Germany's last-ditch effort to keep the advancing American army at bay.

Upon receiving an honorable discharge, Lewis joined the army reserve and rose to the rank of lieutenant. All service personnel were eligible for the G.I. Bill, which enabled them to go to the college of their choice tuition free. Lewis went back to Brooklyn College and continued to study engineering and German literature. Shortly after graduating, he married Elise Wilker, a Viennese émigré whom he had met at college, and he entered the graduate program in German studies at Brown University. In 1952 he began work on his Ph.D. dissertation in linguistics and Middle High German. In the midst of writing it, with the Korean War still going on, Lewis was offered an assignment to the U.S. embassy in Paris. He took that opportunity, and in 1953 he and Elise drove to New York to visit the family before the two of them were to ship out. One night while they were dining out, someone broke into their car and stole their belongings, including Lewis's research notes and the early drafts of his dissertation. He tried at first to reconstruct his work, but he never succeeded in overcoming this setback to his academic career.

After serving as an officer at the embassy, Lewis took on a second assignment in France, as the civilian comptroller of a U.S. air force base in Bar-le-Duc. He eventually became so fond of his life in France and his growing family of five children that he abandoned his plan to return to academic life. He decided to remain in France and became a connoisseur of fine wines and cheeses.

Lewis and Elise's youngest child, Billy, was born in 1961. A few weeks after his birth, Billy developed a high fever from an infection, which frightened Elise greatly. Earlier, she and Lewis had become friends with the Baptist chaplain on the base, whose discussions of Christianity appealed to her search for greater religious involvement. She promised herself that if Billy survived, she would acknowledge the intervention of Christ by converting to Christianity. Billy survived and Elise converted.

When Lewis called to tell of Elise's conversion, my mother failed to appreciate that Elise could be motivated by a search for faith and became extremely upset. To her, this was not a question of accepting a Christian daughter-in-law into our family. Both Lewis and I had had relationships with non-Jewish women, and my mother was prepared to accept the possibility that one of us might marry a non-Jew. But she found Elise's conversion very different. Elise was Jewish. She had been born in Vienna, experienced anti-Semitism, survived, and was now abandoning Judaism. Why did Jews struggle to survive, my mother argued, if not to continue our cultural heritage? To her, the essence of Judaism lay less in the conception of God than in what she saw as the social and intellectual values of Judaism. My mother could not help comparing Elise's actions with those of Denise's mother, who had sacrificed peace of mind and even her daughter's safety in order for Denise to maintain her cultural and historical continuity as a Jew.

Elise and I were on good terms, yet she had never discussed her desire to convert or her search for greater spiritual values with me. I could not grasp what had happened and I worried that this might reflect a psychological crisis in response to Billy's birth, perhaps a postpartum depression. Failing to persuade Elise over the telephone, my mother flew to Bar-le-Duc and spent two weeks with Lewis and Elise, but she did not alter Elise's conviction.

During our sojourn in France, Denise, Paul, and I visited Bar-le-Duc several times, and Elise, Lewis, and their children came to visit us in Paris. These visits gave us the opportunity to discuss Elise's new faith in a more leisurely setting, and I gradually realized that she was searching for a deep belief. In time, Elise also converted their five children, to my mother's deep dismay and my astonishment. Lewis, who had not converted, did not intervene.

By 1965 Lewis and Elise were eager to have their children grow up in the United States. Lewis arranged for a transfer to an air force base in Tobyhanna, Pennsylvania. Two years later he took an administrative position at the Health and Hospitals Administration of the City of New York. He spent the week in New York living with my parents and the weekend in Tobyhanna. In the meantime, Elise moved from being a Baptist to being a Methodist. In the ensuing decade she became a Presbyterian and finally, as I once humorously predicted to her, a Roman Catholic.

From a distance, this progression seemed a search for increasingly greater structure, greater security, by a person who must have been very frightened on a deep level and looked to Christianity to contain her fear. If Elise was frightened, however, it was not apparent to me. I was amazed by her own actions and even more upset by her conversion of the children. Nonetheless, I had gone to a yeshivah and had a vague sense of what a deep religious conviction might mean to someone.

Even more important, I was all too aware that we are all haunted by our own history, our unique problems, our personal demons and that those experiences and fears profoundly influence our actions. During the period we lived in France, my first extended stay in Europe since leaving Vienna in 1939, I was made acutely aware of my own demons. Even while enjoying a productive period of research and a remarkable range of pleasurable cultural experiences, I felt at times intensely isolated and alone. French society and French science are hierarchical, and I was a relatively unknown scientist at the bottom of the ladder.

The year before I went to Paris, I arranged for Tauc to come to Boston to give a series of seminars. He stayed at our house and we gave a welcoming dinner party for him. But once we were in France, the hierarchy kicked in. Neither Tauc nor any of the other senior people at the institute invited us or any of their other postdoctoral fellows to their homes or interacted with us socially. Moreover, I experienced a subtle degree of anti-Semitism—particularly from the technical people in the laboratory, the technicians and secretaries—that I had not experienced since escaping Vienna. My feeling of unease began when I mentioned to Claude Ray, Tauc's technician, that I was Jewish. He looked at me in disbelief and insisted that I did not look Jewish. When I assured him that I was, he quizzed me on whether I participated in the international Jewish conspiracy to control the world. I mentioned this remarkable conversation to Tauc, who pointed out to me that a good part of the French working class shared this belief about the

Jews. This experience led me to wonder whether Elise, during her many years away from the United States, had encountered similar anti-Semitism and whether this demon might have contributed to her conversion.

In 1969 Lewis developed cancer of the kidney. The tumor was removed successfully, apparently without leaving a trace of the disease. Twelve years later, however, the cancer recurred without warning, tragically claiming Lewis's life at age fifty-seven. After my brother's death, my contact with Elise and the children, perhaps predictably, weakened significantly. We continue to see each other but now at intervals measured in years rather than weeks or months.

My brother is an enormous influence on me to this day. My interest in Bach, Mozart, and Beethoven, and classical music in general, my love of Wagner and of opera, and my joy in learning new things were shaped to a great extent by him. In a later phase of my life, as I began to sense the pleasure of the palate, I came to appreciate that even here, in the area of good food and wine, Lewis's efforts on me were not completely wasted.

IN OCTOBER 1963, JUST BEFORE I LEFT PARIS, TAUC AND I HEARD on the radio that Hodgkin, Huxley, and Eccles had won the Nobel Prize in Physiology or Medicine for their work on signaling in the nervous system. We were thrilled. We felt that our field had been recognized in a very important way and that its very best people had been honored. I could not resist saying to Tauc that I thought the problem of learning was of such importance and still so untouched scientifically that whoever solved this problem might someday also get the Nobel Prize.

12

A CENTER FOR NEUROBIOLOGY AND BEHAVIOR

After a very productive fourteen months in Tauc's laboratory, I returned to the Massachusetts Mental Health Center in November 1963 as an instructor, the lowest rung on the faculty ladder. I supervised residents in training in psychotherapy, an exercise I called the blind leading the blind. A resident would discuss with me the various therapy sessions he or she had had with a specific patient, and I would try to give helpful advice.

Three years earlier, when I first arrived at the mental health center to begin my psychiatric residency, I had encountered an unanticipated bonus. Stephen Kuffler, whose thinking had so influenced my own, had been recruited from Johns Hopkins University to build up the neurophysiology faculty in the department of pharmacology at Harvard Medical School. Kuffler brought with him as junior faculty several extraordinarily gifted scientists who had been postdoctoral fellows in his laboratory—David Hubel, Torsten Wiesel, Edwin Furshpan, and David Potter. In one fell swoop, Kuffler had succeeded in forming the premier group of neural scientists in the country. Always a first-class experimentalist, he emerged as the most admired and effective leader of the American neuroscience community.

After my return from Paris, my interactions with Kuffler increased.

He liked the work on *Aplysia* and was very supportive. Until his death in 1980, he proved a friend and counselor of immeasurable strength and generosity. He took an intense interest in people, their careers, and their families. Years after I had left Harvard, he would call on an occasional weekend to discuss a paper of mine he had found interesting or simply to inquire about my family. When he sent me a copy of the book he had written in 1976 with John Nicholls, *From Neuron to Brain*, he inscribed it, "This is meant for Paul and Minouche" (then fifteen and eleven).

DURING THE TWO YEARS I TAUGHT AT HARVARD MEDICAL School, I struggled with three choices that were to have a profound effect on my career. The first arose when I was invited, at age thirty-six, to take on the chairmanship of the department of psychiatry at Beth Israel Hospital in Boston. The psychiatrist who had just retired from that position, Grete Bibring, was a leading psychoanalyst and a former colleague of Marianne and Ernst Kris in Vienna. A few years earlier, such an appointment would have represented my highest aspiration. But by 1965 my thinking had moved in a very different direction, and I decided against it, with Denise's strong encouragement. She summarized it simply: "What, compromise your scientific career by trying to combine basic research with clinical practice and administrative responsibilities!"

Second, I made the even more fundamental and difficult decision not to become a psychoanalyst but to devote myself full-time to biological research. I realized I could not successfully combine basic research and a clinical practice in psychoanalysis, as I had earlier hoped. One problem I encountered repeatedly within academic psychiatry was that young physicians take on much more than they can handle effectively—a problem that only becomes worse with time. I decided that I could not and would not do that.

Finally, I decided to leave Harvard and its clinical environment for an appointment in a basic science department at my alma mater, New York University Medical School. There, I would start a small research group within the department of physiology focused specifically on the neurobiology of behavior.

Harvard—where I had spent my college years and two years of

medical residency, and where I was then a junior faculty member—was wonderful. Boston is an easy place in which to live and bring up young children. Moreover, the university has extraordinary depth in most areas of scholarship. It was not easy to leave this heady intellectual environment. Nevertheless, I did so. Denise and I moved to New York in December 1965, a few months after our daughter, Minouche, was born, completing our family.

During the period in which I was working through these decisions, I was also in the process of terminating a personal psychoanalysis I had undertaken in Boston. Analysis was particularly helpful to me in that difficult and stressful period; it allowed me to dismiss ancillary considerations and focus on the fundamental issues involved in my making a reasonable choice. My analyst, who was extremely supportive, did suggest I consider having a small, specialized practice, one focusing on patients with a particular disorder, and see them once a week. But he readily understood that I was too single-minded at that point to carry off a dual career successfully.

I am often asked whether I benefited from my analysis. To me, there is little doubt. It gave me new insights into my own actions and into the actions of others, and it made me, as a result, a somewhat better parent and a more empathic and nuanced human being. I began to understand aspects of unconscious motivation and connections between some of my actions of which I was previously unaware.

What about giving up a clinical practice? Had I stayed in Boston, I might eventually have followed my analyst's advice and started a small practice. That was still easy for me to do in Boston in 1965. But in New York, where few physicians were sufficiently familiar with my clinical competencies to refer patients to me, it would have been much more difficult. Moreover, one has to know one's self. I am really best when I can focus on one thing at a time. I knew that studying learning in *Aplysia* was all I could handle at this early point in my career.

THE POSITION AT NYU, BEING IN NEW YORK CITY, HAD THREE attractions that proved, in the long run, to be critical. First, it brought Denise and me closer to my parents and to Denise's mother, all of whom were getting on in years, were having medical problems, and would benefit from our being nearby. We also thought it would be wonderful for our children to be closer to our parents. Second, while we were in Paris, Denise and I had spent many weekends visiting art galleries and museums, and in Boston we had begun to collect works on paper by German and Austrian expressionist artists, an interest that was to grow with time. In the mid-1960s, Boston had only a very few galleries, whereas New York was the center of the art world. Moreover, while in medical school I had followed Lewis's lead and fallen in love with the Metropolitan Opera; the return to New York allowed Denise and me to indulge this interest.

In addition, the position at NYU gave me the marvelous opportunity to work with Alden Spencer again. After his stay at NIH, Alden had accepted a position as an assistant professor at the University of Oregon Medical School. He had become frustrated there because teaching was so time-consuming that it crowded out his research. I had tried without success to help him get a position at Harvard. The offer from NYU allowed me to recruit an additional senior neurophysiologist, and Alden agreed to come to New York.

He loved the city. It provided an outlet for his and Diane's love of music, and soon after they arrived, Diane took up the harpsichord, studying with Igor Kipnis, a gifted harpsichordist who also happened to be a classmate of mine from Harvard. Alden occupied the laboratory next to mine. Although we did not collaborate on actual experiments (because Alden was working on the cat and I on *Aplysia*), we talked daily about the neurobiology of behavior and almost everything else, until his untimely death eleven years later. No one else influenced my thinking on matters of science as much as he did.

Within a year Alden and I were joined by James H. Schwartz (figure 12-1), a biochemist who had been recruited to the medical school independently of Alden and me. Jimmy and I had been housemates and friends at Harvard summer school in 1951, and he had been two classes behind me at NYU Medical School, where we renewed our friendship. However, we had not been in touch since I left medical school in 1956.

After graduating from medical school, Jimmy had obtained a Ph.D. at Rockefeller University, where he studied enzyme mechanisms and the chemistry of bacteria. By the time we met again, in the

12-1 James Schwartz (b. 1932), whom I first met in the summer of 1951, obtained both an M.D. at New York University and a Ph.D. in biochemistry at Rockefeller University. He pioneered the biochemistry of *Aplysia* and made major contributions to the molecular underpinnings of learning and memory. (From Eric Kandel's personal collection.)

spring of 1966, Jimmy had established a reputation as an outstanding young scientist. As we talked about science, he mentioned that he was thinking of switching his research from bacteria to the brain. Because the nerve cells of *Aplysia* were so large and uniquely identifiable, they seemed like good candidates for a study of biochemical identity—that is, of how one cell differs from another at the molecular level. Jimmy began by studying the specific chemical transmitters used for signaling by different *Aplysia* nerve cells. He, Alden, and I formed the nucleus of the new Division of Neurobiology and Behavior that I had founded at NYU.

Our group was very much influenced by Stephen Kuffler's group at Harvard—not just by what it had done, but also by what it was not doing. Kuffler had developed the first unified department of neurobiology that brought the electrophysiological study of the nervous system together with biochemistry and cell biology. It was an extraordinarily powerful, interesting, and influential development, a model for modern neural science. Its focus was the single cell and the single synapse. Kuffler shared the view of many good neural scientists that the uncharted territory between the cell biology of neurons and behavior was too great to be mapped and bridged in a reasonable period of time (such as our lifetimes). As a result, the Harvard group did not in its early days recruit anyone who specialized in the study of behavior or learning.

Occasionally, after he had had a glass or two of wine, Steve would talk freely about the higher functions of the brain, about learning and memory, but he told me that in sober moments he thought they were too complex to be tackled on the cellular level at that time. He also felt, unjustifiably I thought, that he did not know much about behavior and did not feel comfortable studying it.

Alden, Jimmy, and I differed with Kuffler on this point. We were not as constrained by what we did not know, and we found enticing the very uncharted nature of the territory and the importance of the problems. We therefore proposed that the new division at NYU examine how the nervous system produces behavior and how behavior is modified by learning. We wanted to merge cellular neurobiology with the study of simple behavior.

In 1967 Alden and I announced this direction in a major review entitled "Cellular Neurophysiological Approaches in the Study of Learning." In that review we pointed out the importance of discovering what actually goes on at the level of the synapse when behavior is modified by learning. The next key step, we noted, was to move beyond analogs of learning, to link the synaptic changes in neurons and their connections to actual instances of learning and memory. We outlined a systematic cellular approach to this challenge and discussed the strengths and weaknesses of a variety of simple systems that lend themselves to such an approach—snails, worms, and insects, and fish and other simple vertebrates. Each of these animals had behaviors that, in principle, should be modifiable by learning, although this had not yet been demonstrated in *Aplysia*. Moreover, delineating the neural circuitry of those behaviors would reveal where learning-induced changes take place. We would then be in a position to use the powerful technique of cellular neurophysiology to analyze the nature of the changes.

By the time Alden and I were writing this review, I was already in transition not only from Harvard to NYU, but also from cellular neurobiology of synaptic plasticity to the cellular neurobiology of behavior and learning.

The impact of our review—perhaps the most influential one I have written—persists to this day. It inspired a number of investigators to take up a reductionist approach to the study of learning and memory, and simple experimental systems devoted to learning began to sprout up all over—in the leech, the snail *Limax*, the marine snails *Tritonia* and *Hermissenda*, the honeybee, the cockroach, the crayfish, and the lobster. These studies supported an idea, first put forth by ethologists studying the behavior of animals in their natural habitat, that learning is conserved through evolution because it is essential for survival. An animal must learn to distinguish prey from predators, food that is nutritious from that which is poisonous, and a place to rest that is comfortable and safe from one that is crowded and dangerous.

The impact of our ideas also extended to vertebrate neurobiology. Per Andersen, whose laboratory pioneered, in 1973, the modern study of synaptic plasticity in the mammalian brain, wrote: "Did such ideas influence scientists working in the field before 1973? To me the answer is obvious."

Alden's and my review convinced David Cohen, a friendly rival who later became a colleague and vice president for arts and sciences at Columbia, of the value of simple systems. Because he was committed to vertebrates, Cohen turned to the pigeon, the favorite experimental animal of Skinner. But whereas Skinner ignored the brain, Cohen focused on brain-controlled changes in heart rate resulting from sensitization and classical conditioning.

Joseph LeDoux, who was also influenced by the review, modified Cohen's protocol for classical conditioning and applied it to the rat, developing what has emerged as the best experimental system for studying the cellular mechanisms of learned fear in mammals. LeDoux focused on the amygdala, a structure that lies deep under the cerebral cortex and is specialized for the detection of fear. Years later, when it proved possible to produce genetically modified mice, I turned to the amygdala and, influenced by LeDoux's work, extended the molecular biology of learned fear in *Aplysia* to learned fear in the mouse.

13

EVEN A SIMPLE BEHAVIOR CAN BE MODIFIED BY LEARNING

When I arrived at NYU in December 1965 I knew the time had come to take a big step. In Tauc's laboratory I had found that in response to different patterns of stimulation modeled on those produced by Pavlovian learning, a synapse can readily undergo long-lasting changes and that these affect the strength of the communication between two nerve cells in an isolated ganglion. But this was an artificial situation. I had no direct evidence that, in a behaving animal, actual learning leads to changes in the effectiveness of synapses. I needed to move beyond modeling learning in individual cells of an isolated ganglion to studying instances of learning and memory in the neural circuit of a behavior in the intact, behaving animal.

I therefore set two goals for the next several years. First, I would develop a detailed catalog of the behavioral repertoire of *Aplysia* and determine which behaviors could be modified by learning. Second, I would select for research one behavior capable of being modified by learning and use it to explore how learning occurs and how memories are stored in the neural circuitry of that behavior. I had this agenda in mind while I was still at Harvard, and I set about finding a postdoctoral fellow with a specific interest in learning in invertebrates to collaborate with me on this problem.

I had the good fortune to recruit Irving Kupfermann, a gifted and idiosyncratic behaviorist trained at the University of Chicago. He joined me in Boston a few months before I left Harvard and then moved with me to NYU. Irving was a typical University of Chicago intellectual. Tall and very thin, extremely bookish, and slightly eccentric, he wore thick glasses and was almost bald despite his youth. One of his students later described him as "a large brain at the end of a long thin rod." Irving was allergic to rodents and cats, so he had worked on a small segmented land invertebrate creature, the pill bug, for his Ph.D. dissertation. He proved an extremely well informed and creative student of behavior who was very clever at designing experiments.

Together, we set about exploring the behavior of *Aplysia*, searching for one behavior that could be used to study learning. We became familiar with almost every feature of the animal's feeding behavior, its daily pattern of locomotor activity (figure 13-1), inking, and egg laying. We were fascinated by its sexual behavior (figure 13-2), the most obvious and impressive social behavior in *Aplysia*. These snails are hermaphrodites; they can be both male and female, with different partners at different times or even simultaneously. By recognizing one another appropriately they can form impressive copulating chains in which each member serves as both male for the animal in front of it and female for the animal behind it in the chain.

As we analyzed and thought about these behaviors, we realized they were all too complex, some involving more than one ganglion of the snail's nervous system. We needed to find a very simple behavior controlled by the cells of one ganglion. We therefore focused on the several behaviors controlled by the abdominal ganglion, the one I had studied in Paris and with which I was most familiar. The abdominal ganglion, which contains only two thousand nerve cells, controls heart rate, respiration, egg laying, inking, release of mucus, and withdrawal of the gill and siphon. In 1968 we settled on the simplest behavior: the gill-withdrawal reflex.

The gill is an external organ that *Aplysia* uses to breathe. It lies in

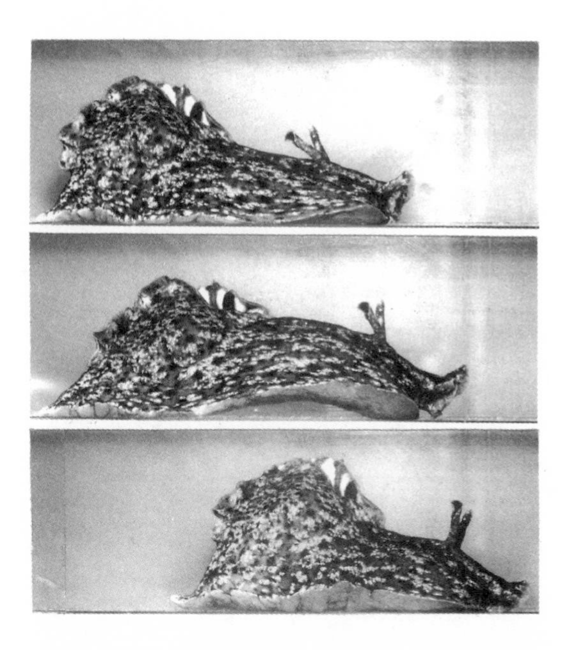

13-1 One full step. *Aplysia* moves by lifting its head and releasing suction to raise the front of its foot, which it then extends for a distance equal to half its body length. The animal brings the front of its foot down, attaches it to a surface, then contracts its anterior end to take up the slack. (Courtesy of Paul Kandel.)

a cavity of the body wall called the mantle cavity and is covered by a sheet of skin called the mantle shelf. The mantle shelf ends in the siphon, a fleshy spout that expels seawater and waste from the mantle cavity (figure 13-3A). Touching the siphon lightly produces a brisk defensive withdrawal of both the siphon and the gill into the mantle cavity (figure 13-3B). The purpose of the withdrawal reflex is clearly to protect the gill, a vital and delicate organ, from possible damage.

Irving and I found that even this very simple reflex can be modified by two forms of learning—habituation and sensitization—and each gave rise to a short-term memory that lasts for a few minutes. An initial light touch to the siphon produces brisk withdrawal of the gill. Repeated light touches produce habituation: the reflex progressively weakens as the animal learns to recognize that the stimulus is trivial. We produced sensitization by applying a strong shock to either the head or the tail. The animal recognized the strong stimu-

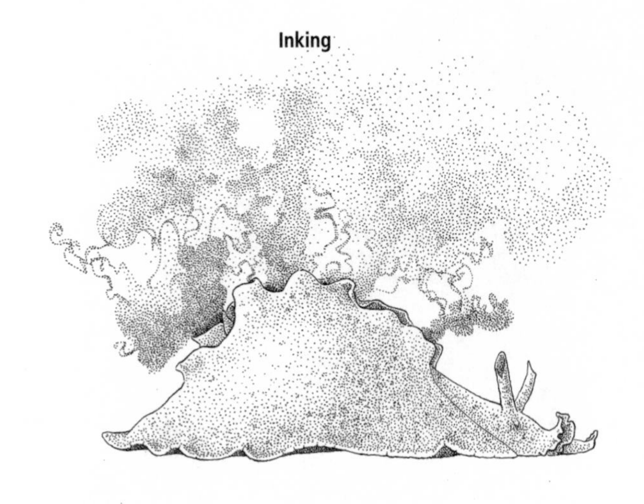

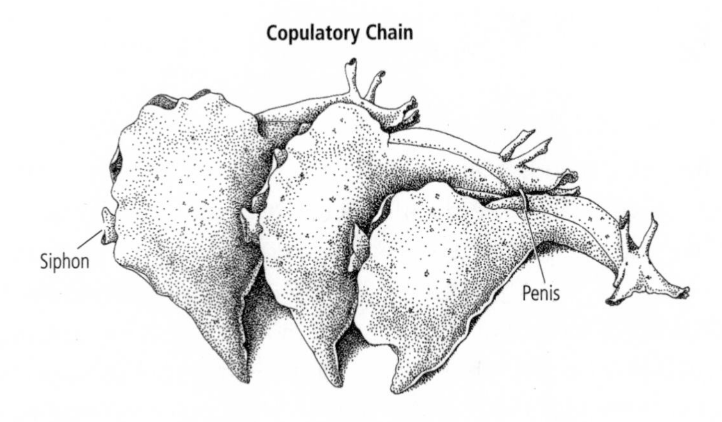

13-2 Simple and complex behaviors in *Aplysia*. Inking (above) is a relatively simple behavior, controlled by cells in a single ganglion (the abdominal ganglion) of the snail's nervous system. Sexual behavior is far more complex and involves nerve cells in several ganglia. *Aplysia* are hermaphrodites, able to be both male and female, and often form copulatory chains such as the one shown (below). (Reprinted from *Cellular Basis of Behavior*, E. R. Kandel, W. H. Freeman and Company, 1976.)

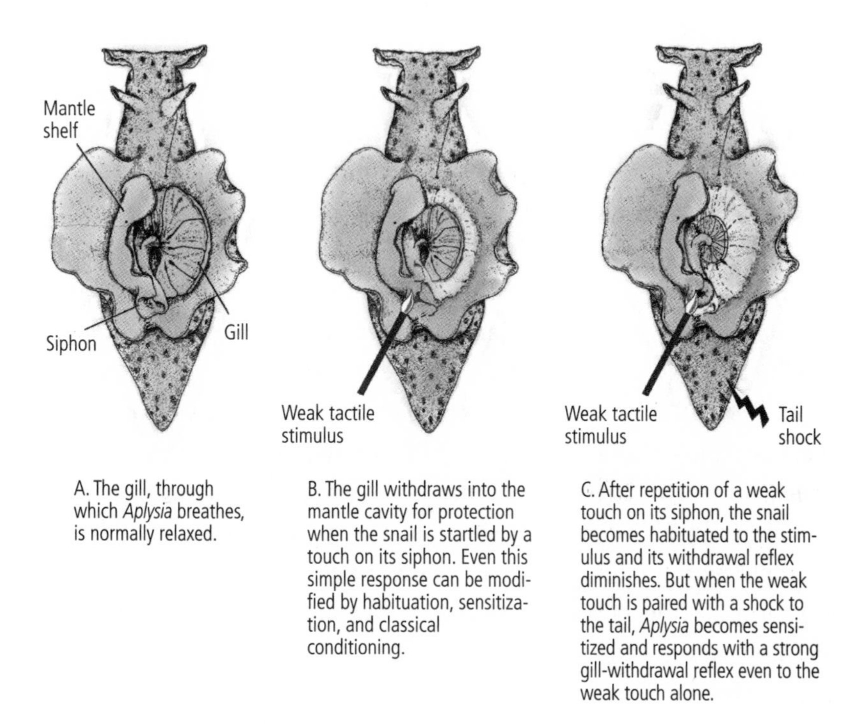

13-3 *Aplysia*'s simplest behavior, the gill-withdrawal reflex.

lus as noxious and subsequently produced an exaggerated gill-withdrawal reflex in response to the same light touch to the siphon (figure 13-3C).

In 1971 we were joined by Tom Carew, a gifted, energetic, and gregarious physiological psychologist from the University of California, Riverside, who opened up our study of long-term memory. Carew simply loved being in the neurobiology and behavior group at NYU. He became good friends with Jimmy Schwartz and Alden Spencer as well as with me. Like a dry sponge, Carew soaked up the culture of the group—not only the science, but also the shared interest in art, music, and scientific gossip. As Carew and I would say to each other, "When other people engage in this talk, it's gossip; when we do it, it's intellectual history."

Carew and I found that long-term memory in *Aplysia*, as in people, requires repeated training interspersed with periods of rest. Practice makes perfect, even in snails. Thus, forty stimuli administered consecutively result in habituation of gill withdrawal that lasts only one day, but ten stimuli every day for four days produce habituation that lasts for weeks. Spacing the training with periods of rest enhances the ability of an *Aplysia* to establish long-term memory.

Kupfermann, Carew, and I had demonstrated that a simple reflex was amenable to two nonassociative forms of learning, each with short- and long-term memory. In 1983 we succeeded in reliably producing classical conditioning of the gill-withdrawal reflex. This was an important advance since it demonstrated that the reflex can also be modified by associative learning.

By 1985, after more than fifteen years of hard work, we had shown that a simple behavior in *Aplysia* could be modified by various forms of learning. This strengthened my hope that some forms of learning had been conserved throughout evolution and would be found even in a simple neural circuit of a very simple behavior. Moreover, I could now foresee the possibility of going beyond the questions of how learning occurs and how memory is stored in the central nervous system to the question of how different forms of learning and memory relate to each other at the cellular level. Specifically, how is short-term memory converted to long-term memory in the brain?

BEHAVIORAL STUDIES OF THE GILL-WITHDRAWAL REFLEX WERE not the sole focus of our work during this time. Indeed, they formed the groundwork for our second and principal interest—devising experiments to explore what happens in the brain of an animal as it learns. Thus, once we had decided to focus our study of learning on the gill-withdrawal reflex in *Aplysia*, we needed to map the neural circuitry of the reflex to learn how the abdominal ganglion brings it about.

Working out the neural circuitry posed its own conceptual challenge. How precise and specialized are the connections among the cells of a neural circuit? In the early 1960s some followers of Karl Lashley argued that the properties of the different neurons in the cerebral cortex are so similar that they are to all intents and purposes identical, and their interconnections random and of roughly equal value.

Other scientists, particularly students of the invertebrate nervous system, championed the idea that many, perhaps all, neurons are unique. This idea was first proposed in 1908 by the German biologist

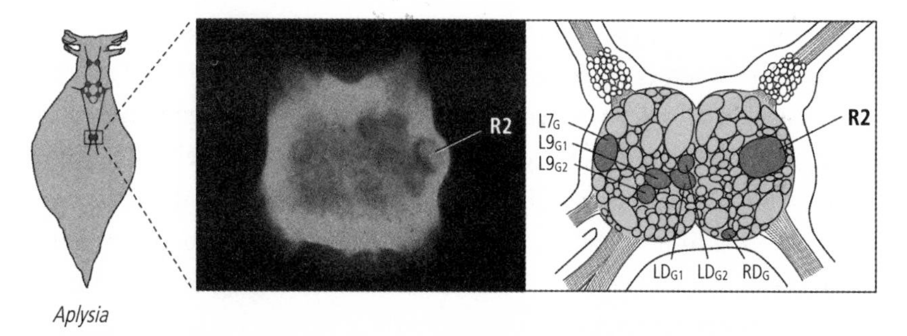

13-4 Identifying specific neurons in the abdominal ganglion of *Aplysia*. Cell R2 can be clearly seen in a photomicrogaph (left) of *Aplysia*'s abdominal ganglion. It measures 1 millimeter in diameter. A drawing (right) shows the position of cell R2 and the six motor neutrons controlling movement of the gill. Once individual neurons had been identified, it was possible to map their connections.

Richard Goldschmidt. Goldschmidt studied a ganglion of the nematode worm *Ascaris*, a primitive intestinal parasite. He found that every animal of the species has the same number of cells in exactly the same position in that ganglion. In a now famous lecture to the German Zoological Association that year, he noted, "the almost startling constancy of the elements of the nervous system: There are 162 ganglion cells in the center, never one more nor less."

Angelique Arvanitaki-Chalazonitis was aware of Goldschmidt's analysis of the worm, and in the 1950s she explored the abdominal ganglion of *Aplysia* in search of identifiable cells. She found that several cells are unique and identifiable in every individual animal on the basis of their location, pigmentation, and size. One such cell was R2, the cell I had focused on in my studies of learning with Ladislav Tauc. First at Harvard and later at NYU, I followed up on this lead, and by 1967 I had found, as had Goldschmidt and Arvanitaki-Chalazonitis earlier, that I could easily identify most of the prominent cells in the ganglion (figure 13-4).

The finding that neurons are unique and that the same cell appears in the same location in every member of the species led to new questions: Are the synaptic connections between these unique neurons also invariant? Does a given cell always signal exactly the same target cell and not others?

To my surprise, I found that I could readily map the synaptic connections between cells. By inserting a microelectrode into a target cell and stimulating action potentials in other cells of the ganglion, one cell at a time, I could identify many of the presynaptic cells that communicate with the target cell. Thus it proved possible for the first time in any animal to map the working synaptic connections between individual cells, which I could use as a method for working out the neural circuit controlling a behavior.

I found the same specificity of connections between individual neurons that Santiago Ramón y Cajal had found between populations of neurons. What's more, just as neurons and their synaptic connections are exact and invariant, so, too, the function of those connections is invariant. This extraordinary invariance would make it easier for me to realize my long-term goal of "trapping" learning in a simple set of neural connections in order to look at how learning gives rise to memory at the cellular level.

By 1969 Kupfermann and I had succeeded in identifying most of the nerve cells that make up the gill-withdrawal reflex. To do this we briefly anesthetized the animal so that we could make a small incision in its neck, and then gently lifted the abdominal ganglion and its attached nerves out through the opening and put them on an illuminated stage. We inserted into various neurons the double-barreled microelectrodes we used for recording and stimulating a cell. Opening up the living animal in this way allowed us to keep its nervous system and all its normal connections intact and thus to observe all of the organs controlled by the abdominal ganglion at the same time. We first set about searching for the motor neurons that control the gill-withdrawal reflex—that is, the motor cells whose axons lead outward from the central nervous system to the gill. We did this by stimulating one cell at a time with the microelectrode and watching to see if that stimulus produced a movement of the gill.

One afternoon in the fall of 1968, working alone, I stimulated a cell and was astonished to see that it produced a powerful contraction of the gill (figure 13-5). For the first time I had identified a motor neuron in *Aplysia* that controlled a specific behavior! I could hardly wait to show

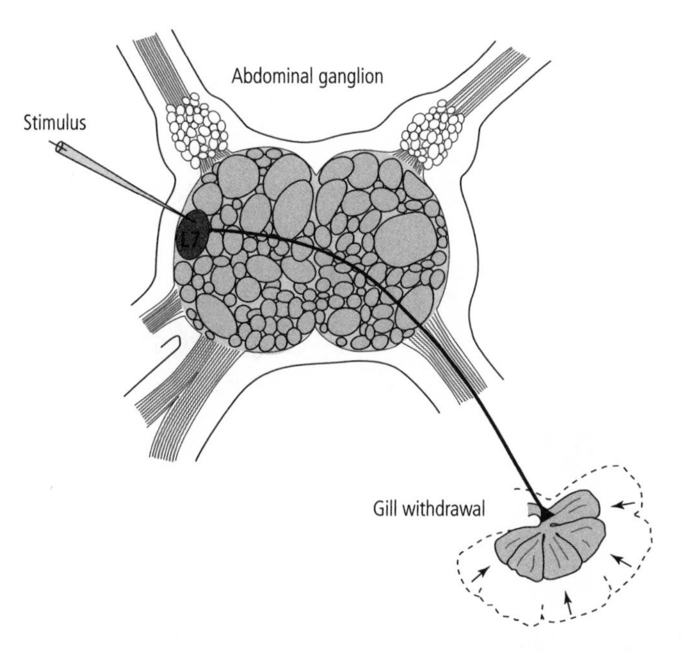

13-5 Discovering a motor neuron that produces a specific behavior in *Aplysia*. Once the individual nerve cells in *Aplysia*'s abdominal ganglion had been identified, it became possible to map their connections. For example, stimulating cell L7 (a motor neuron) produces a sudden contraction of the animal's gill.

Irving. We both were amazed to see the powerful behavioral consequences of stimulating a single cell and knew it boded well for identifying other motor cells. Indeed, within a few months Irving had found five other motor cells. We suspected that those six cells accounted for the motor component of the gill-withdrawal reflex because if we prevented the cells from firing, no reflex response occurred.

In 1969 I was joined by Vincent Castellucci, a delightful and highly cultivated Canadian scientist with an extensive background in biology who regularly trounced me in tennis, and by Jack Byrne, a technically gifted graduate student with training in electrical engineering who brought the rigor of that discipline to bear on our joint work. Together, the three of us identified the sensory neurons of the gill-withdrawal reflex. We then discovered that in addition to their direct connections,

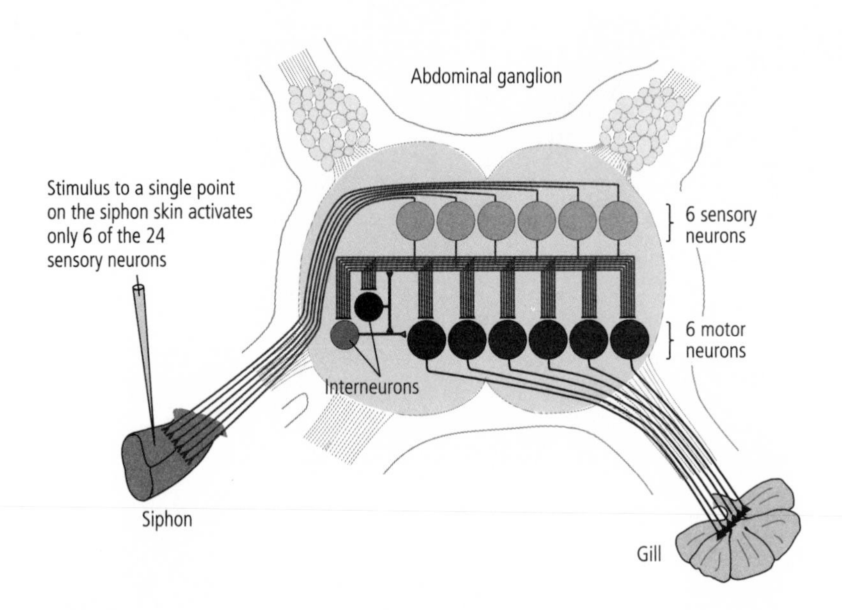

13-6 The neural architecture of *Aplysia's* gill-withdrawal reflex. The siphon system has 24 sensory neurons, but a stimulus applied to any one point on the skin activates only 6 of them. The same 6 sensory neurons relay the sensation of touch to the same 6 motor neurons in every snail, producing the gill-withdrawal reflex.

the sensory neurons formed indirect synaptic connections with motor neurons through interneurons, a type of intermediary neuron. Those two sets of connections—the direct and indirect—relay information about touch to the motor neurons, which actually produce the withdrawal reflex by means of their connections with gill tissue. Moreover, the same neurons were involved in the gill-withdrawal reflex in every snail we studied, and the same cells always formed the same connections with one another. Thus, the neural architecture of at least one behavior of *Aplysia* was amazingly precise (Figure 13-6). In time, we found the same specificity and invariance in the neural circuitry of other behaviors.

Kupfermann and I ended our 1969 paper in *Science*, "Neuronal Controls of a Behavioral Response Mediated by the Abdominal Ganglion of *Aplysia*," on an upbeat note:

In view of its advantages for cellular neurophysiological studies, this preparation may prove useful for analyzing the neuronal mechanisms of learning. Initial experiments indicate that the behavioral reflex response can be modified to show simple learning such as sensitization, habituation. . . . It may also prove possible to study more complex behavioral modifications using either classical or operant conditioning paradigms.

14

SYNAPSES CHANGE WITH EXPERIENCE

Once we had determined that the neural architecture of a behavior is invariant, we were faced with a critical question: How can a behavior that is controlled by a precisely wired neural circuit be changed through experience? One solution had been proposed by Cajal, who suggested that learning could change the strength of the synapses between neurons, thereby strengthening communication between them. Interestingly, Freud's "Project for a Scientific Psychology" outlines a neural model of mind that includes a similar mechanism of learning. Freud postulated that there are separate sets of neurons for perception and memory. The neural circuits concerned with perception form synaptic connections that are fixed, thus assuring the accuracy of our perceptual world. The neural circuits concerned with memory have synaptic connections that change in strength with learning. This mechanism forms the basis of memory and higher cognitive functioning.

The work of Pavlov and the behaviorists and that of Brenda Milner and the cognitive psychologists had led me to the realization that different forms of learning give rise to different forms of memory. I had therefore reformulated Cajal's idea and used that new insight as a basis for developing analogs of learning in *Aplysia*. The results of that work had shown that different patterns of stimulation alter the strength of synaptic connections in different ways. But Tauc and I had not examined how an actual behavior is changed and therefore had no evidence that learning really relies on changes in synaptic strength.

Indeed, the very idea that synapses could be strengthened by learning and thus contribute to memory storage was by no means generally accepted. Two decades after Cajal's proposal, the distinguished Harvard physiologist Alexander Forbes suggested that memory is maintained by dynamic, ongoing changes within a closed loop of self-exciting neurons. To support this idea, Forbes cited a drawing by Rafael Lorente de Nó, a student of Cajal, which showed that neurons connect to each other in closed pathways. The idea was further elaborated by the psychologist D. O. Hebb in his influential 1949 book, *The Organization of Behavior: A Neuropsychological Theory*. Hebb argued that reverberatory circuits are responsible for short-term memory.

Similarly, B. Delisle Burns, a leading student of the biology of the cerebral cortex, challenged the idea that physical changes in synapses can serve as a means of memory storage:

> The mechanisms of synaptic facilitation which have been offered as candidates for an explanation of memory . . . have proven disappointing. Before any of them can be accepted as the cellular changes accompanying conditioned reflex formation, one would have to extend considerably the scale of time on which they have been observed to operate. The persistent failure of synaptic facilitation to explain memory makes one wonder whether neurophysiologists have not been looking for the wrong kind of mechanisms.

Some scholars questioned whether learning could take place at all in fixed neural circuits. For them, learning had to be partially or even totally independent of preestablished neuronal pathways. This view was held by Lashley and by some members of an influential group of early cognitive psychologists, the Gestalt psychologists. A variant of this idea was put forward in 1965 by the neurophysiologist Ross Adey. He began his argument by saying that "no neuron in natural or artificial isolation from other neurons has been shown capable of storing information in the usual notion of memory." He then went on to argue that the flow of current through the space between neurons may carry information that ranks "at least equivalently with neuronal firing in the transaction of information and even more importantly in its deposition and recall." For Adey, as for Lashley, learning was completely mysterious.

Having worked out the neural circuitry of the gill-withdrawal reflex and determined that it could be modified by learning, my colleagues and I were in a position to ask which, if any, of these ideas had merit. In the first of three consecutive papers we published in the journal *Science* in 1970, we outlined the research strategy that we had used and that was to guide our thinking for the next three decades:

> The analysis of the neural mechanisms of learning and similar behavioral modifications requires an animal whose behavior is modifiable and whose nervous system is accessible for cellular analysis. In this and the subsequent two papers, we have applied a combined behavioral and cellular neurophysiological approach to the marine mollusk *Aplysia* in order to study a behavioral reflex that undergoes habituation and dishabituation (sensitization). We have progressively simplified the neural circuit of this behavior so that the action of individual neurons could be related to the total reflex. As a result, it is now possible to analyze the locus and the mechanisms of these behavioral modifications.

In the ensuing papers we established that memory does not depend on self-exciting loops of neurons. For the three simple forms of learning we studied in *Aplysia*, we found that learning leads to a change in the strength of synaptic connections—and therefore in the effectiveness of communication—between specific cells in the neural circuit that mediates the behavior.

Our data spoke clearly and dramatically. We had delineated the anatomical and functional workings of the gill-withdrawal reflex by recording from individual sensory and motor neurons. We had found that touching the skin activates several sensory neurons that together produce a large signal—a large synaptic potential—in each of the motor neurons, causing them to fire several action potentials. These action potentials in the motor neurons produce a behavior—the withdrawal of the gill. We could see that under normal circumstances, the sensory neurons communicate effectively with the motor neurons, sending them an adequate signal to produce the gill-withdrawal reflex.

We now turned our attention to the synapses between the sensory and motor neurons. We observed that when we produced habituation by touching the skin repeatedly, the amplitude of the gill-withdrawal reflex decreased progressively. This learned change in behavior was paralleled by a progressive weakening of the synaptic connections. Conversely, when we produced sensitization by applying a shock to the animal's tail or head, the enhanced gill-withdrawal reflex was accompanied by a strengthening of the synaptic connection. We concluded that during habituation an action potential in the sensory neuron gives rise to a weaker synaptic potential in the motor neuron, leading to less effective communication, while during sensitization it gives rise to a stronger synaptic potential in the motor neuron, leading to more effective communication.

In 1980 we carried our reductionist approach one step further and explored what happens at the synapses during classical conditioning. Carew and I were joined in this endeavor by Robert Hawkins, an insightful young psychologist from Stanford University. The son of an academic family, he did not need New York to broaden his horizons: he was already a devotee of classical music and opera. A fine athlete, Hawkins had played on the varsity soccer team at Stanford, and he proceeded to focus his athletic passion on sailing.

We found that in classical conditioning, the neural signals from the innocuous (conditioned) and noxious (unconditioned) stimuli must occur in a precise sequence. That is, when the siphon is touched just before the tail is—thus predicting the shock to the tail—the sensory neurons will fire action potentials just before they receive signals from the tail. The precisely timed firing of action potentials in the sensory neurons, followed by the precisely timed arrival of the signals from the tail shock, leads to much greater strengthening of the synapse between the sensory and motor neurons than when signals from the siphon or the tail occur separately, as they do in sensitization.

These several results on habituation, sensitization, and classical conditioning led us irresistibly to think about how genetic and developmental processes interact with experience to determine the structure of mental activity. Genetic and developmental processes specify the connections among neurons—that is, which neurons form synaptic connections with which other neurons and when. But they do not specify the strength of those connections. Strength—the long-term effectiveness of synaptic connections—is regulated by experience. This view implies that the *potential* for many of an organism's behaviors is built into the brain and is to that extent under genetic and developmental control; however, a creature's environment and learning alter the effectiveness of the preexisting pathways, thereby leading to the expression of new patterns of behavior. Our findings in *Aplysia* supported this view: in its simplest forms, learning selects among a large repertoire of preexisting connections and alters the strength of a subset of those connections.

In reviewing our results, I could not help being reminded of the two opposing philosophical views of mind that had dominated Western thought from the seventeenth century onward—empiricism and rationalism. The British empiricist John Locke argued that the mind does not possess innate knowledge but is instead a blank slate that is eventually filled by experience. Everything we know about the world is learned, so the more often we encounter an idea, and the more effectively we associate it with other ideas, the more enduring its impact on our minds. Immanuel Kant, the German rationalist philosopher, argued to the contrary, that we are born with certain built-in templates of knowledge. Those templates, which Kant called *a priori* knowledge, determine how sensory experience is received and interpreted.

In choosing between psychoanalysis and biology as a career, I had decided on biology because psychoanalysis, and its predecessor discipline, philosophy, treated the brain as a black box, an unknown. Neither field could resolve the conflict between the empiricist and rationalist views of mind as long as the resolution required a direct examination of the brain. But examining the brain was just what we had begun to do. In the gill-withdrawal reflex of this simplest of organisms, we saw that both views had merit—in fact, they complemented each other. The anatomy of the neural circuit is a simple example of Kantian *a priori* knowledge, while changes in the strength of particular connections in the neural circuit reflect the influence of experience. Moreover, consistent with Locke's notion that practice makes perfect, the persistence of such changes underlies memory.

Whereas the study of complex learning had seemed intractable to Lashley and others, the elegant simplicity of the gill-withdrawal reflex in a snail enabled my colleagues and me to address experimentally a number of the philosophical and psychoanalytical questions that had led me to biology in the first place. This I found to be both amazing and humorous.

In the third of our reports in *Science* in 1970, we concluded with these comments:

> [T]he data indicate that habituation and dishabituation (sensitization) both involve a change in the functional effectiveness of previously existing excitatory connections. Thus, at least in the simple cases, . . . [t]he capability for behavioral modification seems to be built directly into the neural architecture of the behavioral reflex.
>
> Finally, these studies strengthen the assumption . . . that a prerequisite for studying behavioral modification is the analysis of the wiring diagram underlying the behavior. We have, indeed, found that once the wiring diagram of the behavior is known, the analysis of its modification becomes greatly simplified. Thus, although this analysis pertains to only relatively simple and short-term behavioral modifications, a similar approach may perhaps also be applied to more complex as well as longer lasting learning processes.

By sticking with a radically reductionist approach—examining a very simple behavioral reflex and simple forms of learning, delineating cell by cell the neural circuit of the reflex, and then focusing on where change occurs within that circuit—I had reached the long-term goal outlined in my grant application to NIH in 1961. I had "trapped a conditioned response in the smallest possible neural population, the connections made between two cells."

THUS THE REDUCTIONIST APPROACH LED US TO DISCOVER several principles of the cell biology of learning and memory. First, we found that the changes in synaptic strength that underlie the learning of a behavior may be great enough to reconfigure a neural network and its information-processing ability. For example, one particular sensory cell in *Aplysia* communicates with eight different motor cells—five that produce movement of the gill and three that cause contraction of the ink gland and thus inking. Before training, activation of this sensory cell excited the five gill-innervating motor cells moderately, causing them to fire action potentials and thereby causing the gill to contract. Activation of this same sensory cell also excited the three ink gland-innervating motor neurons but only very weakly, not enough to produce action potentials or to elicit inking. Thus, before learning, gill withdrawal would take place in response to stimulation of the siphon but inking would not. After sensitization, however, synaptic communication between the sensory cell and all eight motor cells is enhanced, causing the three ink gland–innervating motor neurons to fire action potentials as well. Thus, as a result of learning, when the siphon is stimulated, inking will occur along with more powerful gill withdrawal.

Second, consistent with my reformulation of Cajal's theory and my earlier work with analogs, we found that a given set of synaptic connections between two neurons can be modified in opposite ways—strengthened or weakened—by different forms of learning. Thus, habituation weakens the synapse, whereas sensitization or classical conditioning strengthens it. These enduring changes in the strength of synaptic connections are the cellular mechanisms underlying learning and short-term memory. Moreover, because the changes occur at several sites in the neural circuitry of the gill-withdrawal reflex, memory is distributed and stored throughout the circuit, not at a single specialized site.

Third, we found that in all three forms of learning, the duration of short-term memory storage depends on the length of time a synapse is weakened or strengthened.

Fourth, we were beginning to understand that the strength of a given chemical synapse can be modified in two ways, depending on which of two neural circuits is activated by learning—a mediating circuit or a modulatory circuit. In *Aplysia*, the mediating circuit is made up of the sensory neurons that innervate the siphon, the interneurons, and the motor neurons that control the gill-withdrawal reflex. The modulatory circuit is made up of sensory neurons that innervate the tail in a completely different part of the body. When the neurons in a mediating circuit are activated, homosynaptic changes in strength occur. This is the case in habituation: the sensory and motor neurons that control the gill-withdrawal reflex fire repeatedly and in a certain pattern in direct response to the repeated sensory stimulus. Heterosynaptic changes in strength occur when the neurons in a modulatory, rather than the mediating, neural circuit are activated. This is the case with sensitization: the strong stimulus to the tail activates a modulatory circuit that controls the strength of synaptic transmission in the mediating neurons.

We later found that classical conditioning recruits both homosynaptic and heterosynaptic changes. Indeed, our studies of the relationship of sensitization to classical conditioning indicate that learning may be a matter of combining various elementary forms of synaptic plasticity into new and more complex forms, much as we use an alphabet to form words.

I now began to realize that the abundance of chemical over electrical synapses in the brains of animals may reflect a fundamental advantage of chemical over electrical transmission: the ability to mediate a variety of forms of learning and of memory storage. Viewed from this perspective, it became clear that the synapses between sensory neurons and motor neurons in the gill-withdrawal circuit—neurons that have evolved to participate in various types of learning—are much more easily changed than synapses that play no role in learning. Our studies showed dramatically that in circuits modified by learning, synapses can undergo large and enduring changes in strength after only a relatively small amount of training.

The Aplisa
by Minouche

An aplisa is like
a squishy snail.
In rain in snow in sleet
in hail.
When it is angry, it shoots
out ink.
The ink is purple, its not
pink.
An aplisa cannot live on
land.
It doesn't have feet so
it can't stand.
It has a very funny
mouth
And in winter it goes to the south.

(Reprinted from *Behavioral Biology of* Aplysia, E. R. Kandel, W. H. Freeman and Company, 1979.)

One of the fundamental features of memory is that it is formed in stages. Short-term memory lasts minutes, while long-term memory lasts many days or even longer. Behavioral experiments suggest that short-term memory grades naturally into long-term memory and, moreover, that it does so through repetition. Practice does make perfect.

How does practice do it? How does training convert a short-term memory into a persistent, self-maintained long-term memory? Does the process take place at the same site—the connection between the sensory and motor cells—or is a new site required? We were now in a position to answer those questions.

In this period, science was again claiming my total concentration, to the exclusion of other activities. In my obsession with *Aplysia*, however, I found an unexpected ally in my daughter, Minouche. In 1970, at age five, as Minouche started to read, she stumbled on a picture of *Aplysia* in *The Larousse Encyclopedia of Animal Life*, a beautiful picture book that I kept in our living room. She simply loved the picture and would squeal, "*Aplysia! Aplysia!*" over and over again, pointing to the picture.

Two years later, at the age of seven, she wrote the following poem on the occasion of my forty-third birthday:

The Aplisa
by Minouche

An aplisa is like
a squishy snail.
In rain, in snow, in sleet,
in hail.
When it is angry, it shoots
out ink.
The ink is purple, it's not
pink.
An aplisa cannot live on
land.
It doesn't have feet so
it can't stand.
It has a very funny
mouth.
And in winter it goes to the south.

Minouche said it all, much better than I could!

15

THE BIOLOGICAL BASIS OF INDIVIDUALITY

I had learned from my research in *Aplysia* that changes in behavior are accompanied by changes in the strength of the synapses between neurons that produce the behavior. But nothing in my research revealed how short-term memory is transformed into long-term memory. Indeed, nothing was known about the cellular mechanisms of long-term memory.

The basis for my early research in learning and memory was the learning paradigms used by behaviorists. The behaviorists focused primarily on how knowledge was acquired and stored in short-term memory. Long-term memory did not particularly interest them. The interest in long-term memory came from studies of human memory by the forerunners of cognitive psychologists.

IN 1885, A DECADE BEFORE EDWARD THORNDIKE BEGAN HIS studies of learning in experimental animals at Columbia University, the German philosopher Hermann Ebbinghaus transformed the analysis of human memory from an introspective study into a laboratory science. Ebbinghaus was influenced by three scientists—the physiologist Ernst Weber, and the physicists Gustav Fechner and Hermann Helmholtz—who introduced rigorous methods to the study of perception. Helmholtz, for example, measured the speed at which a touch on the skin travels to the brain. It was generally thought at the time that conduction along nerves was immeasurably fast, comparable to the speed of light. But Helmholtz found it to be slow—about 90 feet per second. Moreover, the time it took a subject to react to the stimulus—the reaction time—was even slower! This caused Helmholtz to propose that a considerable amount of the brain's processing of perceptual information is carried out unconsciously. He called this processing "unconscious inference" and proposed that it was based on evaluating and transforming the neural signal without conscious awareness of doing so. This processing, he argued, must result from signals being routed and processed at different sites during perception and voluntary movement.

Like Helmholtz, Ebbinghaus held the position that mental processes are biological in nature and can be understood in the same rigorous scientific terms as physics and chemistry. Perception, for example, can be studied empirically as long as the sensory stimuli used to elicit responses are objective and quantifiable. Ebbinghaus conceived the idea of studying memory by means of a similar experimental approach. The techniques he devised for measuring memory are still in use today.

In devising his experiments on how new information is put into memory, Ebbinghaus had to be certain that the people he studied were forming new associations, not relying for help on associations learned earlier. He hit upon the idea of having subjects learn nonsense words, each consisting of two consonants separated by a vowel (RAX, PAF, WUX, CAZ, and so on). Because each word is meaningless, it does not fit into the learner's preestablished network of associations. Ebbinghaus made up about 2000 such words, wrote each on a separate slip of paper, shuffled the slips, and randomly drew slips to create lists that varied in length between 7 and 36 nonsense words. Faced with the arduous task of memorizing the lists, he went to Paris and rented a room in a garret overlooking the roofs of that beautiful city. There, he memorized each list in turn by reading it aloud at the rate of 50 words per minute. As Denise would say, "Only in Paris can one even think of doing such a boring experiment!"

From these self-inflicted experiments, Ebbinghaus generated two principles. First, he found that memory is graded—in other words, practice makes perfect. There was a linear relationship between the number of training repetitions on the first day and the amount of material retained on the following day. Long-term memory therefore appeared to be a simple extension of short-term memory. Second, despite the apparent similarity in mechanism between short- and long-term memory, Ebbinghaus noted that a list of six or seven items could be learned and retained in only one presentation, whereas a longer list required repeated presentations.

Next, he plotted a forgetting curve. He tested himself at various intervals after learning, using different lists for each interval, and determined the amount of time it took to relearn each list with the same degree of accuracy as in the first learning. He found that there were savings: relearning an old list took less time and fewer trials than the original learning. Most interesting of all, he found that forgetting had at least two phases: a rapid initial decline that was sharpest in the first hour after learning and then a much more gradual decline that continued for about a month.

Based on Ebbinghaus's two phases of forgetting and on his own remarkable intuition, William James concluded in 1890 that memory must have at least two different processes: a short-term process, which he called "primary memory," and a long-term process, which he called "secondary memory." He referred to long-term memory as secondary because it involved recalling memory some time after a primary learning event.

IT GRADUALLY BECAME CLEAR TO THE PSYCHOLOGISTS WHO followed Ebbinghaus and James that the next step in understanding long-term memory was to understand how it becomes firmly established, a process now called consolidation. For a memory to persist, the incoming information must be thoroughly and deeply processed. This is accomplished by attending to the information and associating it meaningfully and systematically with knowledge already well established in memory.

The first clue that newly stored information is made more stable for long-term storage came in 1900 from two German psychologists, Georg Müller and Alfons Pilzecker. Using Ebbinghaus's techniques, they asked a group of volunteers to learn a list of nonsense words well enough to remember them twenty-four hours later, which the group did readily. They then asked a second group to learn the same list with the same number of repetitions, but they gave the second group an additional list of words to learn *immediately after* learning the first list. This second group of volunteers failed to remember the first list twenty-four hours later. In contrast, a third group of volunteers, who were given the second list *two hours after* having learned the first list, had little difficulty remembering the first list twenty-four hours later. This result suggested that in the hour after training, when the initial list had been placed into short-term memory and perhaps even into the early stages of long-term memory, memory was still sensitive to disruption. Presumably, a certain amount of time was required for long-term memory to become fixed, or consolidated. Once consolidated, after two or more hours, it was stable for some time and less easily disrupted.

The idea of memory consolidation is supported by two types of clinical observations. First, it has been known since the end of the nineteenth century that head injuries and concussion can lead to a memory loss called retrograde amnesia. A boxer who is hit on the head and sustains a brain concussion in round five of a match will usually remember going to the event, but everything after that will be blank. Undoubtedly, a number of events entered into his short-term memory just before the blow—the excitement of entering the ring, his opponent's movements during the earlier four rounds, perhaps even the moving punch itself and the attempt to avoid it—but the blow to the brain occurred before any of those memory traces could be consolidated. The second clinical observation is that a similar retrograde amnesia often occurs following an epileptic convulsion. People with epilepsy cannot remember events immediately preceding a seizure, even though the seizure has no effect on their memory of earlier events. This suggests that in its early phases memory storage is dynamic and sensitive to disruption.

The first rigorous test of memory consolidation came in 1949, when the American psychologist C. P. Duncan applied electrical stimuli to the brain of animals during or immediately after training, resulting in convulsions that disrupted memory and caused retrograde amnesia. Producing seizures several hours after training had little or no effect on recall. Almost twenty years later, Louis Flexner at the University of Pennsylvania made the remarkable discovery that drugs that inhibit the synthesis of proteins in the brain disrupt long-term memory if given during and shortly after learning, but they do not disrupt short-term memory. This finding suggested that long-term memory storage requires the synthesis of new proteins. Together, the two sets of studies seemed to confirm the idea that memory storage takes place in at least two stages: a short-term memory lasting minutes is converted—by a process of consolidation that requires the synthesis of new protein—into stable, long-term memory lasting days, weeks, or even longer.

Variants of the two-stage model of memory were soon proposed. According to one view, short- and long-term memory take place at different anatomical sites. Alternatively, some psychologists argued that memory was present at one site and simply became progressively stronger with time. The question of whether short- and long-term memory require two separate sites or can be accommodated at one site is central to the analysis of learning, particularly to the analysis of memory on the cellular level. Clearly, it could not be resolved with behavioral analyses alone—cellular analysis was needed. Our studies of *Aplysia* had put us in a position to tackle the question of whether short- and long-term memory are the same or separate neural processes occurring at the same or different sites.

CAREW AND I HAD FOUND IN 1971 THAT, WITH REPEATED TRAINING, habituation and sensitization—the simplest forms of learning—can be sustained for long periods. Thus they could serve as useful tests for differences between long-term and short-term memory. We eventually discovered that the cellular changes accompanying long-term sensitization in *Aplysia* were similar to changes underlying long-term memory in the mammalian brain: long-term memory required the synthesis of new protein.

We wanted to know whether simple forms of long-term memory use the same storage sites—the same group of neurons and the same set of synapses—as short-term memory. I knew from the work of

Brenda Milner on H.M. that, in people, complex, explicit long-term memory—a memory lasting days to years—requires not only the cortex but also the hippocampus. But what about the simpler implicit memory? Carew, Castellucci, and I found that the same synaptic connections between sensory and motor neurons that are altered in short-term habituation and sensitization are also altered in long-term habituation and sensitization. Moreover, in both cases, the synaptic changes parallel the changes in behavior we observed: in long-term habituation, the synapse is depressed for a period of weeks, whereas in long-term sensitization, it is enhanced for weeks. This suggested that, in the simplest cases, the same site can store both short- and long-term memory and that it can do so for different forms of learning.

That left the question of mechanism. Are the mechanisms of short-term and long-term memory the same? If so, what is the nature of the process by which long-term memory is consolidated? Is protein synthesis needed for the long-term synaptic changes associated with long-term memory storage?

I had thought for some time that long-term memory might be consolidated by an anatomical change. Such a change might be one of the reasons that new protein is needed. I sensed that we would soon require an analysis of the structure of memory storage. In 1973 I succeeded in recruiting Craig Bailey, a talented and creative young cell biologist, to explore the structural changes that accompany the transition from short- to long-term memory.

Bailey and his colleague, Mary Chen, and Carew and I found that long-term memory is not simply an extension of short-term memory: not only do the changes in synaptic strength last longer but, more amazingly, the actual number of synapses in the circuit changes. Specifically, in long-term habituation the number of presynaptic connections among sensory neurons and motor neurons decreases, whereas in long-term sensitization sensory neurons grow new connections that persist as long as the memory is retained (figure 15-1). There is in each case a parallel set of changes in the motor cell.

This anatomical change is expressed in several ways. Bailey and Chen found that a single sensory neuron has approximately 1300 presynaptic terminals with which it contacts about 25 different target

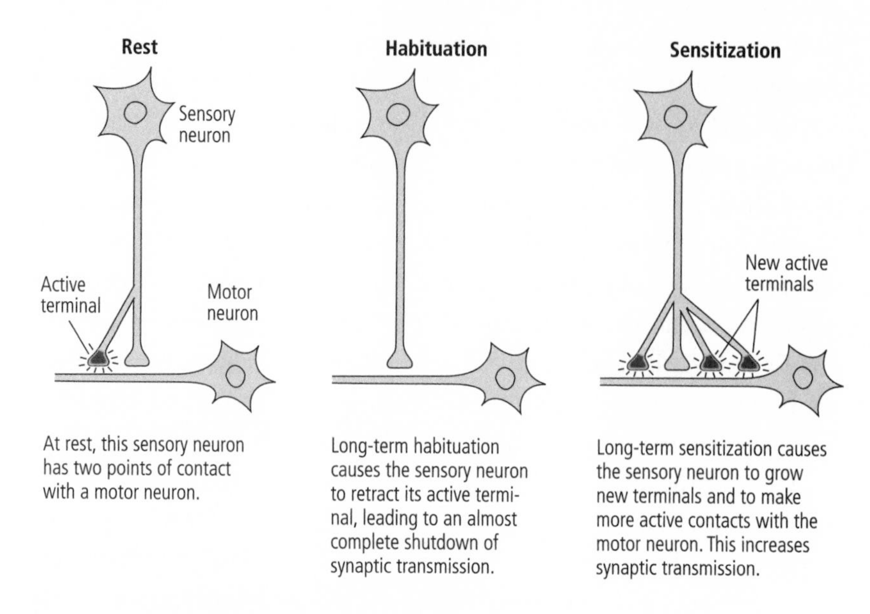

15-1 Anatomical changes accompany long-term memory.

cells—motor neurons, excitatory interneurons, and inhibitory interneurons. Of the 1300 presynaptic terminals, only about 40 percent have active synapses, and only these synapses have the machinery for releasing a neurotransmitter. The remaining terminals are dormant. In long-term sensitization, the number of synaptic terminals more than doubles (from 1300 to 2700), and the proportion of active synapses increases from 40 percent to 60 percent. In addition, there is an outgrowth from the motor neuron to receive some of the new connections. In time, as the memory fades and the enhanced response returns to normal, the number of presynaptic terminals drops from 2700 to about 1500, or slightly more than the initial number. This residual growth presumably is responsible for the fact, first discovered by Ebbinghaus, that an animal can learn a task more readily a second time. In long-term habituation, on the other hand, the number of presynaptic terminals drops from 1300 to about 850, and the number of active terminals diminishes from 500 to about 100—an almost complete shutdown of synaptic transmission (figure 15-1).

Thus in *Aplysia* we could see for the first time that the number of synapses in the brain is not fixed—it changes with learning! Moreover, long-term memory persists for as long as the anatomical changes are maintained.

These studies provided the first clear insights into the two competing theories of memory storage. Both were right, but in different ways. Consistent with the one-process theory, the same site can give rise to both short-term and long-term memory in habituation and sensitization. Moreover, in each case a change in synaptic strength occurs. But consistent with the two-process theory, the mechanisms of short- and long-term change are fundamentally different. Short-term memory produces a change in the function of the synapse, strengthening or weakening preexisting connections; long-term memory requires anatomical changes. Repeated sensitization training (practice) causes neurons to grow new terminals, giving rise to long-term memory, whereas habituation causes neurons to retract existing terminals. Thus, by producing profound structural changes, learning can make inactive synapses active or active synapses inactive.

TO BE USEFUL, A MEMORY HAS TO BE RECALLED. MEMORY retrieval depends on the presence of appropriate cues that an animal can associate with its learning experiences. The cues can be external, such as a sensory stimulus in habituation, sensitization, and classical conditioning, or internal, sparked by an idea or an urge. In the *Aplysia* gill-withdrawal reflex, the cue for memory recall is external: namely, the touch to the siphon that elicits the reflex. The neurons that retrieve the memory of the stimulus are the same sensory and motor neurons that were activated in the first place. But because the strength and number of synaptic connections between these neurons have been altered by learning, the action potential generated by the sensory stimulus to the siphon "reads out" the new state of the synapse when it arrives at the presynaptic terminals and the recall gives rise to a more powerful response.

In long-term memory, as in short-term memory, the number of changed synaptic connections may be great enough to reconfigure a neural circuit, but this time anatomically. For example, prior to training, a stimulus to a sensory neuron in *Aplysia* might be strong enough to cause motor neurons leading to the gill to fire action potentials, but not strong enough to cause motor neurons leading to the ink gland to fire action potentials. Training strengthens not only the synapses between the sensory neuron and the motor neurons to the gill but also the synapses between the sensory neuron and the motor neurons to the ink gland. When the sensory neuron is stimulated after training, it retrieves the memory of the enhanced response, which causes both gill and ink motor neurons to fire action potentials and causes inking as well as gill withdrawal to take place. Thus, the form of *Aplysia*'s behavior is altered. The touch to the siphon elicits not just a change in the magnitude of the behavior—the amplitude of gill withdrawal—but also a change in the animal's behavioral repertory.

Our studies showing that the brain of *Aplysia* is physically changed by experience led us to wonder, Does experience change the primate brain? Does it change the brains of people?

WHEN I WAS A MEDICAL STUDENT IN THE 1950s, WE WERE TAUGHT that the map of the somatosensory cortex discovered by Wade Marshall is fixed and immutable throughout life. We now know that idea is not correct. The map is subject to constant modification on the basis of experience. Two studies in the 1990s were particularly informative in this regard.

First, Michael Merzenich at the University of California, San Francisco discovered that the details of cortical maps vary considerably among individual monkeys. For example, some monkeys have a much more extensive representation of the hand than other monkeys. Merzenich's initial study did not separate the effects of experience from those of genetic endowment, so it was possible that the differences in representation were genetically determined.

Merzenich then carried out additional experiments to determine the relative contributions of genes and experience. He trained monkeys to obtain food pellets by touching a rotating disk with their three middle fingers. After several months, the area of the cortex devoted to the middle fingers—especially the tips of the fingers used for touching the disk—had expanded greatly (figure 15-2). At the same time, the

This drawing shows the relative area that a monkey's somatosensory cortex devotes to its various body parts. Fingers and other particularly sensitive areas take up the most space.

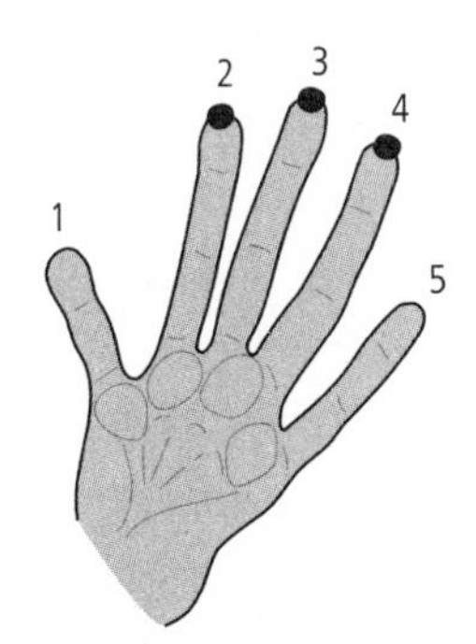

A monkey was trained in a task that requires frequent use of the tips of its middle fingers. After several months of training, these areas became more sensitive.

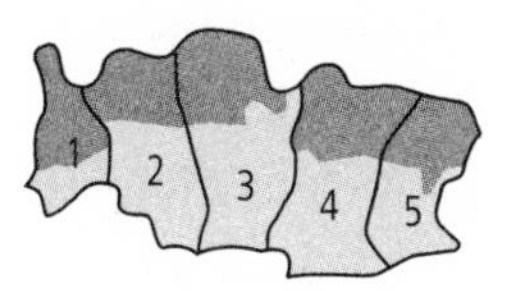

The area of the monkey's somatosensory cortex that corresponds to the monkey's fingertips before training (dark areas).

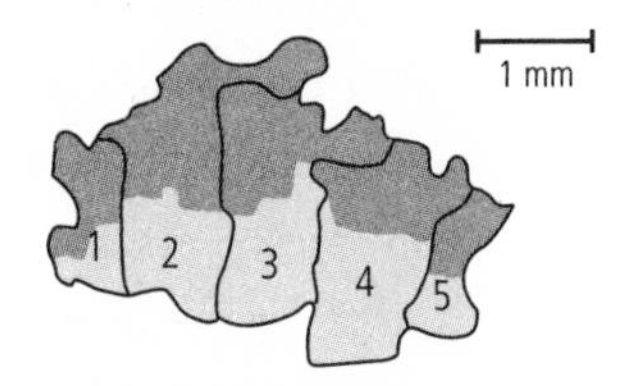

After training, the area that corresponds to the tips of the monkey's middle fingers has expanded.

15-2 Maps of the cortex change with experience. (Adapted from Jenkins et al., 1990.)

tactile sensitivity of the middle fingers increased. Other studies have shown that training in visual discrimination of color or form also leads to changes in brain anatomy and improved perceptual skills.

Second, Thomas Ebert and his colleagues at the University of Konstanz in Germany compared images of violinists' and cellists' brains with images of nonmusicians' brains. Players of stringed instruments use the four fingers of the left hand to modulate the sound of the strings. The fingers of the right hand, which move the bow, are not involved in such highly differentiated movements. Ebert found that the area of the cortex devoted to the fingers of the right hand did not differ in string players and nonmusicians, whereas representations of the fingers of the left hand were much more extensive—by as much as five times—in the brains of string players than in those of nonmusicians. Furthermore, musicians who began playing the instrument before age thirteen had larger representations of the fingers of their left hand than musicians who began playing after that age.

These dramatic changes in cortical maps as a result of learning extended the anatomical insights that our studies in *Aplysia* had revealed: the extent to which a body part is represented in the cortex depends on the intensity and complexity of its use. In addition, as Ebert's study showed, such structural changes in the brain are more readily achieved in the early years of life. Thus, a great musician such as Wolfgang Amadeus Mozart is who he is not simply because he has the right genes (although genes help), but also because he began practicing the skills for which he became famous at a time when his brain was more pliable.

Moreover, our results in *Aplysia* showed that the plasticity of the nervous system—the ability of nerve cells to change the strength and even the number of synapses—is the mechanism underlying learning and long-term memory. As a result, because each human being is brought up in a different environment and has different experiences, the architecture of each person's brain is unique. Even identical twins with identical genes have different brains because of their different life experiences. Thus, a principle of cell biology that first emerged from the study of a simple snail turned out to be a profound contributor to the biological basis of human individuality.

Our finding that short-term memory results from a functional change and long-term memory from an anatomical change raised even more questions: What is the nature of memory consolidation? Why does it require the synthesis of new protein? To find out, we would have to move into the cell and study its molecular makeup. My colleagues and I were ready for that step.

JUST AT THIS POINT, WE HEARD DEVASTATING NEWS. IN THE FALL of 1973 Alden Spencer, my best friend and the cofounder of the Neurobiology and Behavior Division at NYU, began complaining of weakness in his hands, which interfered with his tennis game. Within a few months he was diagnosed as having amyotrophic lateral sclerosis (ALS, or Lou Gehrig's disease), a disease that is invariably fatal. On hearing this diagnosis from one of the country's leading neurologists,

Alden became depressed and began preparing his will, thinking he might be dead within a week. But Alden also had arthritis of the elbow, a feature not typically associated with ALS. I therefore suggested that he see a rheumatologist.

Alden saw a very good doctor who assured him that he did not have ALS but instead had a connective tissue disorder (a collagen disease) related to lupus erythematosus. On hearing this much more optimistic diagnosis, Alden's mood improved. A few months later he went back to his neurologist, who assured him that independent of any arthritis, he clearly had ALS. Alden's mood immediately fell again.

At that point I spoke to the neurologist and told him that Alden was clearly having great difficulty handling his diagnosis, and I asked the neurologist whether he could help Alden by holding out more hope. The neurologist, a thoroughly decent and caring person, insisted that he could not do that because it would deceive Alden about his future, which would not be fair to him. "But," he said, "I have nothing to offer Alden. He simply need not and should not come to see me. Let him continue to see the rheumatologist."

I discussed this course with Alden and independently with his wife, Diane. Both thought it a good idea. Diane was convinced that Alden did not want to confront what she and I had sadly come to agree must be the correct diagnosis, that he had ALS.

During the next two and a half years, Alden slipped slowly and progressively. He first used a cane and then a wheelchair to get around. But at no time did he stop going to his laboratory and doing science. Even though giving lectures became difficult for him, he continued to teach, albeit fewer classes. No one in our group except me knew his true diagnosis, and no one thought, or at least acknowledged, that he did not have a peculiar form of arthritis. He exercised continually and swam on a regular basis at a special swimming pool for the disabled near his house. The day before he died, in November 1977, he was in his laboratory participating in a discussion on sensory processing.

Alden's death was shattering to all of us personally and devastating to our close-knit group. We had talked almost daily for about twenty years, so for a long while afterward the whole rhythm of my working life was disrupted. I still think of Alden frequently.

I was not alone; everyone enjoyed Alden's self-deprecating humor, his modesty and unbounded generosity, and his unending creativity. To honor his memory, we established in 1978 the Alden Spencer Lectureship and Award, given annually to a great scientist under the age of fifty whose best work is still ahead of him or her. The recipient is selected by the entire Center for Neurobiology and Behavior at Columbia—faculty, graduate students, postdoctoral fellows, and professors.

The years following Alden's death were productive and therefore seemed harmonious from the outside, but they were very painful for me personally. Alden's death in 1977 was followed by my father's death the same year and my brother's death in 1981. In each case I was extensively involved in their care, and their deaths left me not only psychologically despondent and depleted but also physically exhausted. I have always been grateful for the serenity I have been able to obtain by focusing hard on my work. At that time, the challenging quest of my work and the surprising insights revealed to me were a particularly welcome retreat from the painful realities of everyday life's irretrievable losses.

This difficult period was made even more painful for me by my son Paul's departure for college in 1979. When Paul was seven years old, I encouraged him to take up chess and to take lessons in tennis, and he subsequently became quite good at both. I played chess and therefore engaged his interest in rooks and knights and checkmates. But I did not play tennis. So at age thirty-nine, I started to take tennis lessons and soon had a mediocre but highly enjoyable game, which I still play regularly. From the time Paul first began playing tennis, he was one of my regular partners. By his last year of high school, he had developed into a wonderfully good player and was my only partner. His leaving home deprived me not just of my son, but also of my tennis and chess partner. I was beginning to feel like Job.

16

MOLECULES AND SHORT-TERM MEMORY

In 1975, twenty years after Harry Grundfest told me that we needed to study the brain one cell at a time, my colleagues and I had begun to explore the cellular basis of memory—how one can remember a social encounter, a natural scene, a lecture, or a medical pronouncement for a lifetime. We had learned that memory derives from changes in the synapses in a neural circuit: short-term memory from functional changes and long-term memory from structural changes. Now we wanted to delve even deeper into the mystery of memory. We wanted to penetrate the molecular biology of a mental process, to know exactly what molecules are responsible for short-term memory. With that question, we were entering completely uncharted territory.

THE JOURNEY WAS MADE LESS DAUNTING BY MY INCREASING confidence that in *Aplysia* we had a simple system in which we could explore the molecular basis of memory storage. We had entered the labyrinth of synaptic connections in *Aplysia*'s nervous system, mapped the neural pathway of its gill-withdrawal reflex, and shown that the synapses forming it could be strengthened by learning. We had quite literally navigated the outer rings of a scientific maze. Now we wanted to determine exactly where along this neural pathway the synaptic changes associated with short-term memory are localized.

WE FOCUSED OUR ATTENTION ON THE CRITICAL SYNAPSE BETWEEN the sensory neuron that transmits information about touch from the snail's siphon and the motor neuron whose action potentials cause the gill to withdraw. We now wanted to know how the two neurons that make up the synapse contribute to the learned change in synaptic strength. Does the sensory neuron change in response to the stimulus, leading its axon terminals to release more or less transmitter? Or does the change occur in the motor neuron, resulting in an increased number of receptors in the cell to the neurotransmitter or an increase in the sensitivity of receptors to the transmitter? We found that the change is quite one-sided: during short-term habituation lasting minutes, the sensory neuron releases less neurotransmitter, and during short-term sensitization it releases more neurotransmitter.

That neurotransmitter, we later discovered, is glutamate, also the major excitatory transmitter in the mammalian brain. By increasing the amount of glutamate a sensory cell sends to a motor cell, sensitization strengthens the synaptic potential elicited in the motor cell, thus making it easier for that neuron to fire an action potential and cause the gill to withdraw.

The synaptic potential between the sensory and motor neurons lasts only milliseconds, yet we had observed that a shock to *Aplysia*'s tail enhances glutamate release and synaptic transmission for many minutes. How does this come about? As my colleagues and I focused on the question, we noticed something curious. The strengthening of the synaptic connection between the sensory and motor neuron is accompanied by a very slow synaptic potential in the sensory cell, one that lasts for minutes rather than the milliseconds typical of synaptic potentials in the motor neuron. We soon found that the shock to *Aplysia*'s tail activates a second class of sensory neurons, one that receives information from the tail. These tail sensory neurons activate a group of interneurons that acts on the sensory neuron from the siphon. It is these interneurons that produce the remarkably slow synaptic potential. We then asked ourselves: What neurotransmitter do the interneurons release? How does this second neurotransmitter lead to the release of more glutamate from the terminals of the sensory neuron, thus creating short-term memory storage?

We found that the interneurons activated by a shock to *Aplysia*'s tail release a neurotransmitter called serotonin. Moreover, the interneurons form synapses not only on the cell body of the sensory neurons but also on the presynaptic terminals, and they not only produce a slow synaptic potential but also enhance the sensory cell's release of glutamate onto the motor cell. In fact, we could simulate the slow synaptic potential, the enhancement of synaptic strength, and the strengthening of the gill-withdrawal reflex simply by applying serotonin to the connections between the sensory and motor neurons.

We called these serotonin-releasing interneurons modulatory interneurons because they do not mediate behavior directly; rather, they modify the strength of the gill-withdrawal reflex by enhancing the strength of the connections between sensory and motor neurons.

These findings caused us to realize that there are two kinds of neural circuits important in behavior and learning: mediating circuits, which we had characterized earlier, and modulating circuits, which we were just beginning to characterize in detail (figure 16-1). Mediating circuits produce behavior directly and are therefore Kantian in nature. These are the genetically and developmentally determined neuronal components of the behavior, the neuronal architecture. The mediating circuit is made up of the sensory neurons that innervate the siphon, the interneurons, and the motor neurons that control the gill-withdrawal reflex. With learning, the mediating circuit becomes the student and acquires new knowledge. The modulating circuit is Lockean in nature; it serves as a teacher. It is not directly involved in producing a behavior but instead fine-tunes the behavior in response to learning by modulating—heterosynaptically—the strength of synaptic connections between the sensory and motor neurons. Activated by a shock to the tail, a completely different part of the body than the siphon, the modulating circuit teaches *Aplysia* to pay attention to a stimulus to the siphon that is important for its safety. Thus the circuit is, in essence,

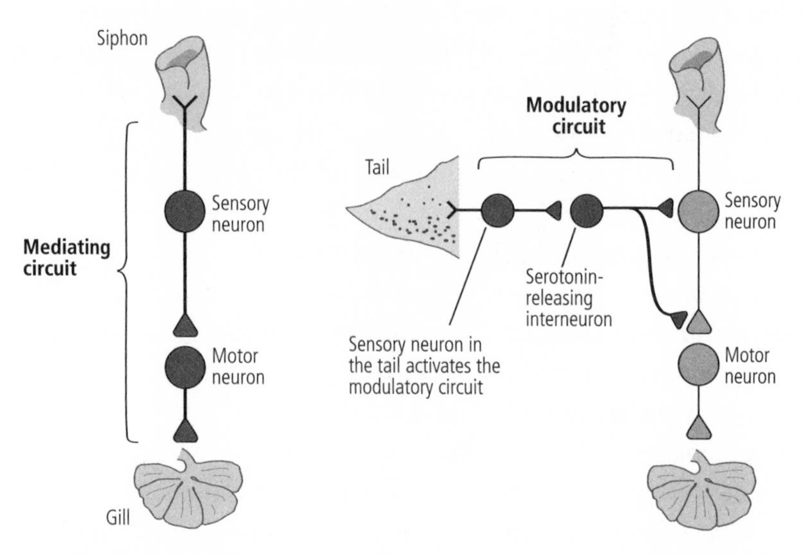

16-1 The two types of circuits in the brain. Mediating circuits produce behaviors. Modulatory circuits act on the mediating circuits, regulating the strength of their synaptic connections.

responsible for arousal or salience in *Aplysia*, just as analogous modulatory circuits are an essential component of memory in more complex animals, as we will see later.

That serotonin was a modulator for sensitization simply amazed me! Some of my first experiments with Dom Purpura in 1956 had focused on the action of serotonin. In fact, on Student Day at NYU Medical School in the spring of 1956, I had given a brief talk entitled "Electrophysiological Patterns of Serotonin and LSD Interaction on Afferent Cortical Pathways." Jimmy Schwartz had been kind enough to listen to a rehearsal of the talk and to help me improve it. I was now beginning to appreciate that life is circular. I had not worked on serotonin for almost twenty years, and here I was returning to it with renewed focus and enthusiasm.

ONCE WE KNEW THAT SEROTONIN ACTED AS A MODULATORY transmitter to enhance the release of glutamate from the presynaptic terminals of the sensory neuron, the stage was set for a biochemical analysis of memory storage. Fortunately, in Jimmy Schwartz I had an excellent guide and fellow traveler on this journey.

Before returning to NYU, Jimmy had worked at Rockefeller University on the bacterium *Escherichia coli*, the single-celled organism in which many fundamental principles of modern biochemistry and molecular biology were first worked out. In 1966 his interest had shifted to *Aplysia*, and he began his research by delineating the chemical transmitters used by a neuron in the abdominal ganglion. In 1971, we joined forces to study the molecular actions that accompany learning.

Jimmy was of inestimable help in this second major stage of my biological education. We were influenced by the work of Louis Flexner, who had shown a few years earlier that long-term memory in mice and rats requires the synthesis of new protein, whereas short-term memory does not. Proteins are the workhorses of the cell. They constitute its enzymes, ion channels, receptors, and transport machinery. Since, as we had found, long-term memory involves the growth of new connections, it is not surprising that the synthesis of new protein constituents is required for that growth.

Jimmy and I set out to test this idea in *Aplysia* and to do so at the level of the siphon sensory cell and its synapses on the motor neurons to the gill. If synaptic changes parallel changes in memory, then the short-term synaptic changes we had delineated should not require the synthesis of new protein. That is exactly what we found. What, then, mediates this short-term change?

Cajal had shown that the brain is an organ constructed of neurons wired to each other in specific pathways. I had seen this remarkable connection specificity in the simple neural circuits that mediate reflex behavior in *Aplysia*. But Jimmy pointed out that this specificity also extends to molecules—to the combinations of atoms that serve as the elementary units of cellular function. Biochemists had found that molecules can interact with one another within a cell and that these chemical reactions are organized in specific sequences known as biochemical signaling pathways. The pathways convey information in the form of molecules from the surface of the cell to the interior, much as one nerve cell conveys information to another. In addition, the pathways are "wireless." Molecules floating within the cell recognize and bind to specific molecular partners and regulate their activity.

My colleagues and I had not only fulfilled my early ambition of trapping a learned response in the smallest possible population of neurons, we had trapped a component of a simple form of memory in a single sensory cell. But even a single *Aplysia* neuron contains thousands of different proteins and other molecules. Which of these molecules are responsible for short-term memory? As Jimmy and I began to discuss the possibilities, we focused in on the idea that the serotonin released in response to a shock to the tail might enhance glutamate release from the sensory neuron by launching a specific sequence of biochemical reactions in the sensory cell.

The sequence of biochemical reactions that Jimmy and I sought would have to serve two fundamental purposes. First, they would have to translate the brief action of serotonin into molecules whose signals would last for minutes within the sensory neuron. Second, those molecules would have to broadcast signals from the cell membrane, where serotonin acts, to the interior of the sensory cell, particularly to the specialized regions of the axon terminal involved in the release of glutamate. We elaborated on these thoughts in our 1971 article in the *Journal of Neurophysiology* and speculated on the possibility that a specific molecule known as cyclic AMP might be involved.

WHAT IS CYCLIC AMP? HOW DID WE HIT UPON IT AS A LIKELY candidate? Cyclic AMP came to mind because this small molecule was known to serve as a master regulator of signaling within muscle and fat cells. Jimmy and I knew that nature is conservative—therefore, a mechanism used in the cells of one tissue is likely to be retained and used in the cells of another tissue. Earl Sutherland at Case Western Reserve University in Cleveland had already shown that the hormone epinephrine (adrenaline) produces a brief biochemical change at the surface membrane of fat and muscle cells that gives rise to a more enduring change inside the cells. That longer lasting change is brought about by an increase in the amount of cyclic AMP inside those cells.

Sutherland's revolutionary findings came to be described as the second-messenger signaling theory. The key to this biochemical signaling theory was his discovery of a new class of receptors at the cell surface of fat and muscle cells that responds to hormones. Earlier, Bernard

Katz had found the neurotransmitter-gated receptors known as ionotropic receptors; on binding a neurotransmitter, these receptors open or close the gate of an ion channel contained within the receptor, thus translating a chemical signal into an electrical signal. But the new class of receptors, called metabotropic receptors, has no ion channel within it to open or close. Instead, one region of these receptors protrudes from the outside surface of the cell membrane and recognizes signals from other cells, while another region protrudes from the inside of the cell membrane and engages an enzyme. When these receptors recognize and bind a chemical messenger on the outside of the cell, they activate an enzyme within the cell called adenylyl cyclase, which makes cyclic AMP.

This process has the advantage of having greatly amplifying the cell's response. When one molecule of chemical messenger binds to a metabotropic receptor, that receptor stimulates adenylyl cyclase to make a thousand molecules of cyclic AMP. Cyclic AMP then binds to key proteins that trigger a whole family of molecular responses throughout the cell. Finally, adenylyl cyclase continues making cyclic AMP for minutes. The actions of metabotropic receptors therefore tend to be more powerful, more widespread, and more persistent than the actions of ionotropic receptors. Whereas ionotropic actions typically last milliseconds, metabotropic actions can last from seconds to minutes—one thousand to ten thousand times longer.

To distinguish between the two spatially distinct functions of metabotropic receptors, Sutherland called the chemical messenger that binds to the metabotropic receptor on the outside of the cell the first messenger, and the cyclic AMP activated inside the cell to broadcast the signal the second messenger. He argued that the second messenger conveys the signal at the cell surface from the first messenger into the interior of the cell and initiates the cell-wide response (figure 16-2). Second-messenger signaling suggested to us that metabotropic receptors and cyclic AMP might be the elusive agents connecting the slow synaptic potential in sensory neurons to the enhanced release of glutamate and consequently to the formation of short-term memory.

In 1968 Ed Krebs at the University of Washington provided the ini-

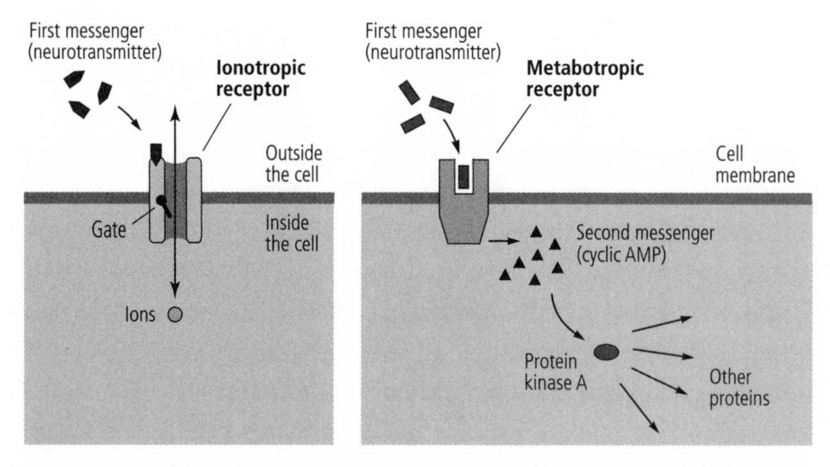

16-2 Sutherland's two classes of receptors. Ionotropic receptors (left) produce changes lasting milliseconds. Metabotropic receptors (e.g., serotonin receptors) act through second messengers (right). They produce changes that last seconds to minutes and are broadcast throughout the cell.

tial insight into how cyclic AMP produces its widespread effects. Cyclic AMP binds to and activates an enzyme that Krebs called cyclic AMP-dependent protein kinase, or protein kinase A (because it was the first protein kinase to be discovered). Kinases modify proteins by adding a phosphate molecule to them, a process known as phosphorylation. Phosphorylation activates some proteins and inactivates others. Krebs found that phosphorylation can be readily reversed and thus can serve as a simple molecular switch, turning the biochemical activity of a protein on or off.

Krebs next went about finding out how this molecular switch works. He discovered that protein kinase A is a complex molecule made up of four units—two regulatory and two catalytic. The catalytic units are designed to carry out phosphorylation, but the regulatory units normally "sit" on them and inhibit them. The regulatory units contain sites that bind cyclic AMP. When the concentration of cyclic AMP in a cell increases, the regulatory units bind the excess molecules. This action changes their shape and causes them to fall off the catalytic units, leaving the catalytic units free to phosphorylate target proteins.

These considerations raised a key issue in our minds: Was the mechanism discovered by Sutherland and Krebs specific to the action of hormones on fat and muscle cells, or could it also involve other transmitters, including those present in the brain? If so, this would represent a previously unknown mechanism of synaptic transmission.

Here we were helped by the work of Paul Greengard, a gifted biochemist who was also trained in physiology and who had recently moved to Yale University from his former position as director of biochemistry at Geigy Pharmaceutical Research Laboratories. On the way to Yale, he stopped off for a year in Sutherland's department. Realizing the importance of a potentially novel signaling mechanism in the brain, Greengard began in 1970 to sort out metabotropic receptors in the brains of rats. Now, a wonderful coincidence emerged that was to link Arvid Carlsson, Paul Greengard, and me on a scientific journey that would lead the three of us to Stockholm in 2000 to share in the Nobel Prize in Physiology or Medicine for signal transformations (transduction) in the nervous system.

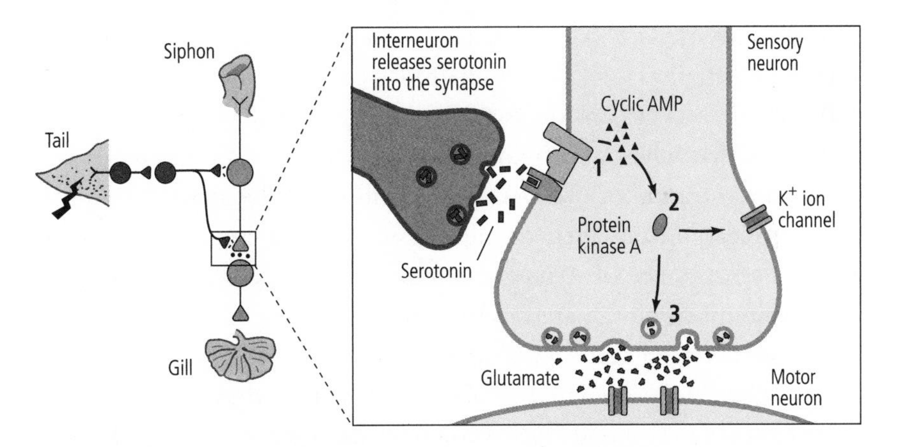

16-3 Biochemical steps in short-term memory. A shock to the tail of *Aplysia* activates an interneuron that releases the chemical messenger serotonin into the synapse. After crossing the synaptic cleft, serotonin binds to a receptor on the sensory neuron, leading to production of cyclic AMP (**1**). Cyclic AMP frees the catalytic unit of protein kinase A (**2**). The catalytic unit of protein kinase A enhances the release of the neurotransmitter glutamate (**3**).

In 1958 Arvid Carlsson, a great Swedish pharmacologist, discovered dopamine to be a transmitter in the nervous system. He then went on to show that when the concentration of dopamine is decreased in a rabbit, the animal develops symptoms that resemble Parkinson's disease. When Greengard started to explore metabotropic receptors in the brain, he started with a receptor for dopamine and found that it stimulates an enzyme that increases cyclic AMP and activates protein kinase A in the brain!

Based on these leads, Jimmy Schwartz and I discovered that cyclic AMP second-messenger signaling is also turned on by serotonin during sensitization. As we have seen, a shock to *Aplysia*'s tail activates modulatory interneurons that release serotonin. Serotonin, in turn, increases the production of cyclic AMP in the presynaptic terminals of the sensory neurons for a few minutes (figure 16-3). Thus it all came together: the increase in cyclic AMP lasts about as long as the slow synaptic potential, the increase in synaptic strength between the sensory and motor neurons, and the animal's enhanced behavioral response to the shock applied to its tail.

THE FIRST DIRECT CONFIRMATION THAT CYCLIC AMP IS INVOLVED in the formation of short-term memory came in 1976, when Marcello Brunelli, an Italian postdoctoral fellow, joined our lab. Brunelli tested the idea that when serotonin signals the sensory neurons to increase the concentration of cyclic AMP, the cells boost the amount of glutamate released from their terminals. We injected cyclic AMP directly into a sensory cell of *Aplysia* and found that it dramatically increased the amount of glutamate released and, therefore, the strength of the synapse between that sensory cell and motor neurons. In fact, injecting cyclic AMP simulated perfectly the increased synaptic strength brought about by applying serotonin to sensory neurons or a shock to the animal's tail. This remarkable experiment not only linked cyclic AMP and short-term memory but also gave us our first insight into the molecular mechanisms of learning. Having begun to trap the basic molecular components of short-term memory, we could now use them to simulate memory formation.

In 1978 Jimmy and I began to collaborate with Greengard. The three of us wanted to know whether cyclic AMP produces its effect on short-term memory through protein kinase A. We pulled the protein apart and injected directly into a sensory neuron only the catalytic unit, which normally carries out phosphorylation. We found that this unit does exactly what cyclic AMP does—it strengthens the synaptic connection by enhancing the release of glutamate. Then, just to make sure we were on the right track, we injected an inhibitor of protein kinase A into a sensory neuron and found that it indeed blocked the ability of serotonin to enhance glutamate release. In finding that cyclic AMP and protein kinase A are both necessary and sufficient for strengthening the connections between sensory and motor neurons, we were able to identify the first links in the chain of biochemical events leading to short-term memory storage (figure 16-4).

That did not tell us how serotonin and cyclic AMP bring about the slow synaptic potential or how this synaptic potential relates to the enhanced release of glutamate, however. In 1980 I met Steven Siegelbaum in Paris, where I was giving a series of seminars at the College of France. Steve was a technically gifted young biophysicist who specialized in studying the properties of single ion channels. We hit it off extremely well and, as fate would have it, he had recently accepted a position in the pharmacology department at Columbia. We therefore decided to join forces upon his arrival in New York and explore the biophysical nature of the slow synaptic potential.

Steve discovered one of the targets of cyclic AMP and protein kinase A: a potassium ion channel in sensory neurons that responds to serotonin. We called this channel the S channel because it responds to serotonin and because it was discovered by Steve Siegelbaum. The channel is open when the neuron is at rest and contributes to its resting membrane potential. Steve found that the channel is present in the presynaptic terminals and that he could cause it to close either by applying serotonin (the first messenger) to the outside of the cell membrane or by applying cyclic AMP (the second messenger) or protein kinase A to the inside. Closing the potassium ion channel causes the slow synaptic potential that drew our attention to cyclic AMP in the first place.

Closing the channel also helps to enhance the release of glutamate. When the channel is open, it contributes, along with other potassium channels, to the resting membrane potential and to the outward movement of potassium during the descending stroke of the action potential. But when it is closed by serotonin, the ions move out of the cell less rapidly, slightly increasing the duration of the action potential by slowing the descending stroke. Steve showed that slowing the action potential allows more time for calcium to flow into the presynaptic terminals—and calcium, as Katz had shown in the squid's giant synapse, is essential for the release of glutamate. In addition, cyclic AMP and protein kinase A act directly on the machinery that releases the synaptic vesicles, thus stimulating the release of glutamate even further.

These exciting results regarding cyclic AMP were soon complemented by important genetic studies of learning in fruit flies, a research favorite for over half a century. In 1907 Thomas Hunt Morgan at Columbia began to use the fruit fly, *Drosophila*, as a model organism for genetic studies because of its small size and short reproductive cycle (twelve days). This proved to be a happy choice because *Drosophila* has only four pairs of chromosomes (compared with twenty-three pairs in humans), making it a relatively easy animal to study genetically. It had long been obvious that many physical characteristics of animals—the shape of the body, eye color, and speed, among others—are inherited. If external physical characteristics can be inherited, can mental characteristics produced by the brain also be inherited? Do genes have a role in a mental process such as memory?

The first person to address this question with modern techniques was Seymour Benzer, at the California Institute of Technology. In 1967 he began a brilliant series of experiments in which he treated flies with chemicals designed to produce random mutations, or changes, in single genes. He then examined the effects of these mutations on learning and memory. To study memory in the fruit fly, Benzer's students Chip Quinn and Yadin Dudai used a classical conditioning procedure. They placed the flies in a small chamber and exposed them to

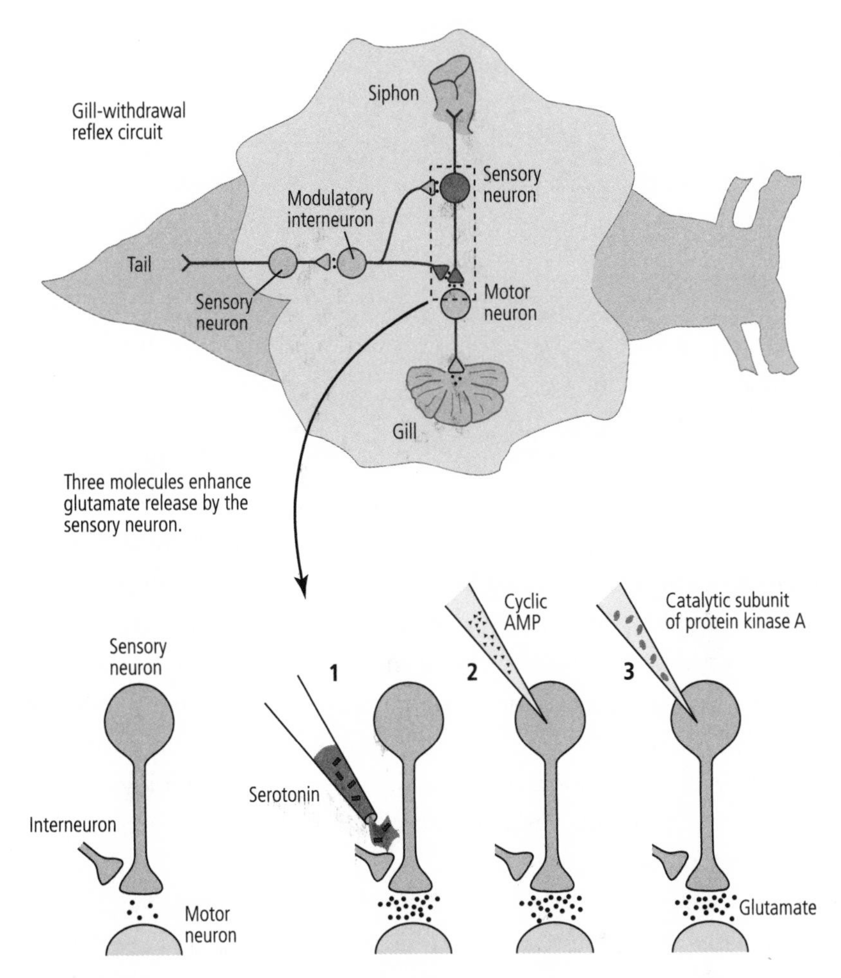

16-4 Molecules involved in short-term memory. Applying serotonin to the terminal of a sensory neuron (**1**), injecting cyclic AMP into the neuron (**2**), and injecting the catalytic part of protein kinase A (**3**) all lead to increased release of the neurotransmitter glutamate. This suggests that each of these three substances participates in the pathway for short-term memory.

two odors in sequence. The flies were then given an electric shock in the presence of odor 1, teaching them to avoid that odor. Later, the flies were placed in another chamber with the sources of the two odors at opposite ends. The conditioned flies avoided the end containing odor 1 and streamed to the end containing odor 2.

This training procedure gave Quinn and Dudai a way of identifying the flies that lacked the ability to remember that odor 1 is accompanied by a shock. By 1974 they had screened thousands of flies and isolated the first mutant with a defect in short-term memory. Benzer called the mutant *dunce*. In 1981 Benzer's student Duncan Byers, following up on the work in *Aplysia*, began to examine the cyclic AMP pathway in *dunce* and found a mutation in the gene responsible for disposing of cyclic AMP. As a result, the fly accumulates too much of the substance; its synapses presumably become saturated, making them insensitive to further change and preventing them from functioning optimally. Other mutations in memory genes were subsequently identified. They, too, involve the cyclic AMP pathway.

THE MUTUALLY REINFORCING RESULTS IN *APLYSIA* AND *Drosophila*—two very different experimental animals examined for different types of learning using different approaches—were vastly reassuring. Together, they made it clear that the cellular mechanisms underlying simple forms of implicit memory are likely to be the same in many animal species, including in people, and in many different forms of learning because those mechanisms have been conserved through evolution. Biochemistry and, later, molecular biology would be powerful tools for revealing common features in the biological machinery of different organisms.

The discoveries in *Aplysia* and *Drosophila* also reinforced an important biological principle: evolution does not require new, specialized molecules to produce a new adaptive mechanism. The cyclic AMP pathway is not unique to memory storage. As Sutherland had shown, it is not even unique to neurons: the gut, the kidney, and the liver all make use of the cyclic AMP pathway to produce persistent metabolic changes. In fact, of all the known second messengers, the cyclic AMP system is probably the most primitive. It is the most important, and in some cases the only second-messenger system found in single-celled organisms such as the bacterium *E. coli*, in which it signals hunger. Thus the biochemical actions underlying memory did not arise specifically to support memory. Rather, neurons simply recruited an efficient signaling system employed for other purposes in other cells and used it to produce the changes in synaptic strength required for memory storage.

As the molecular geneticist François Jacob has pointed out, evolution is not an original designer that sets out to solve new problems with completely new sets of solutions. Evolution is a tinkerer. It uses the same collection of genes time and again in slightly different ways. It works by varying existing conditions, by sifting through random mutations in gene structure that give rise to slightly different variations of a protein or to variations in the way that protein is deployed in cells. Most mutations are neutral or even detrimental and do not survive the test of time. Only the rare mutation that enhances an individual's survival and reproductive capacities is likely to be retained. As Jacob writes:

> The action of natural selection has often been compared to that of an engineer. This comparison, however, does not seem suitable. First . . . the engineer works according to a preconceived plan. Second, an engineer who prepares a new structure does not necessarily work from older ones. The electric bulb does not derive from the candle, nor does the jet engine descend from the internal combustion engine. . . . Finally, the objects thus produced de novo by the engineer, at least by the good engineer, reach the level of perfection made possible by the technology of the time.
>
> In contrast to the engineer, evolution does not produce innovations from scratch. It works on what already exists, either transforming a system to give it a new function or combining several systems to produce a more complex one. If one wanted to use a comparison, however, one would have to say that this process resembles not engineering but tinkering, *bricolage* we say in French. While the engineer's work relies on his having the raw materials and the tools that exactly fit his project, the tinkerer manages with odds and ends. . . . He uses whatever he finds around him, old cardboards, pieces of string, fragments of wood or metal, to make some kind of workable object. The tinkerer

> picks up an object that happens to be in his stock and gives it an unexpected function. Out of an old car wheel, he will make a fan; from a broken table a parasol.

In living organisms, new capabilities are achieved by modifying existing molecules slightly and adjusting their interaction with other existing molecules. Because human mental processes have long been thought to be unique, some early students of the brain expected to find many new classes of proteins lurking in our gray matter. Instead, science has found surprisingly few proteins that are truly unique to the human brain and no signaling systems that are unique to it. Almost all of the proteins in the brain have relatives that serve similar purposes in other cells of the body. This is true even of proteins used in processes that are unique to the brain, such as the proteins that serve as receptors for neurotransmitters. All life, including the substrate of our thoughts and memories, is composed of the same building blocks.

I SUMMARIZED THE FIRST COHERENT INSIGHTS INTO THE CELL biology of short-term memory in a book entitled *Cellular Basis of Behavior*, published in 1976. In it I spelled out my belief—almost as a manifesto—that to understand behavior, one had to apply to it the same type of radical reductionist approach that had proved so effective in other areas of biology. At about the same time, Steve Kuffler and John Nicholls published *From Neuron to Brain*, a book that emphasizes the power of the cellular approach. They used cell biology to explain how nerve cells work and how they form circuits in the brain, and I used cell biology to connect the brain to behavior. Steve also sensed that connection and saw that the field of neurobiology was poised to take another big step.

I was therefore particularly pleased that in August 1980 Steve and I had a chance to travel together. We were both invited to Vienna to be inducted as honorary members of the Austrian Physiological Society. Steve had fled Vienna in 1938. We were introduced to the medical faculty of the University of Vienna by Wilhelm Auerwald, a pretentious academic who had accomplished little scientifically and who acted as if nothing out of the ordinary had caused these two sons of Vienna to flee the country. The professor blithely remarked that Kuffler had attended medical school in Vienna and that I had lived in Severingasse, literally around the corner from the university. His silence regarding our actual experiences in Vienna spoke volumes. Neither Steve nor I responded to his comments.

Two days later, we took a boat down the Danube from Vienna to Budapest, where we attended the International Meeting of Physiologists. It was the last important meeting Steve attended. He gave a superb lecture. Shortly thereafter, in October 1980, he died of a heart attack at his weekend home in Woods Hole, Massachusetts, having just come back from a long swim.

Like most of the neural science community, I was shattered when I heard the news. We were all indebted to and in some ways dependent upon him. Jack McMahan, one of Steve's most devoted students, described the reaction many of us felt: "How could he do this to us?"

I was president of the Society for Neuroscience that year and was responsible with the program committee for organizing the annual meeting in November. The meeting was held in Los Angeles just a few weeks after Steve's death, and about ten thousand neuroscientists attended. David Hubel delivered a remarkable eulogy. Accompanied by slides, he illustrated how prescient, insightful, and generous Steve had been and how much he had meant to us all. I don't think anyone on the American scene since then has been as influential or as beloved as Steve Kuffler. Jack McMahan organized a posthumous volume in his honor, and in my contribution I stated, "In writing this piece, I sense how much he is still here. Next to Alden Spencer, there is no colleague in science that I have lost that I think of and miss more."

The death of Steve Kuffler marked the end of an era, an era in which the neural science community was still relatively small and focused on the cell as the unit of brain organization. Steve's death coincided with the merger of molecular biology and neural science, a step that expanded dramatically both the scope of the field and the number of scientists in it. My own work reflects this change: to a large degree I ended my cellular and biochemical studies of learning and memory in 1980. By that time it was becoming clear to me that the increase in cyclic AMP and the enhancement of transmitter release produced by serotonin in response to a single learning trial lasts only minutes. Long-term facilitation lasting days and weeks must involve something more, perhaps changes in the expression of genes as well as anatomical changes. So I turned to the study of genes.

I was ready for this step. Long-term memory was beginning to fire my imagination. How can one remember events from childhood for the whole of one's life? Denise's mother, Sara Bystryn, who imbued Denise and her brother, Jean-Claude, as well as their spouses and children, with her taste in decorative arts—art nouveau furniture, vases, and lamps—rarely spoke to me about my science. But she must have somehow sensed that I was ready to tackle genes and long-term memory.

On my fiftieth birthday, November 7, 1979, she bought me a beautiful Viennese vase by Teplist (figure 16-5) and gave it to me with the following note:

Dear Eric,

This vase by Teplist
The look of the Viennese forest
The nostalgia which emanates from
The trees
The flowers
The light
The sunset
Will bring memories to you
from other times
Reminiscences of your childhood.
And while you are jogging
along the trees of Riverdale forest,
the nostalgia of the Viennese

forest will envelop you.
And for a short moment
make you forget the events
of your daily life.

Love,
Sara

Sara Bystryn had defined my task.

16-5 The Teplist Vase. (From Eric Kandel's personal collection.)

17

LONG-TERM MEMORY

In reflecting on his genetic studies of bacteria, François Jacob distinguished between two categories of scientific investigation: day science and night science. Day science is rational, logical, and pragmatic, carried forward by precisely designed experiments. "Day science employs reasoning that meshes like gears, and achieves results with the force of certainty," Jacob wrote. Night science, on the other hand, "is a sort of workshop of the possible, where are elaborated what will become the building materials of science. Where hypotheses take the form of vague presentiments, of hazy sensations."

By the mid-1980s, I felt that our studies of short-term memory in *Aplysia* were edging toward the threshold of day science. We had succeeded in tracing a simple learned response in *Aplysia* to the neurons and synapses that mediate it and had found that learning gives rise to short-term memory by producing transient changes in the strength of existing synaptic connections between sensory and motor neurons. Those short-term changes are mediated by proteins and other molecules already present at the synapse. We had discovered that cyclic AMP and protein kinase A enhance the release of glutamate from the terminals of the sensory neurons, and that this enhanced release is a key element in short-term memory formation. In brief, we had in *Aplysia* an experimental system whose molecular components we could manipulate experimentally in a logical way.

But a central mystery in the molecular biology of memory storage remained: How are short-term memories transformed into enduring, long-term memories? This mystery became for me a subject of night science: of romantic musings and unconnected ideas, of months of considering how we might pursue the solution through day science experiments.

Jimmy Schwartz and I had found that long-term memory formation depends upon the synthesis of new proteins. I had a hunch that long-term memory, which involves enduring changes in synaptic strength, could be tracked to changes in the genetic machinery of sensory neurons. Pursuing this vague idea meant carrying our analysis of memory formation even deeper into the molecular labyrinth of the neuron: to the nucleus of the cell, where genes reside and where their activity is controlled.

In my late-night musings, I dreamed of taking the next step, of using the newly developed techniques of molecular biology to listen in on the dialogue between sensory neurons' genes and their synapses. This next step could not have come at a more opportune time. By 1980, molecular biology had become the dominant and unifying force within biology. It would soon extend its influence to neural science and help create a new science of mind.

HOW DID MOLECULAR BIOLOGY, PARTICULARLY MOLECULAR genetics, get to be so important? The emergence of molecular biology and its initial influence can be traced to the 1850s, when Gregor Mendel first realized that hereditary information is passed from parent to offspring by means of discrete biological units we now call genes. In about 1915 Thomas Hunt Morgan discovered in fruit flies that each gene resides at a specific site, or locus, on the chromosomes. In flies and other higher organisms, the chromosomes are paired: one comes from the mother, the other from the father. The offspring thus receives one copy of each gene from each of its two parents. In 1942 the Austrian-born theoretical physicist Erwin Schrödinger gave a series of lectures in Dublin that was later published in a little volume entitled *What Is Life?* In that book he noted that it is the differences in their genes that distinguish one animal species from another and human beings from other animals. Genes, Schrödinger wrote, endow organisms with their distinctive features; they code biological information in a stable form so that it can be copied and transmitted reliably from generation to generation. Thus when a paired chromosome separates, as it does during cell division, the genes on each chromosome must be copied exactly into genes on the new chromosome. Life's key processes—the storing and passing on of biological information from one generation to the next—are carried out through the replication of chromosomes and the expression of genes.

Schrödinger's ideas caught the attention of physicists and brought a number of them into biology. In addition, his ideas helped transform biochemistry, one of the core areas of biology, from a discipline concerned with enzymes and the transformation of energy (that is, with how energy is produced and utilized in the cell) to a discipline concerned with the transformation of information (how information is copied, transmitted, and modified within the cell). Viewed from this new perspective, the importance of chromosomes and genes is that they are carriers of biological information. By 1949 it was already clear that a number of neurological diseases, such as Huntington's and Parkinson's, as well as several mental illnesses, including schizophrenia and depression, had genetic components. The nature of the gene therefore became the central question for all of biology, including, ultimately, the biology of the brain.

What is the nature of the gene? Of what is it made? In 1944 Oswald Avery, Maclyn McCarty, and Colin MacLeod at the Rockefeller Institute made the breakthrough discovery that genes are not proteins as many biologists had thought, but instead are made of deoxyribonucleic acid (DNA).

Nine years later, in the April 25, 1953, issue of *Nature*, James Watson and Francis Crick described their now historic model of the structure of DNA. With the help of X-ray photographs taken by the structural biologists Rosalind Franklin and Maurice Wilkins, Watson and Crick were able to infer that DNA is composed of two long strands wound around each other in the form of a spiral, or helix. Knowing that each strand in this double helix is made up of four small, repeating units called nucleotide bases—adenine, thymine, guanine, and cytosine—Watson and Crick assumed that the four nucleotides are the information-carrying elements of the gene. This led them to the striking discovery that the two strands of DNA are complementary and that the nucleotide bases on one strand of DNA form pairs with specific nucleotide bases on the other strand: adenine (A) in one stand pairing and binding only to thymine (T) in the other, and guanine (G) in one strand pairing and binding only to cytosine (C) in the other. The pairing of nucleotide bases at multiple points along their length holds the two strands together.

Watson and Crick's discovery put Schrödinger's ideas into a molecular framework, and molecular biology took off. The essential operation of genes, as Schrödinger pointed out, is replication. Watson and Crick ended their classic paper with the now famous sentence, "It has not escaped our notice that the specific pairing we have postulated immediately suggests a copying mechanism for genetic material."

The double helix model illustrates how gene replication works. When the two strands of DNA unwind during replication, each parent strand acts as a template for the formation of a complementary daughter strand. Since the sequence of the information-containing nucleotides on the parent strand is given, it follows that the sequence on the daughter strand will also be given: A will bind to T and G to C. The daughter strand can then serve as a template for the formation of still another strand. In this way, multiple copies of DNA can be replicated faithfully as a cell divides and the copies can be distributed to daughter cells. This pattern extends to all the cells of an organism, including the sperm and the egg, thus enabling the organism as a whole to be replicated from generation to generation.

Taking their cue from gene replication, Watson and Crick further suggested a mechanism for protein synthesis. Since each gene directs the production of a particular protein, they reasoned that the sequence of nucleotide bases in each gene carries the code for protein production. As in gene replication, the genetic code for proteins is "read out" by making a complementary copy of the nucleotide bases in a strand of DNA. But in protein synthesis, later work showed, the code is carried by an intermediary molecule called messenger RNA (ribonucleic acid). Like DNA, messenger RNA is a nucleic acid made up of four nucleotides. Three of them—adenine, guanine, and cytosine—are identical to the nucleotides in DNA, but the fourth, uracil, is unique to RNA and replaces thymine. When the two strands of DNA in a gene separate, one of the strands is copied into messenger RNA. The sequence of nucleotides in messenger RNA is later translated into protein. Watson and Crick thus formulated the central dogma of molecular biology: DNA makes RNA, and RNA makes protein.

The next step was to crack the genetic code, the rules whereby the nucleotides in messenger RNA are translated into the amino acids of protein, including proteins important for memory storage. Attempts to do this began in earnest in 1956, when Crick and Sydney Brenner focused on how the four nucleotides in DNA could code for the twenty amino acids that combine to form proteins. A one-to-one system, with each nucleotide coding for a single amino acid, would yield only four amino acids. A code using different pairs of nucleotides would yield only sixteen amino acids. To produce twenty unique amino acids, Brenner argued, the system would have to be based on triplets—that is, on combinations of three nucleotides. However, triplets of nucleotides yield not twenty, but sixty-four combinations. Brenner therefore suggested that a code based on triplets is degenerate (redundant), meaning that more than one triplet of nucleotides encodes the same amino acid.

In 1961 Brenner and Crick proved that the genetic code consists of a series of nucleotide triplets, each of which contains the instructions for forming a unique amino acid. But they did not show which triplets code for which amino acids. That was revealed later in the same year by Marshall Nirenberg at NIH and by Har Gohind Khorana at the University of Wisconsin. They tested Brenner and Crick's idea biochemically and cracked the genetic code by describing the specific combinations of nucleotides that code for each amino acid.

In the late 1970s Walter Gilbert at Harvard and Frederick Sanger in Cambridge, England, developed a new biochemical technique that made it possible to sequence DNA rapidly, that is, to read segments of the nucleotide sequences in DNA with relative ease and thus to determine what protein a given gene encodes. This proved a remarkable advance. It enabled scientists to observe that the same stretches of DNA occur in different genes and encode identical or similar regions in a variety of proteins. These recognizable regions, called domains, mediate the same biological function, regardless of the protein in which they occur. Thus, by merely looking at some of the nucleotide sequences that make up a gene, scientists could determine important aspects of how the protein encoded by that gene would work, whether the protein was a kinase, an ion channel, or a receptor, for example. Furthermore, by comparing the sequence of amino acids in different proteins, they could recognize similarities between proteins encountered in very different contexts, such as in different cells of the body or even in vastly different organisms.

From these sequences and comparisons of them, a blueprint emerged of how cells work and how they signal one another, forming a conceptual framework for understanding many of life's processes. In particular, these studies revealed once again that different cells—indeed, different organisms—are made out of the same material. All multicellular organisms have the enzyme that synthesizes cyclic AMP; they all have kinases, ion channels, and on and on. In fact, half of the genes expressed in the human genome are present in much simpler invertebrate animals such as the worm *C. elegans*, the fly *Drosophila*, and the snail *Aplysia*. The mouse has more than 90 percent and the higher apes 98 percent of the coding sequences of the human genome.

A KEY ADVANCE IN MOLECULAR BIOLOGY THAT FOLLOWED DNA sequencing, and the one that brought me into the field, was the emergence of recombinant DNA and gene cloning, techniques that make it possible to identify genes, including those expressed in the brain, and to determine their function. The first step is to isolate from a person, a mouse, or a snail the gene one wishes to study—that is, the segment of DNA that codes for a particular protein. One does this by locating the gene on the chromosome and then snipping it out with molecular scissors—enzymes that cut the DNA at appropriate spots.

The next step is to make many copies of the gene, a process known as cloning. In cloning, the ends of the excised gene are stitched to stretches of DNA from another organism, such as a bacterium, creating what is known as recombinant DNA—recombinant because a gene snipped from one organism's DNA is recombined with the genome from another organism. The genome of a bacterium divides every twenty minutes or so, producing large numbers of identical copies of the original gene. The final step is to decipher the protein that the gene encodes. This is done by reading the sequence of nucleotides or molecular building blocks, in the gene.

In 1972 Paul Berg of Stanford University succeeded in creating the first recombinant DNA molecule, and in 1973 Herbert Boyer of the University of California, San Francisco and Stanley Cohen of Stanford University elaborated on Berg's technique to develop gene cloning. By 1980 Boyer had spliced the human insulin gene into a bacterium, a feat that gave rise to an unlimited amount of human insulin and thereby created the biotechnology industry. Jim Watson, the co-discoverer of the structure of DNA, would write of these achievements as playing God:

> We wanted to do the equivalent of what a word processor can now achieve: to cut, paste, and copy DNA . . . after we cracked the genetic code. . . . A number of discoveries made in the late sixties and seventies, however, serendipitously came together in 1973 to give us so-called "recombinant DNA" technology—the capacity to edit DNA. This was no ordinary advance in lab technique. Scientists were suddenly able to tailor DNA molecules, creating ones that had never before been seen in nature. We could "play God" with the molecular underpinning of all of life.

Before long, the remarkable tools and molecular insights that had been used to dissect gene and protein function in bacteria, yeast, and non-neuronal cells were eagerly seized upon by neuroscientists, especially by me, to study the brain. I had no experience with any of these methods—it was all night science for me. But even at night I understood the power of molecular biology.

18

MEMORY GENES

Three events conspired to transform my plan to apply molecular biology to the study of memory from night science to day science. The first was my move in 1974 to Columbia University's College of Physicians and Surgeons to replace my mentor Harry Grundfest, who was retiring. Columbia was attractive to me because it was a great university with a wonderful tradition in scientific medicine and was particularly strong in neurology and psychiatry. Founded as King's College in 1754, it was the fifth oldest college in the United States and the first to grant a medical degree. The decisive factor was that Denise was on the faculty of the College of Physicians and Surgeons and we had bought our house in Riverdale because it was convenient to the campus. My move from NYU to Columbia therefore shortened my commute dramatically and made it possible for both of us to have independent careers yet participate in a common faculty.

Moving to Columbia led to the second event, my collaboration with Richard Axel (figure 18-1). Just as Grundfest had been my mentor in the first stage of my biological career, spurring me to study brain functions at the cellular level, and Jimmy Schwartz had been my guide in the second stage, exploring the biochemistry of short-term memory, Richard Axel would prove to be the collaborator who guided

18-1 Richard Axel (b. 1946) and I became friends during our early years at Columbia University. Through our scientific interactions, I learned molecular biology and Richard began to work on the nervous system. In 2004 Richard and his colleague Linda Buck (b. 1947), who had been his postdoctoral fellow, won the Nobel Prize in Physiology or Medicine for their classic work on the sense of smell. (From Eric Kandel's personal collection.)

me into the third stage of my biological career, one centered on the dialogue between a neuron's genes and its synapses in the formation of long-term memory.

Richard and I met in 1977 at a tenure committee meeting. At the end of the meeting, he walked up to me and said, "I'm getting tired of all of this gene cloning. I want to do something on the nervous system. We should talk and maybe do something on the molecular biology of walking." This proposal was nowhere near as naïve and grandiose as my proposal to Harry Grundfest that I study the biological basis of the ego, superego, and id. Nevertheless, I felt obliged to tell Richard that as of the moment, walking was probably out of the reach of molecular biology. Perhaps a simple behavior in *Aplysia*, such as gill withdrawal, inking, or egg laying, might be more tractable.

As I got to know Richard, I quickly appreciated how remarkably interesting, intelligent, and generous he is. In his book on the origins of cancer, Robert Weinberg gives an excellent description of Richard's curiosity and his incisive intellect:

> Tall, lanky, stoop-shouldered, Axel had an intense, angular face made even more intense by the shiny steel-rimmed glasses he always wore. Axel . . . was the source of the "Axel syndrome,"

> which I had discovered through careful observation and then described on occasion to members of my lab. I first recognized its existence at several scientific meetings where Axel was in attendance.
>
> Axel would sit in the front row of a lecture audience, listening intently to every word from the podium. Afterwards he would ask penetrating, perceptive questions that came out in slow, well-measured words, each syllable pronounced with care and clarity. His questions invariably reached straight to the heart of the lecture, uncovering a weak point in the speaker's data or arguments. The prospect of a probing question from Axel was extremely unsettling for those not entirely comfortable with their own science.

Richard's glasses have actually always been gold-rimmed, but otherwise the description is right on target. Besides having added the "Axel syndrome" to the annals of academic discomfort, Richard had made important contributions to recombinant DNA technology. He had developed a general method of transferring any gene into any cell in tissue culture. The method, called co-transfection, is widely used both by scientists in their research and by the pharmaceutical industry in generating drugs.

Richard was also an opera addict, and soon after we became friends, we went to the opera together on a number of occasions, always without tickets. The first time we went, we caught a performance of Wagner's *Walküre*. Richard insisted that we enter the opera house through the lower entrance that connects to the garage. The usher who collected tickets at this entrance recognized Richard immediately and let us in. We went into the orchestra and stood in the back until the lights were dimmed. Then another usher who had recognized Richard as we entered came to us and pointed to two empty seats. Richard slipped him money, the exact amount of which he refused to reveal to me. The performance was marvelous, but periodically I would break out in a cold sweat as I worried about reading a headline in the next day's *New York Times*: "Two Columbia Professors Discovered Sneaking into the Metropolitan Opera."

Shortly after we began our collaboration, Richard asked the people in his laboratory, "Does anyone want to learn neurobiology?" Only Richard Scheller stepped forward, and he became our joint postdoctoral student. Scheller proved a most fortunate addition—creative and bold, as his volunteering to explore the brain indicated. Scheller also knew a great deal about genetic engineering; he had contributed important technical innovations while still a graduate student, and he was generous in helping me learn molecular biology.

When Irving Kupfermann and I were investigating the behavioral function of various cells and cell clusters in *Aplysia*, we had found two symmetrical clusters of neurons, each containing about two hundred identical cells, which we called bag cells. Irving found that the bag cells release a hormone that initiates egg laying, an instinctive, fixed pattern of complex behavior. *Aplysia*'s eggs are packaged in long gelatinous strings, each of which contains a million or more eggs. In response to the egg-laying hormone, the animal extrudes an egg string from an opening in its reproductive system, which is located near its head. As it does so, its heart rate increases and it breathes more rapidly. It then grabs the emerging egg string with its mouth and waves its head back and forth to draw the string out of the reproductive duct, kneads the egg string into a ball, and deposits it on a rock or an alga.

Scheller succeeded in isolating the gene that controls egg laying and showed that it encodes a peptide hormone, or short string of amino acids, that is expressed in the bag cells. He synthesized the peptide hormone, injected it into *Aplysia*, and watched as it set off the animal's whole egg-laying ritual. This was an extraordinary accomplishment for its day because it showed that a single short string of amino acids could trigger a complex sequence of behavioral actions. My work with Axel and with Scheller on the molecular biology of a complex behavior—egg laying—sparked both men's long-term interest in neurobiology and fueled my desire to move even further into the maze of molecular biology.

Our studies of learning and memory in the early 1970s had linked cellular neurobiology to learning in a simple behavior. My studies with Scheller and Axel, beginning in the late 1970s, convinced me, as they did

Axel, that molecular biology, brain biology, and psychology could be merged to create a new molecular science of behavior. We spelled this conviction out in the introduction to our first paper on the molecular biology of egg laying: "We describe a useful experimental system in *Aplysia* for examining the structure, expression, and modulation of genes that code for a peptide hormone of known behavioral function."

This shared project exposed me to the technique of recombinant DNA, which became crucial to my subsequent work on long-term memory. In addition, my collaboration with Axel laid the foundation for an important scientific and personal friendship. I therefore was delighted and not at all surprised when I learned on October 10, 2004, four years after I was recognized by the Nobel Prize committee, that Richard and one of his former postdoctoral fellows, Linda Buck, had been awarded the Nobel Prize in Physiology or Medicine for their extraordinary work in molecular neurobiology. Together, Richard and Linda made the astonishing discovery that there are about a thousand different receptors for smell in the nose of a mouse. This vast array of receptors—completely unpredicted—explains why we can detect thousands of specific odorants and indicates that a significant aspect of the brain's analysis of odors is carried out by receptors in the nose. Richard and Linda then used tnese receptors in independent studies to demonstrate the precision of connections between neurons in the olfactory system.

The third and final event that promoted my goal of learning molecular biology and using it to study memory occurred in 1983, when Donald Fredrickson, the newly appointed president of the Howard Hughes Medical Institute, asked Schwartz, Axel, and me to form the nucleus of a group devoted to this new science of mind—molecular cognition. Each group of scientists the medical institute supports at universities and other research institutions around the country is named by its location. We thus became the Howard Hughes Medical Institute at Columbia.

Howard Hughes was a creative and eccentric industrialist who also produced movies and designed and raced airplanes. He inherited from his father a major interest in the Hughes Tool Company and used it to build a large business empire. Within the tool company he established an aircraft division, the Hughes Aircraft Company, which became a major defense contractor. In 1953 he gave the aircraft company in its entirety to the Howard Hughes Medical Institute, a medical research organization that he had just founded. By 1984, eight years after Hughes's death, the institute had become the largest private supporter of biomedical research in the United States. By 2004 the institute's endowment had risen to over $11 billion, and it supported 350 investigators in numerous universities in the United States. About 100 of those scientists belonged to the National Academy of Sciences, and 10 had Nobel Prizes.

The motto of the Howard Hughes Medical Institute is "People, not projects." It believes that science flourishes when outstanding researchers are provided both the resources and the intellectual flexibility to carry out bold, cutting-edge work. In 1983 the institute started three new initiatives—in neural science, in genetics, and in metabolic regulation. I was invited to be senior investigator of the neural science initiative, an opportunity that had an extraordinary impact on my career, as it did on Axel's.

The newly formed institute gave us the chance to recruit Tom Jessell and Gary Struhl from Harvard and to ask Steven Siegelbaum, who was about to leave Columbia, to remain. These were marvelous additions to the Hughes group at Columbia and to the Center for Neurobiology and Behavior. Jessell rapidly emerged as the leading scientist working on the development of the vertebrate nervous system. In a brilliant series of studies, he pinpointed the genes that endow different nerve cells in the spinal cord with their identity (the same cells that Sherrington and Eccles had studied). He went on to show that those genes also control the outgrowth of the axon and the formation of synapses. Siegelbaum brought his remarkable insights into ion channels to bear on how channels control the excitability of nerve cells and the strength of synaptic connections and how these are modulated by activity and by various modulatory neurotransmitters. Struhl developed an imaginative genetic approach in *Drosophila* to explore how the fruit fly develops its body form.

WITH THE TOOLS OF MOLECULAR BIOLOGY AND THE SUPPORT of the Howard Hughes Medical Institute in hand, we could now address questions about genes and memory. Since 1961 my experimental strategy had been to trap a simple form of memory in the smallest possible neural population and to use multiple microelectrodes to track the activity of participating cells. We could record signals from single sensory and motor cells for several hours in the intact animal, which was more than adequate for the study of short-term memory. But for long-term memory we needed to be able to record for one or more days. This required a new approach, so I turned to tissue cultures of the sensory and motor cells.

One cannot simply remove sensory and motor cells from adult animals and grow them, because adult cells do not survive well in culture. Instead, cells must be taken from the nervous system of very young animals and provided with an environment in which they can grow into adult cells. The crucial advance toward this goal was made by Arnold Kriegstein, an M.D.-Ph.D. student. Just before our lab moved to Columbia, Kriegstein succeeded in rearing *Aplysia* in the laboratory from the embryonic stage of the egg mass to adulthood, a feat that had eluded biologists for almost a century.

As it grows, *Aplysia* changes from a transparent, free-swimming larva that feeds on single-celled algae into a crawling, seaweed-eating juvenile slug, a small version of the adult. To achieve this radical change in body form, the larva must rest on a particular species of seaweed and be exposed to a specific chemical. No one had ever observed the metamorphosis in nature, so no one knew what the process entailed. Kriegstein observed immature *Aplysia* in the wild and noticed that they frequently rested on a particular species of seaweed. When he tested that seaweed by exposing larvae to it, he found that the larvae were transformed into juvenile slugs (figure 18-2). Most of us who were at Kriegstein's extraordinary seminar in December 1973 will not readily forget his description of how the larvae seek out a red seaweed called *Laurencia pacifica*, rest on it, and extract from it the chemicals needed to trigger metamorphosis. When Kriegstein showed the first pictures of the tiny juvenile snail, I remember saying to myself, "Babies are always so beautiful!"

After Kriegstein's discovery, we began to grow the seaweed and soon had all the juvenile animals we needed to culture cells of the

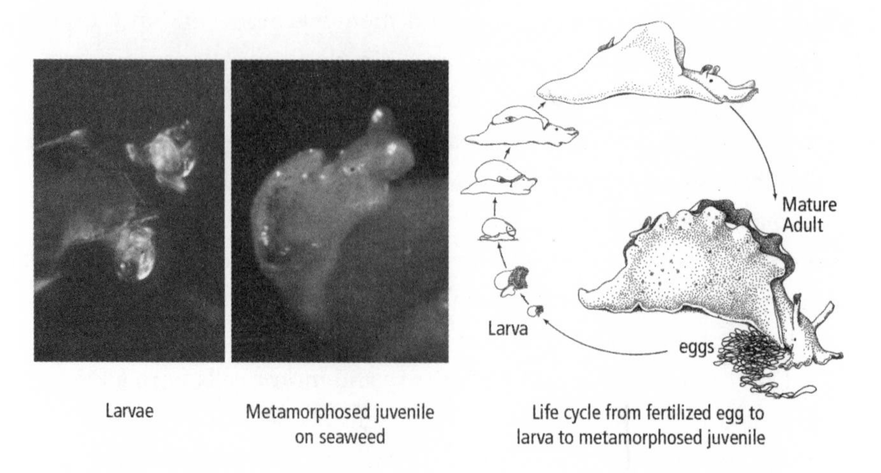

18-2 The life cycle of *Aplysia*. *Aplysia* larvae rest on a particular red seaweed (*Laurencia pacifica*) and extract from it the chemicals needed to trigger metamorphosis into a juvenile snail. (Drawing reprinted from *Cellular Basis of Behavior*, E. R. Kandel, W. H. Freeman and Company, 1976.)

nervous system. The next major task—how to grow individual nerve cells in culture and have them form synapses—was taken on by a former student of mine, Samuel Schacher, a cell biologist. With the help of two postdoctoral fellows, Schacher soon succeeded in culturing the individual sensory neurons, motor neurons, and interneurons involved in the gill-withdrawal reflex (figure 18-3).

We now had the elements of a learning circuit in tissue culture. This circuit enabled us to study a component of memory storage by focusing on a single sensory neuron and a single motor neuron. Our experiments showed that these isolated sensory and motor neurons form the same precise synaptic connections and exhibit the same physiological behavior in culture as they do in the intact animal. In nature, a shock to the tail activates modulatory interneurons that release serotonin, thereby strengthening the connections between sensory neurons and motor neurons. Since we already knew that these modulatory interneurons release serotonin, we found after a few experiments that we did not even need to culture them. We simply injected serotonin near the synapses between the sensory neuron and the motor neurons—that is, at the site in the intact animal where the modulatory interneurons terminate on the sensory neurons and

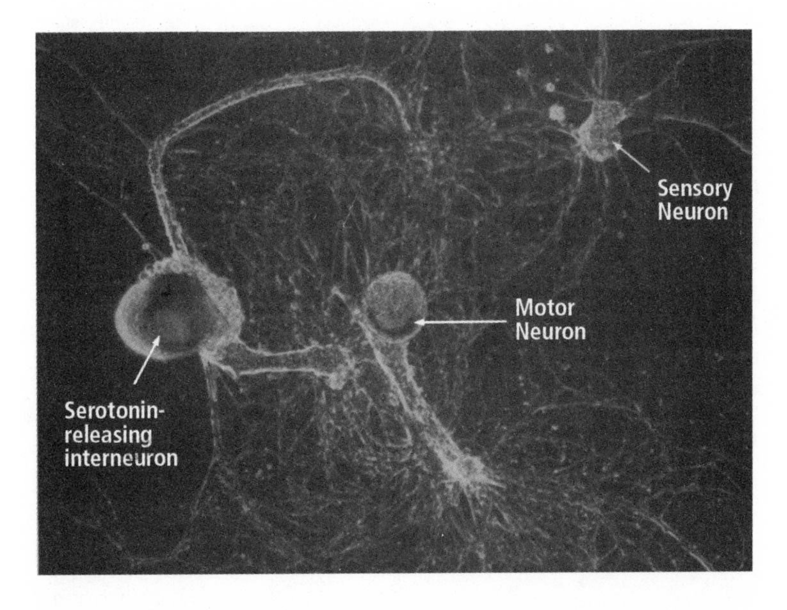

18-3 Using individual nerve cells grown in the lab to study long-term memory. Single sensory neurons, motor neurons, and serotonin-releasing modulatory interneurons grown in culture form synapses that reproduce the simplest form of the circuit mediating and modulating the gill-withdrawal reflex. This simple learning circuit—the first available in tissue culture—made it possible to investigate the molecular biology of long-term memory. (Courtesy of Sam Schacher.)

release serotonin. One of the great pleasures of working on a biological system over a long period is seeing today's discoveries become tomorrow's experimental tools. Our years of study of this neural circuit, our ability to isolate the key chemical signals being transmitted between and within its cells, enabled us to use these same signals to manipulate the system and probe more deeply.

We found that one brief pulse of serotonin strengthened the synaptic connection between the sensory and motor neuron for a few minutes by enhancing the release of glutamate from the sensory cell. As in the intact animal, this short-term enhancement of synaptic strength is a functional change: it does not require the synthesis of new proteins. In contrast, five separate pulses of serotonin, designed to simulate five shocks to the tail, strengthened the synaptic connection for

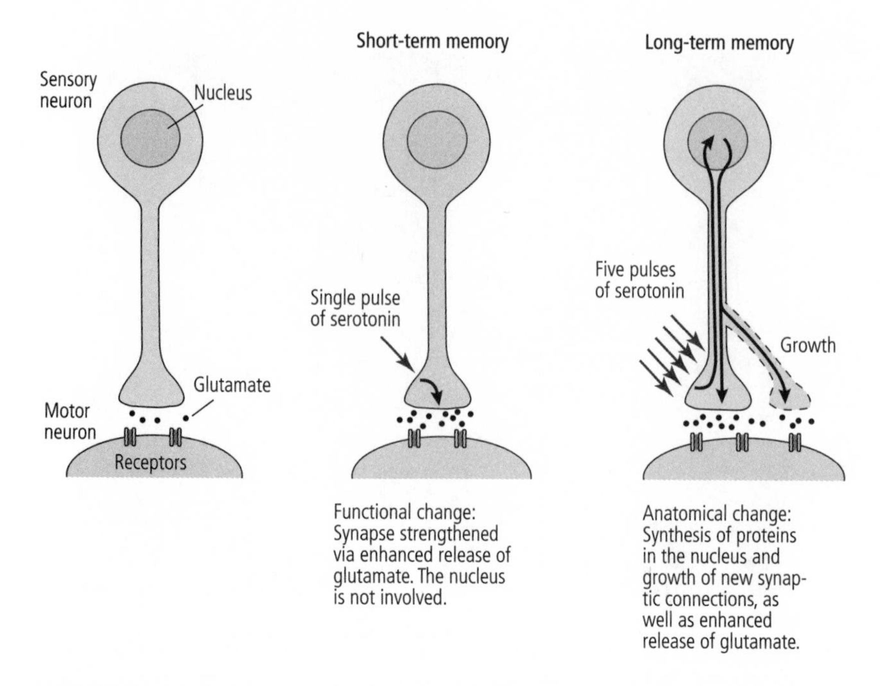

18-4 Changes underlying short- and long-term memory in a single sensory and motor neuron.

days and led to the growth of new synaptic connections, an anatomical change that did involve the synthesis of new protein (figure 18-4). This showed us that we could initiate new synaptic growth in the sensory neuron in tissue culture, but we still needed to find out what proteins are important for long-term memory.

My career in neurobiology now intersected with one of the great intellectual adventures of modern biology: the unraveling of the molecular machinery for regulating genes, the coded hereditary information at the heart of every life form on earth.

THIS ADVENTURE BEGAN IN 1961 WHEN FRANÇOIS JACOB AND Jacques Monod of the Institut Pasteur in Paris published a paper entitled "Genetic Regulatory Mechanisms in the Synthesis of Protein." Using bacteria as a model system, they made the remarkable discovery that genes can be regulated—that is, they can be switched on and off like a water faucet.

Jacob and Monod inferred what we now know to be a fact: that even in a complex organism like a human being, almost every gene of the genome is present in every cell of the body. Every cell has in its nucleus all of the chromosomes of the organism and therefore all of the genes necessary to form the entire organism. This inference raised a serious question for biology: Why do not all genes function in the same way in every cell of the body? Jacob and Monod proposed what ultimately proved to be the case—namely, that a liver cell is a liver cell, and a brain cell is a brain cell because in each cell type only some of those genes are turned on, or expressed; all of the other genes are shut off, or repressed. Thus, each cell type contains a unique mix of proteins—a subpopulation of all the proteins available to the cell. This mix of proteins enables the cell to perform its specific biological functions.

Genes are switched on and off as needed to achieve optimal functioning of the cell. Some genes are repressed for most of the lifetime of the organism; other genes, such as those involved in the production of energy, are always expressed because the proteins they encode are essential for survival. But in every cell type some genes are expressed only at certain times, whereas others are turned on and off in response to signals from within the body or from the environment. This set of arguments caused a lightbulb to go on in my brain one night: What is learning but a set of sensory signals from the environment, with different forms of learning resulting from different types or patterns of sensory signals?

What sort of signals regulate the activity of genes? Just how are genes turned on and off? Jacob and Monod found that in bacteria, genes are switched on and off by other genes. This led them to distinguish between effector genes and regulatory genes. Effector genes encode effector proteins such as enzymes and ion channels, which mediate specific cellular functions. Regulatory genes encode proteins called gene regulatory proteins, which switch the effector genes on or off. Jacob and Monod then asked: How do the proteins of the regulatory genes act on the effector genes? They postulated that every effector gene has in its DNA not only a coding region that encodes a particular protein but also a control region, a specific site now known as the promoter. Regulatory proteins bind to the promoter of effector sites and thereby determine whether the effector genes are going to be switched on or off.

Before an effector gene can be switched on, regulatory proteins must assemble on its promoter and help to separate the two strands of DNA. One of the exposed strands is then copied into messenger RNA in a process known as transcription. Messenger RNA carries the gene's instructions for protein synthesis from the nucleus of the cell to the cytoplasm, where structures known as ribosomes translate the messenger RNA into protein. Once the gene has been expressed, the two strands of DNA zip up again, and the gene is shut off until the next time regulatory proteins initiate transcription.

Jacob and Monod not only outlined a theory of gene regulation, they also discovered the first regulators of gene transcription. These regulators come in two forms—repressors, genes that encode the regulatory proteins that shut genes off, and as later work showed, activators, genes that encode the regulatory proteins that turn genes on. Through brilliant reasoning and insightful genetic experiments, Jacob and Monod found that when the common intestinal bacterium *E. coli* has a plentiful supply of a food source, the sugar lactose, the bacterium turns on a gene for an enzyme that breaks down lactose for consumption. When no lactose is present, the gene for this digestive enzyme is suddenly turned off. How does this occur?

The two scientists found that in the absence of lactose, the repressor gene encodes a protein that binds to the promoter of the gene for the digestive enzyme, thereby preventing the gene's DNA from being transcribed. When they reintroduced lactose into the medium in which the bacteria were grown, the lactose moved into the cell and bound to the repressor proteins, causing them to fall off the promoter. The promoter was then free to bind proteins encoded by an activator gene. The activator proteins turn the effector gene on, resulting in production of the enzyme that metabolizes lactose.

These studies showed that *E. coli* adjusts the rate of transcription of particular genes in response to environmental cues. Later studies revealed that when a bacterium finds itself in the presence of a low concentration of glucose, it responds by synthesizing cyclic AMP, which sets off a process that enables the cell to consume an alternative sugar.

The finding that gene function can be regulated up and down in response to environmental needs by signaling molecules from outside the cell (such as different sugars) as well as from inside the cell (second messenger signals such as cyclic AMP) was revolutionary to me. It caused me to rephrase in molecular terms the question of how short-term memory is converted to long-term memory. I now asked: What is the nature of the regulatory genes that respond to a specific form of learning, that is, to cues from the environment? And how do these regulatory genes switch a short-term synaptic change that is critical to a specific short-term memory into a long-term synaptic change that is critical to a specific long-term memory?

Our studies in invertebrates, as well as several studies in vertebrates, had demonstrated that long-term memory requires the synthesis of new protein, indicating that the mechanisms of memory storage are likely to be quite similar in all animals. Moreover, Craig Bailey had made the remarkable discovery that long-term memory in *Aplysia* endures because sensory neurons grow new axon terminals that strengthen their synaptic connections with motor neurons. Yet exactly what it takes to throw the switch for any form of long-term memory remained a mystery. Does the pattern of learning that produces long-term sensitization activate certain regulatory genes, and do the proteins encoded by those genes prompt effector genes to direct the formation of new axon terminals?

By studying living sensory and motor cells in culture we had reduced our behavioral system sufficiently to address these questions. We had located a critical component of long-term memory in the synaptic connection between only two cells. Now we could use the techniques of recombinant DNA to ask: Do regulatory genes switch on and maintain long-term strengthening of this connection?

AT ABOUT THIS TIME, I BEGAN TO RECEIVE FORMAL RECOGNITION for my work. In 1983 I shared with Vernon Mountcastle the Lasker Award in basic medical sciences, the most important scientific recognition awarded in the United States, and I received my first honorary degree, from the Jewish Theological Seminary in New York. I was thrilled that they would even know of my work. I suspect they learned of it from my colleague Mortimer Ostow, one of the psychoanalysts who first stirred my interest in psychoanalysis and the brain.

My father had died by then, but my mother came to the graduation ceremony and, in his introductory remarks, Gerson D. Cohen, the chancellor of the seminary, referred to my having received a good Hebrew education at the Yeshivah of Flatbush, an acknowledgment that filled my mother's Jewish heart with pride. I think the recognition that her father, my grandfather, had tutored me well in Hebrew may have meant more to her than the distinction of the Lasker Award a few months later.

19

A DIALOGUE BETWEEN GENES AND SYNAPSES

In 1985 I finally began to apply the insights I had gained from night science—from months of thinking about the proteins that regulate gene expression—to a daytime framework for working on gene expression and long-term memory. This thinking had become more focused with the arrival at Columbia of Philip Goelet, a postdoctoral student who had trained with Sydney Brenner at the Medical Research Council Laboratory in Cambridge, England. Goelet and I reasoned as follows: long-term memory requires that new information be encoded and then consolidated, put into more permanent storage. With the finding that long-term memory requires the growth of new synaptic connections, we had gained some insight into what form this more permanent storage took. But we still did not understand the intervening molecular genetic steps—the nature of memory consolidation. How is a fleeting short-term memory converted to a stable long-term memory?

In the Jacob-Monod model, signals from a cell's environment activate gene regulatory proteins that switch on the genes encoding particular proteins. This led Goelet and me to wonder whether the crucial step in switching on long-term memory in sensitization might involve similar signals and similar gene regulatory proteins. We wondered if the repeated learning trials required for sensitization are important because they send signals to the nucleus, telling it to activate regulatory genes that encode regulatory proteins that in turn switch on the effector genes needed for the growth of new synaptic connections. If so, the consolidation phase of memory might be the interval during which the regulatory proteins switch on effector genes. Our thinking would provide a genetic explanation for the finding that blocking the synthesis of new protein during a critical period—that is, during and shortly after learning—blocks both the growth of new synaptic connections and the conversion from short- to long-term memory. In blocking protein synthesis, we reasoned, we were actually preventing the expression of the genes that initiate protein synthesis essential for synaptic growth and long-term memory storage.

We summarized our views in "The Long and Short of Long-Term Memory," a conceptual review published in 1986 in *Nature*. In this paper, we proposed that if gene expression was required to convert short-term memory at a synapse into long-term memory, then the synapse stimulated by learning somehow had to send a signal to the nucleus telling it to turn on certain regulatory genes. In short-term memory, synapses use cyclic AMP and protein kinase A inside the cell to call for the release of more neurotransmitter. Goelet and I hypothesized that in long-term memory this kinase moves from the synapse to the nucleus, where it somehow activates proteins that regulate gene expression.

To test our hypothesis, we would have to identify the signal sent from the synapse to the nucleus, find the regulatory genes activated by the signal, and then identify the effector genes switched on by the regulator—that is, the genes responsible for the new synaptic growth underlying long-term memory storage.

THE SIMPLIFIED NEURAL CIRCUIT WE HAD CREATED IN TISSUE culture—a single sensory neuron connected to a single motor neuron—gave us a complete biological system in which to test these ideas. In our culture dish, serotonin acted as an arousal signal initiated by sensitization. One pulse—the equivalent of one shock, one training trial—alerted the cell that a stimulus was of momentary, short-term interest, while five pulses—the equivalent of five training trials—signaled a stimulus of lasting, long-term interest. We found that injecting a high concentration of cyclic AMP into a sensory neuron produced not only a short-term but also a long-term increase in the strength of the synapse. We now collaborated with Roger Tsien at the University of California, San Diego and used a method developed by him that allowed us to visualize the location of the cyclic AMP and protein kinase A in the neuron. We found that whereas a single pulse of serotonin increases cyclic AMP and protein kinase A primarily at the synapse, repeated pulses of serotonin produce even higher concentrations of cyclic AMP, causing protein kinase A to move into the nucleus, where it activates genes. Later studies found that protein kinase A recruits another kinase, called MAP kinase, which is also associated with synaptic growth and also migrates to the nucleus. Thus we confirmed our idea that one of the functions of repeated sensitization training—why practice makes perfect—is to cause the appropriate signals in the form of kinases to move into the nucleus.

Once in the nucleus, what do these kinases do? We knew from recently published studies of non-neuronal cells that protein kinase A can activate a regulatory protein called CREB (cyclic AMP response element-binding protein), which binds to a promoter (the cyclic AMP response element). This suggested to us that CREB might be a key component of the switch that converts short-term facilitation of synaptic connections to long-term facilitation and the growth of new connections.

In 1990, joined by two postdoctoral students, Pramod Dash and Benjamin Hochner, we found that CREB is present in the sensory neurons of *Aplysia* and is indeed essential to the long-term strengthening of synaptic connections underlying sensitization. By blocking the action of CREB in the nucleus of a sensory neuron in culture, we prevented long-term, but not short-term, strengthening of these synaptic connections. This was astonishing: blocking this one regulatory protein blocked the entire process of long-term synaptic change! Dusan Bartsch, a creative and technically brilliant postdoctoral fellow, later found that simply injecting CREB that had been phosphorylated by protein kinase A into the nucleus of sensory neurons was sufficient to turn on the genes that produce long-term facilitation of these connections.

Thus, even though I had long been taught that the genes of the brain are the governors of behavior, the absolute masters of our fate, our work showed that, in the brain as in bacteria, genes also are servants of the environment. They are guided by events in the outside world. An environmental stimulus—a shock to an animal's tail—activates modulatory interneurons that release serotonin. The serotonin acts on the sensory neuron to increase cyclic AMP and to cause protein kinase A and MAP kinase to move to the nucleus and activate CREB. The activation of CREB, in turn, leads to the expression of genes that changes the function and the structure of the cell.

In 1995 Bartsch found that there are in fact two forms of the CREB protein, much as the model of Jacob and Monod might have predicted: one that activates gene expression (CREB-1), and one that suppresses gene expression (CREB-2). Repeated stimulation causes protein kinase A and MAP kinase to move to the nucleus, where protein kinase A activates CREB-1 and MAP kinase inactivates CREB-2. Thus long-term facilitation of synaptic connections requires not only a switching on of some genes, but also the switching off of others (figure 19-1).

As these exciting findings were emerging in the laboratory, I was struck by two things. First, we were seeing the Jacob-Monod model of gene regulation applied to the process of memory storage. Second, we were seeing Sherrington's discovery of the integrative action of the neuron carried to the level of the nucleus. I was amazed by the parallels: on the cellular level, excitatory and inhibitory synaptic signals converge on a nerve cell, while on the molecular level, one CREB regulatory protein facilitates gene expression and the other inhibits it. Together, the two CREB regulators integrate opposing actions.

Indeed, CREB's opposing regulatory actions provide a threshold for memory storage, presumably to ensure that only important, life-serving experiences are learned. Repeated shocks to the tail are a significant learning experience for an *Aplysia*, just as, say, practicing the piano or conjugating French verbs are to us: practice makes perfect, repetition is necessary for long-term memory. In principle, however, a highly emotional state, such as that brought about by a car crash, could bypass the normal restraints on long-term memory. In such a situation, enough MAP kinase molecules would be sent into the nucleus rapidly enough to

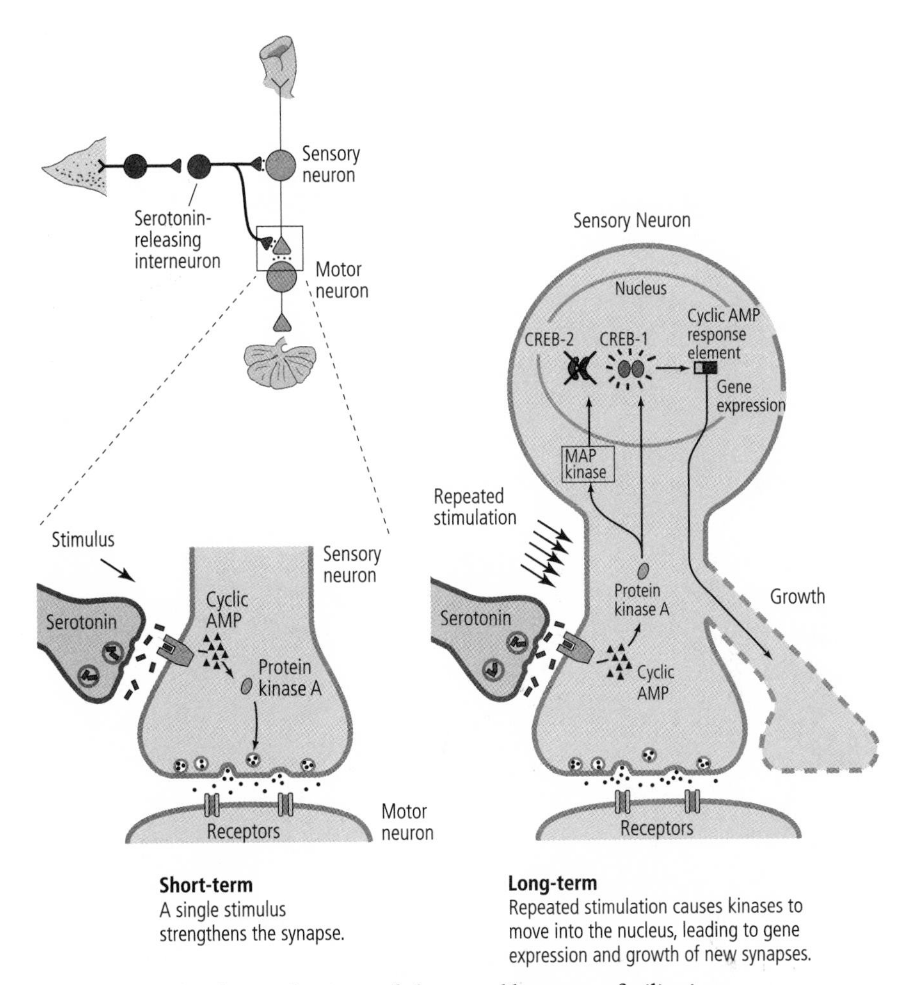

19-1 The molecular mechanisms of short- and long-term facilitation.

inactivate all of the CREB-2 molecules, thereby making it easy for protein kinase A to activate CREB-1 and put the experience directly into long-term memory. This might account for so-called flashbulb memories, memories of emotionally charged events that are recalled in vivid detail—like my experience with Mitzi—as if a complete picture had been instantly and powerfully etched on the brain.

Similarly, the exceptionally good memory exhibited by some people may stem from genetic differences in CREB-2 that limit the activity of this repressor protein in relation to CREB-1. Although long-term memory typically requires repeated, spaced training with intervals of rest, it occasionally occurs following a single exposure that is not emotionally charged. One-trial learning was particularly well developed in the famous Russian memorist S. V. Shereshevski, who seemed never to forget anything he had learned following a single exposure, even after more than a decade. Usually, memorists have more restricted capabilities: they may be exceptionally good at remembering certain types of knowledge but not others. Some people have an astonishing memory for visual images, for musical scores, for chess games, for poetry, or for faces. Some Talmudic memorists from Poland can recall, from visual memory, every word on every page of the twelve volumes of the Babylonian Talmud as if that one page (out of several thousand) were in front of their eyes.

Conversely, a characteristic of age-related memory loss (benign senescent forgetfulness) is the inability to consolidate long-term memories. This defect of aging may represent not only a weakening of the ability to activate CREB-1, but also an insufficiency of signals to remove the braking action of CREB-2 on memory consolidation.

The CREB switch for long-term memory, like the cellular mechanisms of short-term memory, proved to be the same in several species of animals, indicating that it has been conserved through evolution. In 1993 Tim Tully, a behavioral geneticist working at the Cold Spring Harbor Laboratory in Long Island, New York, developed an elegant protocol for examining long-term memory of learned fear in the fly. In 1995 Tully joined forces with the molecular geneticist Jerry Yin, and together they discovered that CREB proteins are essential for long-term memory in *Drosophila*. As in *Aplysia*, CREB activators and repressors played critical roles. The CREB repressor blocked the conversion of short-term memory to long-term memory. Even more fascinating, mutant flies bred to produce more copies of the CREB activator had the equivalent of flashbulb memories. A few training trials in which a specific odor was paired with a shock produced only a short-term memory of fear of the odor in normal flies, but the same number of trials led to long-term memory of fear in the mutant flies. In time, it would become clear that the same CREB switch is important for many forms of implicit memory in a variety of other species, from bees to mice to people.

Thus, by combining behavioral analysis first with cellular neural science and then with molecular biology, we were able, collectively, to help lay the foundation of a molecular biology of elementary mental processes.

THE FACT THAT THE SWITCH FOR CONVERTING SHORT-TERM memory to long-term memory is the same in a variety of simple animals that are learning simple tasks was heartening, confirming our belief that the core mechanisms of memory storage are conserved across different species. But it posed a considerable problem for the cell biology of neurons. A single sensory neuron has 1200 synaptic terminals and makes contact with about 25 target cells: gill motor neurons, siphon motor neurons, inking motor neurons, and excitatory and inhibitory interneurons. We had found that short-term changes occur in just some of those synapses and not others. That made sense, since a single shock to the tail or a single pulse of serotonin increases cyclic AMP locally, at a particular set of synapses. But long-lasting synaptic change requires gene transcription, which takes place in the nucleus and leads to the production of new proteins. One would expect the newly synthesized proteins to be shipped to all of the neuron's synaptic terminals. Thus, unless some special mechanism in the cell limits the changes to specific synapses, all of the neuron's synaptic terminals would be affected by long-term facilitation. If that were so, each long-term change would be stored at all the synapses of a neuron. This creates a paradox: How are long-term learning and memory processes localized to specific synapses?

Goelet and I thought about this question a great deal and outlined a scheme in our 1986 review in *Nature* that has become known as "synaptic marking." We hypothesized that the transient modification of a given synapse produced by short-term memory would somehow mark that synapse. Marking would allow proteins to be recognized by and stabilized at that synapse.

How the cell targets proteins to specific synapses was a question ideally suited for Kelsey Martin, an extremely gifted cell biologist who had obtained a combined M.D.-Ph.D. at Yale. After graduating from Harvard College, she and her husband joined the Peace Corps and worked in Africa. By the time they came to Columbia, they already had a son, Ben. While she was in our laboratory, they had a daughter, Maya. Kelsey proved a special presence in the laboratory, not only doing first-class science with extraordinary skill, but also lifting all of our spirits by turning our little conference-lunchroom into a joyful kindergarten for gifted children from 4:00 to 6:00 P.M.

The tracking of protein kinase A to the nucleus and the discovery in the nucleus of the CREB regulators had led us along a molecular pathway from the synapse to the nucleus. Now we had to begin the reverse journey. Kelsey and I needed to explore, in a single sensory cell, how a stimulated synapse that is undergoing long-term structural changes differs from an unstimulated synapse. We did this by developing an elegant new cell culture system.

We grew a single sensory neuron with a branched axon that formed synaptic connections with two separate motor neurons. We simulated behavioral training as before by applying pulses of serotonin, but now we could apply them selectively to one or the other set of synaptic connections. A single pulse of serotonin applied to one set of synapses produced short-term facilitation in those synapses only, as expected. Yet five pulses of serotonin applied to one set of synapses produced long-lasting facilitation and growth of new synaptic terminals only at the stimulated synapses. This result was surprising because long-term facilitation and growth require the activation of genes by CREB, an action that takes place in the nucleus of the cell and should theoretically affect all of the cell's synapses. When Kelsey blocked the action of CREB in the nucleus of the cell, it suppressed both facilitation and growth at the stimulated synapse (figure 19-2).

This finding gave us tremendous insight into the computational power of the brain. It illustrates that even though a neuron may make one thousand or more synaptic connections with different target cells, the individual synapses can be modified independently, in long-term as well as short-term memory. The synapses' independence of long-term action gives the neuron extraordinary computational flexibility.

How does this extraordinary selectivity come about? We considered two possibilities: Do the neurons ship messenger RNA and proteins only to synapses marked for long-term memory storage? Or are messenger RNA and proteins shipped to all of the neuron's

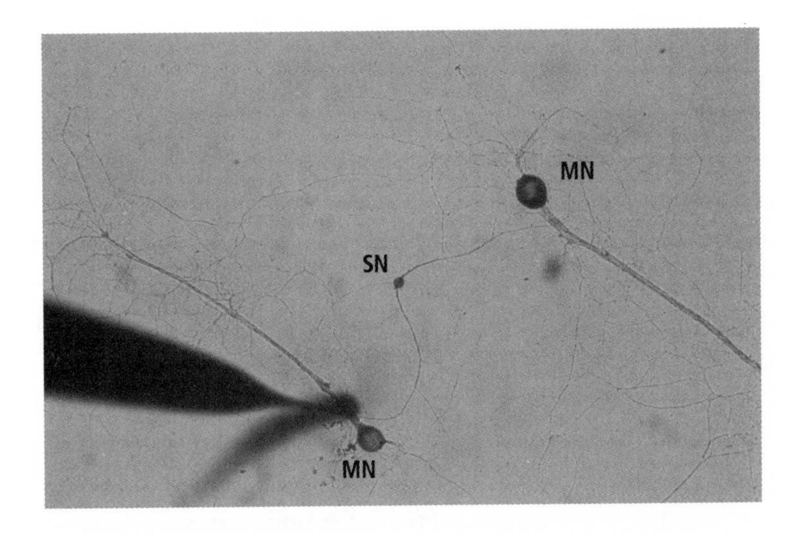

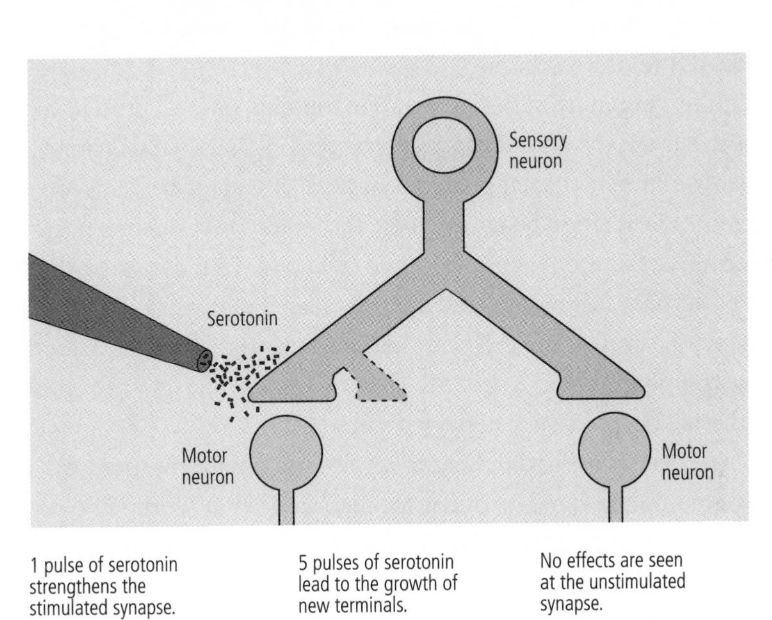

19-2 Setup for studying the role of serotonin in synaptic change. A sensory neuron (SN in photo, above) with a branched axon forms synapses with two motor neurons (MN). Serotonin is applied to just one of the synapses. Only that synapse undergoes short- and long-term change. (Image courtesy of Kelsey Martin.)

synapses and only the marked synapses are able to utilize them for growth? We first began by testing the second hypothesis because it was easy to explore.

What makes this "marked for growth" process possible? Kelsey found that two things must occur at the marked synapse. The first is simply the activation of protein kinase A. If protein kinase A is not activated at the synapse, no facilitation occurs at all. The second is the activation of machinery that regulates *local* protein synthesis. This was a very surprising finding, and it made new sense out of a fascinating area of nerve cell biology that had not been fully appreciated and therefore had been largely ignored. In the early 1980s Oswald Steward, now at the University of California, Irvine, had discovered that even though the vast majority of protein synthesis takes place in the cell body of the neuron, some also occurs locally, at the synapses themselves.

Our findings now indicated that one function of local protein synthesis is to sustain the long-term strengthening of the synaptic connection. When we inhibited local protein synthesis at a synapse, the process of long-term facilitation began and new terminals grew, making use of the proteins sent to the synapse from the cell body. That new growth could not be sustained, however, and after one day it regressed. Thus the proteins synthesized in the cell body and shipped to the terminals are sufficient to initiate synaptic growth, but to sustain that growth, proteins synthesized locally are necessary (figure 19-3).

These results opened up a new window on long-term memory. They suggested that two independent mechanisms are at work. One process initiates long-term synaptic facilitation by sending protein kinase A to the nucleus to activate CREB, thereby turning on the effector genes that encode the proteins needed for the growth of new synaptic connections. The other process perpetuates memory storage by maintaining the newly grown synaptic terminals, a mechanism that requires local protein synthesis. Thus we realized that there are separate processes for initiation and for maintenance. How does that second mechanism work?

IT WAS AT THIS POINT IN 1999 THAT KAUSIK SI, A REMARKABLY original and effective scientist, joined the laboratory. Kausik came

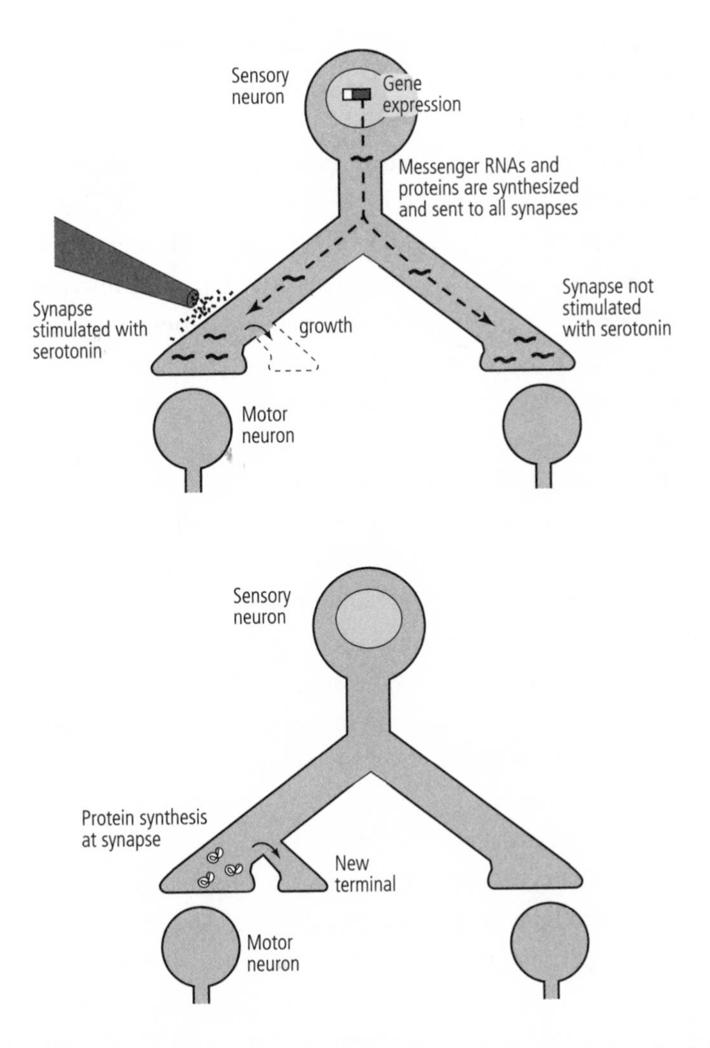

19-3 Two mechanisms of long-term change. New proteins are sent to all of the synapses (above), but only synapses stimulated with serotonin use them to initiate the growth of new axon terminals. Proteins synthesized locally (below) are needed to sustain the growth initiated by gene expression.

from a small town in India, where his father taught in the local high school. When Kausik's father realized that Kausik was interested in biology, he asked a colleague, the local biology teacher, to take the boy under his wing. This biology teacher taught Kausik a great deal and engendered his interest in genetic mechanisms. The teacher also encouraged Kausik to seek graduate training in biology in the United States, which ultimately led to his taking his postdoctoral training with me at Columbia.

Kausik had done his Ph.D. research on protein synthesis in yeast, and after arriving at Columbia, he began to think about the problem of local protein synthesis in *Aplysia*. We knew that messenger RNA molecules are synthesized in the nucleus and translated into protein at specific synapses. So the question was this: Is the messenger RNA sent to the terminals in an active state? Or is it sent in a dormant state, like Sleeping Beauty waiting to be kissed at the marked synapses by some molecular Prince Charming?

Kausik favored the Sleeping Beauty hypothesis. He argued that the dormant messenger RNA molecules become activated only if they reach an appropriately marked synapse and encounter a particular signal. He pointed out that an interesting example of this sort of regulation occurs in the development of the frog. As the frog's egg becomes fertilized and matures, dormant messenger RNA molecules are awakened and activated by a novel protein that regulates local protein synthesis. This protein is known as CPEB (cytoplasmic polyadenylation element-binding protein).

As we journeyed deeper into the maze of molecular processes underlying memory, Kausik found that a novel form of CPEB in *Aplysia* was indeed the Prince Charming we had sought. The molecule is present only in the nervous system, is located at all the synapses of a neuron, is activated by serotonin, and is required at the activated synapses to maintain protein synthesis and the growth of new synaptic terminals. But Kausik's finding advanced the question just one step further. Most proteins are degraded and destroyed in a period of hours. What maintains growth over longer periods of time? What, in principle, has sustained my memory of Mitzi for a lifetime?

As Kausik looked carefully at the amino acid sequence of the novel CPEB, he noticed something very peculiar. One end of the protein had all the characteristics of a prion.

Prions are probably the weirdest proteins known to modern biology. They were first discovered by Stanley Prusiner of the University of California, San Francisco as the causal agents of several mysterious neurodegenerative diseases, such as mad cow disease (bovine spongiform encephalopathy) in cattle and Creutzfeldt-Jakob disease in people (this is the disease that tragically killed Irving Kupfermann in 2002, at the prime of his scientific career). Prions differ from other proteins in that they can fold into two functionally distinct shapes, or conformations; one of these is dominant, the other recessive. The genes that encode prions give rise to the recessive form, but the recessive form can be converted to the dominant form, either by chance alone, as may have happened with Irving, or by eating food that contains an active form of the protein. In the dominant form, prions can be lethal to other cells. The second way in which prions differ from other proteins is that the dominant form is self-perpetuating; it causes the recessive conformation to change its shape and thereby become dominant and self-perpetuating as well (figure 19-4).

I remember it was a beautiful New York afternoon in the spring of 2001, the bright sunlight rippling off the Hudson River outside my office windows, when Kausik walked into my office and asked, "What would you say if I told you that the CPEB has prion-like properties?"

A wild idea! But if true, it could explain how long-term memory is maintained in synapses indefinitely, despite constant protein degradation and turnover. Clearly, a self-perpetuating molecule could remain at a synapse indefinitely, regulating the local protein synthesis needed to maintain newly grown synaptic terminals.

In my late-night thoughts about long-term memory, I had once briefly entertained the idea that prions might somehow be involved in long-term memory storage. Moreover, I was familiar with Prusiner's groundbreaking work on prions and prion diseases, work for which he would receive the 1997 Nobel Prize in Medicine or Physiology. Thus, while I had never anticipated that the novel form of CPEB might be a prion, I was immediately enthusiastic about Kausik's ideas.

Prions were a major field of study in yeast, but no one had identified a normal function of these proteins until Kausik's discovery of the novel form of CPEB in neurons. Thus his discovery not only offered deep new insights into learning and memory, it broke new ground in biology. We soon found that in the sensory neurons of the gill-withdrawal reflex, the conversion of CPEB from the inactive, non-

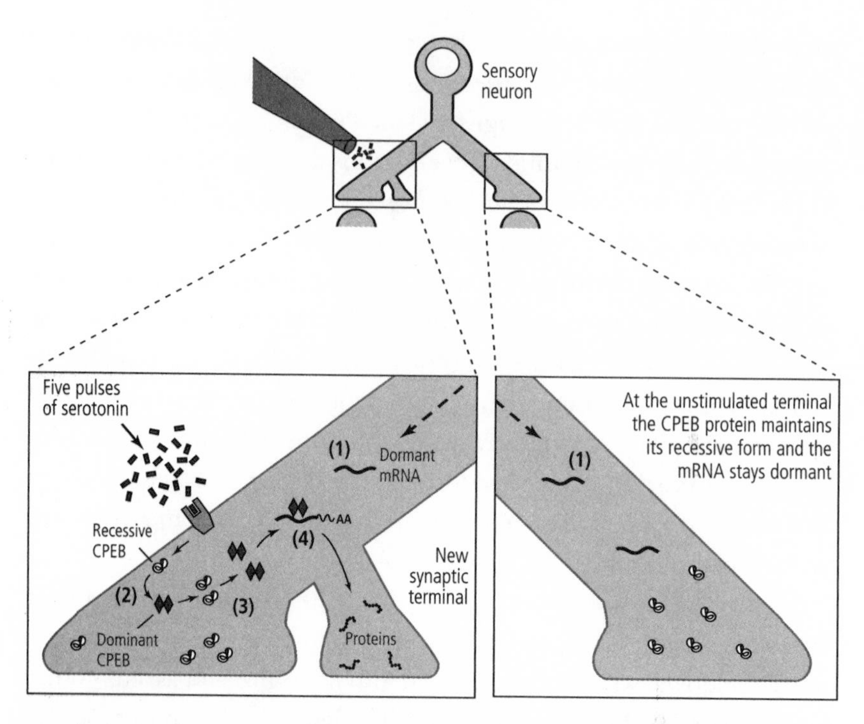

19-4 Long-term memory and the prion-like CPEB protein. As a result of a prior stimulus, the sensory cell's nucleus has sent dormant messenger RNA (mRNA) to all axon terminals (**1**). Five pulses of serotonin at one terminal convert a prion-like protein (CPEB) that is present at all synapses into a dominant, self-perpetuating form (**2**). Dominant CPEB can convert recessive CPEBs to the dominant form (**3**). Dominant CPEB activates dormant messenger RNA (**4**). The activated messenger RNA regulates protein synthesis at the new synaptic terminal, stabilizes the synapse, and perpetuates the memory.

propagating form to the active, propagating form is controlled by serotonin, the transmitter that is required for converting short- to long-term memory (figure 19-4). In its self-perpetuating form, CPEB maintains local protein synthesis. Moreover, the self-perpetuating state is not easily reversed.

These two features make the new variant of the prion ideally designed for memory storage. Self-perpetuation of a protein that is critical for local protein synthesis allows information to be stored selectively and in perpetuity at one synapse, and not, Kausik soon discovered, at the many others that a neuron makes with its target cells.

Beyond discovering the new prion's relevance to the persistence of memory or even to the functioning of the brain, Kausik and I had found two new biological features of prions. First, a normal physiological signal—serotonin—is critical for converting CPEB from one form to another. Second, CPEB is the first self-propagating form of a prion known to serve a physiological function—in this case, perpetuation of synaptic facilitation and memory storage. In all other cases previously studied, the self-propagating form either causes disease and death by killing nerve cells or, more rarely, is inactive.

We have come to believe that Kausik's discovery may be just the tip of a new biological iceberg. In principle, this mechanism—activation of a nonheritable, self-perpetuating change in a protein—could operate in many other biological contexts, including development and gene transcription.

This exciting finding in my laboratory illustrates that basic science can be like a good mystery novel with surprising twists: some new, astonishing process lurks in an undiscovered corner of life and is later found to have wide significance. This particular finding was unusual in that the molecular processes underlying a group of strange brain diseases are now seen also to underlie long-term memory, a fundamental aspect of healthy brain function. Usually, basic biology contributes to our understanding of disease states, not vice versa.

IN RETROSPECT, OUR WORK ON LONG-TERM SENSITIZATION AND the discovery of the prionlike mechanism brought to the forefront three new principles that relate not only to *Aplysia* but to memory storage in all animals, including people. First, activating long-term memory requires the switching on of genes. Second, there is a biological constraint on what experiences get stored in memory. To switch on the genes for long-term memory, CREB-1 proteins must be activated and CREB-2 proteins, which suppress the memory-enhancing genes, must be inactivated. Since people do not remember everything they have learned—nor would anyone want to—it is clear that the genes that encode suppressor proteins set a high threshold for converting short-term to long-term memory. It is for this reason that we remember only certain events and experiences for the long run. Most things we simply forget. Removing that biological constraint triggers the switch to long-term memory. The genes activated by CREB-1 are required for new synaptic growth. The fact that a gene must be switched on to form long-term memory shows clearly that genes are not simply determinants of behavior but are also responsive to environmental stimulation, such as learning.

Finally, the growth and maintenance of new synaptic terminals makes memory persist. Thus, if you remember anything of this book, it will be because your brain is slightly different after you have finished reading it. This ability to grow new synaptic connections as a result of experience appears to have been conserved throughout evolution. As an example, in people, as in simpler animals, the cortical maps of the body surface are subject to constant modification in response to changing input from sensory pathways.

FOUR

These scenes. . . . why do they survive undamaged year after year unless they are made of something comparatively permanent?

—Virginia Woolf, "Sketch of the Past" (1953)

20

A RETURN TO COMPLEX MEMORY

When I first began to study the biological basis of memory, I focused on the memory storage that ensues from the three simplest forms of learning: habituation, sensitization, and classical conditioning. I found that when a simple motor behavior is modified by learning, those modifications directly affect the neural circuit responsible for the behavior, altering the strength of preexisting connections. Once stored in the neural circuit, that memory can be recalled immediately.

This finding gave us our first insight into the biology of implicit memory, a form of memory that is not recalled consciously. Implicit memory is responsible not only for simple perceptual and motor skills but also, in principle, for the pirouettes of Margot Fonteyn, the trumpeting technique of Wynton Marsalis, the accurate ground strokes of Andre Agassi, and the leg movements of an adolescent riding a bicycle. Implicit memory guides us through well-established routines that are not consciously controlled.

The more complex memory that had inspired me initially—the explicit memory for people, objects, and places—is consciously recalled and can typically be expressed in images or words. Explicit memory is far more sophisticated than the simple reflex I had studied in *Aplysia*. It depends on the elaborate neural circuitry of the hippocampus and the medial temporal lobe, and it has many more possible storage sites.

Explicit memory is highly individual. Some people live with such memories all the time. Virginia Woolf falls into this category. Her memories of childhood were always at the edge of her consciousness, ready to be summoned up and incorporated into everyday moments, and she had an exquisite ability to describe the details of her recalled experiences. Thus, years after the death of her mother, Woolf's memory of her was still fresh:

> . . . there she was, in the very center of that great Cathedral space which was childhood; there she was from the very first. My first memory is of her lap. . . . Then I see her in her white dressing gown on the balcony. . . . It is perfectly true that she obsessed me in spite of the fact that she died when I was thirteen, until I was forty-four.
>
> . . . these scenes . . . why do they survive undamaged year after year unless they are made of something comparatively permanent?

Other people call up their past life only occasionally. Periodically, I think back and recall the two police officers coming to our apartment and ordering us to leave on the day of Kristallnacht. When this memory enters my consciousness, I can once again see and feel their presence. I can visualize the worried expression on my mother's face, feel the anxiety in my body, and perceive the confidence in my brother's actions while retrieving his coin and stamp collections. Once I place these memories in the context of the spatial layout of our small apartment, the remaining details emerge in my mind with surprising clarity.

Remembering such details of an event is like recalling a dream or watching a movie in which we play a part. We can even recall past emotional states, though often in a much simplified form. To this day I remember some of the emotional context of my romantic encounter with our housekeeper Mitzi.

As Tennessee Williams wrote in *The Milk Train Doesn't Stop Here Anymore*, describing what we now call explicit memory, "Has it ever struck you . . . that life is all memory, except for the one present moment that goes by you so quickly you hardly catch it going? It's really all memory . . . except for each passing moment."

For all of us, explicit memory makes it possible to leap across space and time and conjure up events and emotional states that have vanished into the past yet somehow continue to live on in our minds. But recalling a memory episodically—no matter how important the memory—is not like simply turning to a photograph in an album. Recall of memory is a creative process. What the brain stores is thought to be only a core memory. Upon recall, this core memory is then elaborated upon and reconstructed, with subtractions, additions, elaborations, and distortions. What biological processes enable me to review my own history with such emotional vividness?

ON REACHING MY SIXTIETH BIRTHDAY, I FINALLY GATHERED THE courage to return to the study of the hippocampus and explicit memory. I had long been curious to see whether some of the basic molecular principles we had learned from a simple reflex circuit in *Aplysia* also applied to the complex neural circuits of the mammalian brain. By 1989, three major breakthroughs had made it feasible to explore this question in the laboratory.

The first was the discovery that the pyramidal cells of the hippocampus play a critical role in an animal's perception of its spatial environment. The second was the discovery of a remarkable synaptic strengthening mechanism in the hippocampus called long-term potentiation. Many researchers thought this mechanism might underlie explicit memory. The third breakthrough, and the one most immediately relevant to my own molecular approach to learning, was the invention of powerful new methodologies for modifying mice genetically. My colleagues and I would adapt those methods to the brain in an attempt to explore explicit memory in the hippocampus in the same molecular detail as we had studied implicit memory in *Aplysia*.

The new era of research in the hippocampus began in 1971, when

John O'Keefe at University College, London made an amazing discovery about how the hippocampus processes sensory information. He found that neurons in the hippocampus of the rat register information not about a single sensory modality—sight, sound, touch, or pain—but about the space surrounding the animal, a modality that depends on information from several senses. He went on to show that the hippocampus of rats contains a representation—a map—of external space and that the units of that map are the pyramidal cells of the hippocampus, which process information about place. In fact, the pattern of action potentials in these neurons is so distinctively related to a particular area of space that O'Keefe referred to them as "place cells." Soon after O'Keefe's discovery, experiments with rodents showed that damage to the hippocampus severely compromises the animals' ability to learn a task that relies on spatial information. This finding indicated that the spatial map plays a central role in spatial cognition, our awareness of the environment around us.

Since space involves information acquired through several sensory modalities, it raised the questions: How are these modalities brought together? How is the spatial map established? Once established, how is the spatial map maintained?

The first clue to the answers emerged in 1973, when Terje Lømo and Tim Bliss, postdoctoral students in Per Andersen's laboratory in Oslo, discovered that the neuronal pathways leading to the hippocampus of rabbits can be strengthened by a brief burst of neural activity. Lømo and Bliss were unaware of O'Keefe's work and did not attempt to examine the functioning of the hippocampus in the context of memory or a specific behavior, as we had done with *Aplysia*'s gill-withdrawal reflex. Instead, they adopted an approach similar to the one Ladislav Tauc and I had first taken in 1962: they developed a neural analog of learning. Rather than basing their neural analog on conventional behavioral paradigms, such as habituation, sensitization, or classical conditioning, they based it on neuronal activity per se. They applied a very rapid train of electrical stimuli (100 impulses per second) to a neuronal pathway leading to the hippocampus and found that the synaptic connections in that pathway were strengthened for several hours to one or more days. Lømo and Bliss called this form of synaptic facilitation long-term potentiation.

It soon emerged that long-term potentiation occurs in all three of the pathways within the hippocampus and that it is not a unitary process. Instead, long-term potentiation describes a family of slightly different mechanisms, each of which increases the strength of the synapse in response to different rates and patterns of stimulation. Long-term potentiation is analogous to long-term facilitation of the connections between sensory and motor neurons in *Aplysia* in that it enhances the strength of synaptic connections. But whereas long-term facilitation in *Aplysia* strengthens synapses heterosynaptically, by means of a modulatory transmitter acting on the homosynaptic pathway, many of long-term potentiation can be initiated merely by means of homosynaptic activity. As we and others found later, however, neuromodulators are usually recruited to switch short-term homosynaptic plasticity into long-term heterosynaptic plasticity.

In the early 1980s, Andersen simplified Lømo and Bliss's research methodology greatly by removing the hippocampus from the brain of a rat, cutting it into slices, and placing the slices in an experimental dish. This enabled him to observe the several neural pathways in a specific segment of the hippocampus. Amazingly, such brain slices can function for hours when prepared properly. With this advance, researchers could analyze the biochemistry of long-term potentiation and observe the effects of drugs blocking various signaling components.

Key molecules involved in long-term potentiation began to emerge from such studies. In the 1960s David Curtis collaborating with Geoffrey Watkins discovered that glutamate, a common amino acid, is the major excitatory transmitter in the vertebrate brain (just as it is in the invertebrate brain, we later discovered). Watkins and Graham Collingridge then found that glutamate acts on two different types of ionotropic receptors in the hippocampus, the AMPA receptor and the NMDA receptor. The AMPA receptor mediates normal synaptic transmission and responds to an individual action potential in the presynaptic neuron. The NMDA receptor, on the other hand, responds only to extraordinarily rapid trains of stimuli and is required for long-term potentiation.

When a postsynaptic neuron is stimulated repeatedly, as in Bliss and Lømo's experiments, the AMPA receptor generates a powerful synaptic potential that depolarizes the cell membrane by as much as 20 or 30 millivolts. This depolarization causes an ion channel in the NMDA receptor to open, allowing calcium to flow into the cell. Roger Nicoll at the University of California, San Francisco and Gary Lynch at the University of California, Irvine discovered independently that the flow of calcium ions into the postsynaptic cell acts as a second messenger (much as cyclic AMP does), triggering long-term potentiation. Thus the NMDA receptor can translate the electrical signal of the synaptic potential into a biochemical signal.

These biochemical reactions are important because they trigger molecular signals that can be broadcast throughout the cell and thus contribute to long-lasting synaptic modifications. Specifically, calcium activates a kinase (called the calcium, calmodulin-dependent protein kinase) that increases synaptic strength for about an hour. Nicoll went on to show that the calcium influx and the activation of this kinase lead to the strengthening of synaptic connections by causing additional AMPA receptors to be assembled and inserted into the membrane of the postsynaptic cell.

The analysis of how the NMDA receptor functions elicited great excitement among neuroscientists, for it showed that the receptor acts as a coincidence detector. It allows calcium ions to flow through its channel if and only if it detects the coincidence of two neural events, one presynaptic and the other postsynaptic. The presynaptic neuron must be active and release glutamate, *and* the AMPA receptor in the postsynaptic cell must bind glutamate and depolarize the cell. Only then will the NMDA receptors become active and allow calcium to flow into the cell, thereby triggering long-term potentiation. Interestingly, in 1949 the psychologist D. O. Hebb had predicted that some kind of neural coincidence detector would be present in the brain during learning: "When an axon of cell A . . . excites cell B and repeatedly or persistently takes part in its firing, some growth process or metabolic changes take place in one or both cells so that A's efficiency is increased."

Aristotle, and subsequently the British empiricist philosophers and many other thinkers, had proposed that learning and memory are somehow the result of mind's ability to associate and form some lasting mental connection between two ideas or stimuli. With the discovery of the NMDA receptor and long-term potentiation, neuroscientists had unearthed a molecular and cellular process that could well carry out this associative process.

21

SYNAPSES ALSO HOLD OUR FONDEST MEMORIES

The new discoveries in the hippocampus—place cells, the NMDA receptor, and long-term potentiation—raised exciting prospects for neuroscience. But it was not at all clear how the spatial map and long-term potentiation were related to each other or to explicit memory storage. To begin with, although long-term potentiation in the hippocampus was a fascinating and widespread phenomenon, it was a highly artificial way of producing changes in synaptic strength. This artificiality caused even Lømo and Bliss to wonder "whether or not the intact animal makes use in real life of a property which has been revealed by synchronous, repetitive volleys. . . ." Indeed, it seemed unlikely that the same pattern of firing occurs in the course of learning. Many scientists questioned whether the changes in synaptic strength produced by long-term potentiation play any role at all in spatial memory or the formation and maintenance of the spatial map.

I began to realize that the ideal way to explore these relationships would be through genetics, much as Seymour Benzer had used genetics to study learning in *Drosophila*. In the 1980s biologists began to combine selective breeding and the tools of recombinant DNA to produce genetically modified mice. These techniques made it possible to manipulate the genes underlying long-term potentiation and thus to answer some pressing questions that interested me. Does long-term potentiation, like long-term facilitation in *Aplysia*, have different phases? Do those phases correspond to short- and long-term storage of spatial memory? If they do correspond, we could interfere with one or the other phase of long-term potentiation and thereby determine what actually happens to the spatial map in the hippocampus when an animal learns and remembers a new environment.

It was exhilarating for me to return to the hippocampus, an old love found again. I had kept up with the advances in research, so it did not seem as though thirty years had passed. Per Andersen was a good friend, as was Roger Nicoll. But most of all I was motivated by memories of my experiments with Alden Spencer when we were both at NIH. I was feeling once again the excitement of being on the edge of something new—but this time armed with molecular genetic techniques whose power and specificity Alden and I could not have imagined in our wildest dreams.

THESE ADVANCES IN MOLECULAR GENETICS HAD THEIR intellectual roots in the selective breeding of mice. Experiments at the turn of the twentieth century showed that various lines of mice differ not only in their genetic makeup but also in their behavior. Some lines proved extremely gifted at learning a variety of tasks, while others were exceptionally dull at those tasks. Such observations showed that genes contribute to learning. Animals differ similarly in their degree of fearfulness, sociability, and parenting ability. By inbreeding and creating some lines that are abnormally fearful and others that are not, behavioral geneticists overcame the randomness of natural selection. Selective breeding was thus the first step in isolating the genes responsible for particular behaviors. Recombinant DNA now made it possible both to try to identify the specific genes needed and to examine the role of those genes in the alteration of the synapses that underlie each behavior, emotional state, or learning capability.

Until 1980 molecular genetics in the mouse relied on a classical analysis known as forward genetics, which is the technique Benzer used in *Drosophila*. It begins by exposing mice to a chemical that usually damages only one of the 15,000 genes in the animal's genome. The damage occurs at random, however, so which gene is affected is anyone's guess. The animals are given a variety of tasks to see which, if any, are affected by the randomly altered gene. Because mice must be bred for several generations, forward genetics is very demanding and time-consuming, but it has the great advantage of being unbiased. There are no hypotheses and therefore no biases involved in screening for genes in this manner.

The recombinant DNA revolution enabled molecular biologists to develop a less demanding, less time-consuming strategy, reverse genetics. In reverse genetics, a specific gene is either removed from or introduced into the mouse's genome and the effects on synaptic change and learning examined. Reverse genetics is biased—it is designed to test a specific hypothesis, such as whether a particular gene and the protein it encodes are involved in a particular behavior.

Two methods of modifying individual genes made reverse genetics in mice possible. The first, transgenesis, involves the introduction of a foreign gene, called a transgene, into the DNA of a mouse egg. Once the egg is fertilized, the transgene becomes part of the genome of the baby mouse. Adult transgenic mice are then bred to obtain a genetically pure strain of mice, all of which express the transgene. The second method of genetically modifying mice involves "knocking out" one gene in the mouse genome. This is done by inserting a segment of genetic material into the mouse's DNA that renders the chosen gene dysfunctional and thus eliminates the protein encoded by that gene from the mouse's body.

IT WAS BECOMING EVIDENT TO ME THAT WITH THESE ADVANCES in genetic engineering, the mouse would be a superb experimental animal for identifying the genes and proteins responsible for the various forms of long-term potentiation. One could then relate those genes and proteins to the storage of spatial memory. Although mice are relatively simple mammals, they have a brain that is anatomically similar to that of humans and, as in humans, the hippocampus is involved in storing memories of places and objects. Moreover, mice breed much faster than larger mammals such as cats, dogs, monkeys, and people. As a result, large populations with identical genes, including identical transgenes or knockout genes, can be bred within months.

These revolutionary new experimental techniques also had major biomedical ramifications. Almost every gene in the human genome exists in several different versions, called alleles, which are present in different members of the human population. Genetic studies of human neurological and psychiatric disorders had made it possible to identify alleles that account for behavioral differences in normal people as well as alleles that underlie many neurological disorders, such as amyotrophic lateral sclerosis, early onset Alzheimer's disease, Parkinson's disease, Huntington's disease, and several forms of epilepsy. The ability to insert disease-causing alleles into the mouse genome and then study how they wreak havoc on the brain and on behavior revolutionized neurology.

The final stimulus for me to move into studying genetically engineered mice was the presence in our lab of several talented postdoctoral fellows, among them Seth Grant and Mark Mayford. Grant and Mayford knew mouse genetics much better than I, and they greatly influenced the direction of our research. Grant was the driving force behind my beginning to work on genetically modified mice, while Mayford's critical thinking became important later, as we began to improve on the methodology we and others had used in our first generation of behavioral studies in mice.

Our original methods for producing transgenic mice affected every cell in the mouse's body. We needed to find a way to restrict our genetic manipulations to the brain, specifically to the regions that form the neural circuits of explicit memory. Mayford developed ways of limiting the expression of newly implanted genes to certain regions of the brain. He also developed a method for controlling the timing of gene expression in the brain, thereby making it possible to turn the gene on and off. These two feats inaugurated a new stage in our studies and were widely adopted by other researchers; they remain cornerstones of the modern analysis of behavior in genetically modified mice.

THE FIRST ATTEMPT TO RELATE LONG-TERM POTENTIATION TO spatial memory was made in the late 1980s. Richard Morris, a physiologist at the University of Edinburgh, had shown that by blocking the NMDA receptor pharmacologically, one could block long-term potentiation and thus interfere with spatial memory. In independent experiments, Grant and I at Columbia and Susumu Tonegawa and his postdoctoral fellow Alcino Silva at the Massachusetts Institute of Technology carried this analysis one important step further. We each created a different line of genetically modified mice that lacked a key protein thought to be involved in long-term potentiation. We then observed how learning and memory were affected in the genetically modified mice, compared with normal mice.

We tested the animals' performance on several well-established spatial tasks. For example, we placed a mouse in the center of a large, white, well-illuminated circular platform surrounded by a rim of forty holes. Only one of the holes led to an escape chamber. The platform was in a small room, each of whose walls was decorated with a different, distinctive marking. Mice do not like being in an open space, especially a well-lit one. They feel defenseless and try to escape. The only way a mouse could escape from the platform was to find the hole that led to the escape chamber. Ultimately, the mouse found that hole by learning the spatial relationship between the hole and the markings on the wall.

In trying to escape, mice use three strategies in sequence: random, serial, and spatial. Each strategy allows the animals to find the escape hatch, but each varies greatly in efficiency. The mice first go to any hole at random and quickly learn that this strategy is not efficient. Next, they begin with one hole and then try each consecutive hole until they find the right one. This is a better strategy but still not optimal. Neither strategy is spatial—neither requires the mice to have an internal map of the spatial organization of the environment stored in their brain—and neither strategy requires the hippocampus. Finally, the mice use a spatial strategy that does require the hippocampus. They learn to look to see which marked wall is aligned with the target hole, and then they make a beeline for that hole, using the marking on the wall as a guide. Most mice go through the first two strategies quickly and soon learn to use the spatial strategy.

We then focused on long-term potentiation in an area of the hippocampus called the Schaffer collateral pathway. Larry Squire of the University of California, San Diego had shown that damage to this one pathway produces a memory deficit similar to that experienced by H.M., Brenda Milner's patient. We found that by knocking out a particular gene encoding a protein that is important for long-term potentiation, we could compromise synaptic strengthening in the Schaffer collateral pathway. Moreover, the genetic defect was correlated with a defect in the mouse's spatial memory.

Each year the Cold Spring Harbor Laboratory holds a meeting devoted solely to one major topic in biology. The topic for 1992 was "The Cell Surface," but because Susumu Tonegawa's and our work relating genes to memory in the mouse was deemed so interesting, a new slot, unrelated to the cell surface, was created for the two of us so we could give back-to-back talks. Tonegawa and I presented our separate experiments on how knocking out a single gene inhibits both long-term potentiation in a pathway of the hippocampus and spatial memory. At the time, it was the most direct correlation that had been made between long-term potentiation and spatial memory. Soon thereafter, both of us went one step further and examined how long-term potentiation relates to the spatial map of the external environment represented in the hippocampus.

By the time of that meeting, Tonegawa and I already knew each other a bit. In the 1970s he had determined the genetic basis of antibody diversity, an extraordinary contribution to immunology for which he received the Nobel Prize in Physiology or Medicine in 1987. With this accomplishment behind him, he wanted to turn to the brain for new scientific worlds to conquer. He was a good friend of Richard Axel, who suggested that he talk to me.

The problem that most interested Tonegawa when he came to see me in 1987 was consciousness. I tried to encourage his enthusiasm for brain research while dissuading him from taking on consciousness, which was too difficult and too poorly defined for a molecular approach at that time. Susumu had started to use genetically modified mice to study the immune system, so it was natural and much more realistic for him to turn to learning and memory, which he did when Silva joined his lab.

Since 1992 many other research groups have obtained results parallel to our own. Although there are occasional, important exceptions to the link between disrupted long-term potentiation and deficient spatial memory, it has nevertheless proven to be a good place to start examining the molecular mechanisms of long-term potentiation and the role of these molecules in memory storage.

I KNEW THAT SPATIAL MEMORY IN MICE, LIKE THE IMPLICIT memory studied in *Aplysia* and *Drosophila*, has two components: a short-term memory that does not require protein synthesis and a long-term memory that does. Now I wanted to find out whether storage of explicit short- and long-term memory also has distinctive synaptic and molecular mechanisms. In *Aplysia*, short-term memory requires short-term synaptic changes that rely only on second-messenger signaling. Long-term memory requires more persistent synaptic changes that are based on alterations in gene expression as well.

My colleagues and I examined slices of the hippocampus taken from genetically modified mice and found that in each of the three major pathways of the hippocampus, long-term potentiation has two phases similar to those of long-term facilitation in *Aplysia*. A single train of electrical stimuli produces a transient, early phase of long-term potentiation that lasts only one to three hours and does not require the synthesis of new protein. The reaction of neurons to those stimuli was just as Roger Nicoll had described: NMDA receptors in the postsynaptic cell are activated, leading to the flow of calcium ions into the postsynaptic cell. Here calcium acts as a second messenger; it triggers long-term potentiation by enhancing the existing AMPA receptors' response to glutamate and by stimulating the insertion of new AMPA receptors into the membrane of the postsynaptic cell. In response to certain patterns of stimulation, the postsynaptic cell also sends a signal back to the presynaptic cell, calling for more glutamate.

Repeated trains of electrical stimuli produce a late phase of long-term potentiation that lasts for more than a day. We found the properties of this phase, which previously had not been extensively explored, to be very similar to long-term facilitation of synaptic strength in *Aplysia*. In both *Aplysia* and mice, the late phase of long-term potentiation is strongly affected by modulatory interneurons, which in mice are recruited to switch a short-term, homosynaptic into long-term, heterosynaptic change. In mice those neurons release dopamine, a neurotransmitter commonly recruited in the mammalian brain for attention and reinforcement. Like serotonin in *Aplysia*, dopamine prompts a receptor in the hippocampus to activate an enzyme that increases the amount of cyclic AMP. However, an important part of the increase in cyclic AMP in the mouse hippocampus occurs in the postsynaptic cell, whereas in *Aplysia* the increase occurs in the presynaptic sensory neuron. In each case, the cyclic AMP recruits protein kinase A and other protein kinases, which leads to the activation of CREB and the turning on of effector genes.

One of the striking things we had found in studying memory in *Aplysia* was the existence of the memory-suppressor gene that produces the CREB-2 protein. Blocking the expression of that gene in *Aplysia* enhances both the strengthening and the increase in number of synapses associated with long-term facilitation. In the mouse, we found that blocking this and similar memory-suppressor genes enhances both long-term potentiation in the hippocampus and spatial memory.

In the course of these studies, I found myself once again in an enjoyable collaboration with Steven Siegelbaum. We were interested in a particular ion channel that inhibits synaptic strengthening, especially in certain dendrites. Alden Spencer and I had studied those dendrites in 1959 and inferred that they produce action potentials in response to activity in the perforant pathway, which leads from the entorrhinal cortex to the hippocampus. Steve and I bred mice in which the gene for this particular ion channel was lacking. We found that long-term potentiation in response to stimulation of the perforant pathway was greatly enhanced in those mice, in part by dendritic action potentials. As a result, these mice were brilliant; they had a much stronger spatial memory than normal mice!

My colleagues and I also discovered that explicit memory in the mammalian brain, unlike implicit memory in *Aplysia* or *Drosophila*, requires several gene regulators in addition to CREB. Although the evidence is less complete, it appears that in mice, expression of genes also gives rise to anatomical changes—specifically, to the growth of new synaptic connections.

Despite the significant behavioral differences between implicit and explicit memory, some aspects of implicit memory storage in invertebrates have been conserved over millions of years of evolutionary time in the mechanisms by which explicit memory is stored in vertebrates. Although the great neurophysiologist John Eccles had urged me early in my career not to abandon research on the splendid mammalian brain for work on a slimy, brainless sea snail, it is now clear that several key molecular mechanisms of memory are shared by all animals.

22

THE BRAIN'S PICTURE OF THE EXTERNAL WORLD

The study of the explicit memory for space in the mouse drew me ineluctably to the larger questions that had attracted me to psychoanalysis at the beginning of my career. I was starting to think about the nature of attention and consciousness, mental states not associated with simple reflex actions but with complex psychological processes. I wanted to focus on how space—the internal environment in which the mouse navigates—is represented in the brain and how this representation is modified by attention. In doing so, I was moving from a system in *Aplysia* that was reasonably well understood to systems in the mammalian brain that had yielded (and to some degree still yield) only a few fascinating results and many unresolved questions. Nevertheless, the time had come to try to help move the molecular biology of cognition a step forward.

In examining implicit memory in *Aplysia*, I had built a neurobiological and molecular approach to elementary mental processes on a foundation laid by Pavlov and the behaviorists. Their methods were rigorous, but they reflected a narrow and limited definition of behavior, one that focused on motor acts. In contrast, our research on the explicit memory and the hippocampus posed enormous new intellectual challenges, in no small part because the encoding and recall of spatial memory requires conscious attention.

As a first step in thinking about complex memory of space and its internal representation in the hippocampus, I turned from the behaviorists to the cognitive psychologists, the scientific successors to the psychoanalysts and the first group of scientists to think systematically about how the outside world is re-created and represented in our brain.

COGNITIVE PSYCHOLOGY EMERGED IN THE EARLY 1960s IN response to the limitations of behaviorism. While attempting to retain the experimental rigor of behaviorism, cognitive psychologists focused on mental processes that were more complex and closer to the domain of psychoanalysis. Much like the psychoanalysts before them, the new cognitive psychologists were not satisfied with simply describing motor responses elicited by sensory stimuli. Rather, they were interested in investigating the mechanisms in the brain that intervene between a stimulus and a response—the mechanisms that convert a sensory stimulus into an action. The cognitive psychologists set up behavioral experiments that allowed them to infer how sensory information from the eyes and the ears is transformed in the brain into images, words, or actions.

The thinking of cognitive psychologists was driven by two underlying assumptions. The first was the Kantian notion that the brain is born with *a priori* knowledge, "knowledge that is . . . independent of experience." That idea was later advanced by the European school of Gestalt psychologists, the forerunners, together with psychoanalysis, of modern cognitive psychology. The Gestalt psychologists argued that our coherent perceptions are the end result of the brain's built-in ability to derive meaning from the properties of the world, only limited features of which can be detected by the peripheral sensory organs. The reason that the brain can derive meaning from, say, a limited analysis of a visual scene is that the visual system does not simply record a scene passively, as a camera does. Rather, perception is creative: the visual system transforms the two-dimensional patterns of light on the retina of the eye into a logically coherent and stable interpretation of a three-dimensional sensory world. Built into neural pathways of the brain are complex rules of guessing; those rules allow the brain to extract information from relatively impoverished patterns of incoming neural signals and turn it into a meaningful image. The brain is thus the ambiguity-resolving machine par excellence!

Cognitive psychologists demonstrated this ability with studies of illusions, that is, misreadings of visual information by the brain. For example, an image that does not contain the complete outline of a triangle is nevertheless seen as a triangle because the brain expects to form certain images (figure 22-1). The brain's expectations are built into the anatomical and functional organization of the visual pathways; they are derived in part from experience but in large part from the innate neural wiring for vision.

To appreciate these evolved perceptual skills, it is useful to compare the computational abilities of the brain with those of artificial computational or information-processing devices. When you sit at a sidewalk café and watch people go by, you can, with minimal clues, readily distinguish men from women, friends from strangers. Perceiving and recognizing objects and people seem effortless. However, computer

22-1 The brain's reconstruction of sensory information. The brain resolves ambiguities by creating shapes from incomplete data—for example, filling in the missing lines of these triangles. If you hide parts of these pictures, your brain is deprived of some clues it uses to form conclusions and the triangles vanish.

scientists have learned from constructing intelligent machines that these perceptual discriminations require computations that no computer can begin to approach. Merely recognizing a person is an amazing computational achievement. All of our perceptions—seeing, hearing, smelling, and touching—are analytical triumphs.

The second assumption developed by cognitive psychologists was that the brain achieves these analytic triumphs by developing an internal representation of the external world—a cognitive map—and then using it to generate a meaningful image of what is out there to see and to hear. The cognitive map is then combined with information about past events and is modulated by attention. Finally, the sensory representations are used to organize and orchestrate purposeful action.

The idea of a cognitive map proved an important advance in the study of behavior and brought cognitive psychology and psychoanalysis closer together. It also provided a view of mind that was much broader and more interesting than that of the behaviorists. But the concept was not without problems. The biggest problem was the fact that the internal representations inferred by cognitive psychologists were only sophisticated guesses; they could not be examined directly and thus were not readily accessible to objective analysis. To see the internal representations—to peer into the black box of mind—cognitive psychologists had to join forces with biologists.

FORTUNATELY, AT THE SAME TIME THAT COGNITIVE PSYCHOLOGY was emerging in the 1960s, the biology of higher brain function was maturing. During the 1970s and 1980s, behaviorists and cognitive psychologists began collaborating with brain scientists. As a result, neural science, the biological science concerned with brain processes, began to merge with behaviorist and cognitive psychology, the sciences concerned with mental processes. The synthesis that emerged from these interactions gave rise to the field of cognitive neural science, which focused on the biology of internal representations and drew heavily on two lines of inquiry: the electrophysiological study of how sensory information is represented in the brains of animals, and the imaging of sensory and other complex internal representations in the brains of intact, behaving human beings.

Both of these approaches were used to examine the internal representation of space, which I wanted to study, and they revealed that space is indeed the most complex of sensory representations. To make any sense of it, I needed first to take stock of what had already been learned from the study of simpler representations. Fortunately for me, the major contributors to this field were Wade Marshall, Vernon Mountcastle, David Hubel, and Torsten Wiesel, four people I knew very well, and whose work I was intimately familiar with.

THE ELECTROPHYSIOLOGICAL STUDY OF SENSORY REPRESENTATION was initiated by my mentor, Wade Marshall, the first person to study how touch, vision, and hearing were represented in the cerebral cortex. Marshall began by studying the representation of touch. In 1936 he discovered that the somatosensory cortex of the cat contains a map of the body surface. He then collaborated with Philip Bard and Clinton Woolsey to map in great detail the representation of the entire body surface in the brain of monkeys. A few years later Wilder Penfield mapped the human somatosensory cortex.

These physiological studies revealed two principles regarding sensory maps. First, in both people and monkeys, each part of the body is represented in a systematic way in the cortex. Second, sensory maps are not simply a direct replica in the brain of the topography of the body surface. Rather, they are a dramatic distortion of the body form. Each part of the body is represented in proportion to its importance in sensory perception, not to its size. Thus the fingertips and the mouth, which are extremely sensitive regions for touch perception, have a disproportionately larger representation than does the skin of the back, which although more extensive is less sensitive to touch. This distortion reflects the density of sensory innervation in different areas of the body. Woolsey later found similar distortions in other experimental animals; in rabbits, for example, the face and nose have the largest representation in the brain because they are the primary means through which the animal explores its environment. As we have seen, these maps can be modified by experience.

In the early 1950s Vernon Mountcastle at Johns Hopkins extended the analysis of sensory representation by recording from single cells. Mountcastle found that individual neurons in the somatosensory cortex respond to signals from only a limited area of the skin, an area he called the receptive field of the neuron. For example, a cell in the hand region of the somatosensory cortex of the left hemisphere might respond only to stimulation of the tip of the middle finger of the right hand and to nothing else.

Mountcastle also discovered that tactile sensation is made up of several distinct submodalities; for example, touch includes the sensation produced by hard pressure on the skin as well as that produced by a light brush against it. He found that each distinct submodality has its own private pathway within the brain and that this segregation is maintained at each relay in the brain stem and in the thalamus. The most fascinating example of this segregation is evident in the somatosensory cortex, which is organized into columns of nerve cells extending from its upper to its lower surface. Each column is dedicated to one submodality and one area of skin. Thus all the cells in one column might receive information on superficial touch from the end of the index finger. Cells in another column might receive information on deep pressure from the index finger. Mountcastle's work revealed the extent to which the sensory message about touch is deconstructed; each submodality is analyzed separately and reconstructed and combined only in later stages of information processing. Mountcastle also proposed the now generally accepted idea that these columns form the basic information-processing modules of the cortex.

OTHER SENSORY MODALITIES ARE ORGANIZED SIMILARLY. THE analysis of perception is more advanced in vision than in any other sense. Here we see that visual information, relayed from one point to another along the pathway from the retina to the cerebral cortex, is also transformed in precise ways, first being deconstructed and then reconstructed—all without our being in any way aware of it.

In the early 1950s, Stephen Kuffler recorded from single cells in the retina and made the surprising discovery that those cells do not signal absolute levels of light; rather, they signal the contrast between light and dark. He found that the most effective stimulus for exciting retinal cells is not diffuse light but small spots of light. David Hubel and Torsten Wiesel found a similar principle operating in the next relay stage, located in the thalamus. However, they made the astonishing discovery that once the signal reaches the cortex, it is transformed. Most cells in the cortex do not respond vigorously to small spots of light. Instead, they respond to linear contours, to elongated edges between lighter and darker areas, such as those that delineate objects in our environment.

Most amazingly, each cell in the primary visual cortex responds only to a specific orientation of such light-dark contours. Thus if a square block is rotated slowly before our eyes, slowly changing the angle of each edge, different cells will fire in response to these different angles. Some cells respond best when the linear edge is oriented vertically, others when the edge is horizontal, and still other cells when the axis is at an oblique angle. Deconstructing visual objects into line segments of different orientation appears to be the initial step in encoding the forms of objects in our environment. Hubel and Wiesel next found that in the visual system, as in the somatosensory system, cells with similar properties (in this case, cells with similar axes of orientation) are grouped together in columns.

I found this work enthralling. As a scientific contribution to brain science, it stands as the most fundamental advance in our understanding of the organization of the cerebral cortex since the work of Cajal at the turn of the last century. Cajal revealed the precision of the interconnections between populations of individual nerve cells. Mountcastle, Hubel, and Wiesel revealed the functional significance of those patterns of interconnections. They showed that the connections filter and transform sensory information on the way to and within the cortex, and that the cortex is organized into functional compartments, or modules.

As a result of the work of Mountcastle, Hubel, and Wiesel, we can begin to discern the principles of cognitive psychology on the cellular level. These scientists confirmed the inferences of the Gestalt psychologists by showing us that the belief that our perceptions are precise and direct is an illusion—a perceptual illusion. The brain does not simply take the raw data that it receives through the senses and reproduce it faithfully. Instead, each sensory system first analyzes and deconstructs, then restructures the raw, incoming information according to its own built-in connections and rules—shades of Immanuel Kant!

The sensory systems are hypothesis generators. We confront the world neither directly nor precisely, but as Mountcastle pointed out:

> . . . from a brain linked to what is "out there" by a few million fragile sensory nerve fibers, our only information channels, our lifelines to reality. They provide also what is essential for life itself: an afferent excitation that maintains the conscious state, the awareness of self.
>
> Sensations are set by the encoding functions of sensory nerve endings, and by the integrating neural mechanics of the central nervous system. Afferent nerve fibers are not high-fidelity recorders, for they accentuate certain stimulus features, neglect others. The central neuron is a story-teller with regard to the nerve fibers, and it is never completely trustworthy, allowing distortions of quality and measure. . . . *Sensation is an abstraction, not a replication, of the real world.*

SUBSEQUENT WORK ON THE VISUAL SYSTEM SHOWED THAT IN addition to dissecting objects into linear segments, other aspects of visual perception—motion, depth, form, and color—are segregated from one another and conveyed in separate pathways to the brain, where they are brought together and coordinated into a unified perception. An important part of this segregation occurs in the primary visual area of the cortex, which gives rise to two parallel pathways. One pathway, the "what" pathway, carries information about the form of an object: what the object looks like. The other, the "where" pathway, carries information about the movement of the object in space: where the object is located. These two neural pathways end in higher regions of the cortex that are concerned with more complex processing.

The idea that different aspects of visual perception might be handled in separate areas of the brain was predicted by Freud at the end of the nineteenth century, when he proposed that the inability of certain patients to recognize specific features of the visual world was due not to a sensory deficit (resulting from damage to the retina or the optic nerve), but to a cortical defect that affected their ability to combine aspects of vision into a meaningful pattern. These defects, which Freud called *agnosias* (loss of knowledge), can be quite specific. For example, there are specific defects caused by lesions in either the "where" or the "what" pathway. A person with depth agnosia due to a defect in the "where" system is unable to perceive depth but has otherwise intact vision. One such person was unable "to appreciate depth or thickness of objects seen. . . . The most corpulent individual might be a moving cardboard figure; everything is perfectly flat." Similarly, persons with motion agnosia are unable to perceive motion, but all other perceptual abilities are normal.

Striking evidence indicates that a discrete region of the "what" pathway is specialized for face recognition. Following a stroke, some people can recognize a face as a face, the parts of the face, and even specific emotions expressed on the face but are unable to identify the face as belonging to a particular person. People with this disability (prosopagnosia) often cannot recognize close relatives or even their own face in the mirror. They have not lost the ability to recognize a person's identity, they have lost the connection between a face and an identity. To recognize a close friend or relative, these patients must rely on the person's voice or other nonvisual clues. In his classic essay "The Man Who Mistook His Wife for a Hat," the gifted neurologist-neuropsychologist Oliver Sacks describes a patient with prosopagnosia who failed to recognize his wife sitting next to him and, thinking she was his hat, tried to pick her up and put her on his head as he was about to leave Sacks's office.

How is information about motion, depth, color, and form, which is carried by separate neural pathways, organized into a cohesive perception? This problem, called the binding problem, is related to the unity of conscious experience: that is, to how we see a boy riding a bicycle not by seeing movement without an image or an image that is stationary, but by seeing in full color a coherent, three-dimensional, moving version of the boy. The binding problem is thought to be resolved by bringing into association temporarily several independent neural pathways with discrete functions. How and where does this binding occur? Semir Zeki, a leading student of visual perception at University College, London, put the issue succinctly:

> At first glance, the problem of integration may seem quite simple. Logically it demands nothing more than that all the signals from the specialized visual areas be brought together, to 'report' the results of their operations to a single master cortical area. This master area would then synthesize the information coming from all these diverse sources and provide us with the final image, or so one might think. But the brain has its own logic. . . . If all the visual areas report to a single master cortical area, who or what does that single area report to? Put more visually, who is 'looking' at the visual image provided by that master area? The problem is not unique to the visual image or the visual cortex. Who, for example, listens to the music provided by a master auditory area, or senses the odour provided by the master olfactory cortex? It is in fact pointless pursuing this grand design. For here one comes across an important anatomical fact, which may be less grand but perhaps more illuminating in the end: *there is no single cortical area to which all other cortical areas report exclusively, either in the visual or in any other system. In sum, the cortex must be using a different strategy for generating the integrated visual image.*

WHEN A COGNITIVE NEUROSCIENTIST PEERS DOWN AT THE brain of an experimental animal, the scientist can see which cells are firing and can read out and understand what the brain is perceiving. But what strategy does the brain use to read itself out? This question, which is central to the unitary nature of conscious experience, remains one of the many unresolved mysteries of the new science of mind.

An initial approach was developed by Ed Evarts, Robert Wurtz, and Michael Goldberg at NIH. They pioneered methods for recording the activity of single nerve cells in the brains of intact, behaving monkeys focusing on cognitive tasks that require action and attention. Their new research techniques enabled investigators such as Anthony Movshon at NYU and William Newsome at Stanford to correlate the action of individual brain cells with complex behavior—that is, with perception and action—and to see the effects on perception and action of stimulating or reducing activity in small groups of cells.

These studies also made it possible to examine how the firing of individual nerve cells involved in perceptual and motor processing is modified by attention and decision making. Thus unlike behaviorism, which focused only on the behavior stemming from an animal's response to a stimulus, or cognitive psychology, which focused on the abstract notion of an internal representation, the merger of cognitive psychology and cellular neurobiology revealed an actual physical representation—an information-processing capability in the brain—that leads to a behavior. This work demonstrated that the unconscious inference described by Helmholtz in 1860, the unconscious information processing that intervenes between a stimulus and a response, could also be studied on the cellular level.

The cellular studies of internal representation in the cerebral cortex of the sensory and motor world were extended in the 1980s with the introduction of brain imaging. These techniques, such as positron emission tomography (PET) and functional magnetic resonance imaging (fMRI), carried the work of Paul Broca, Carl Wernicke, Sigmund Freud, the British neurologist John Hughlings Jackson, and Oliver Sacks a giant step forward by revealing the locale in the brain of a variety of complex behavioral functions. With these new technologies, investigators could look into the brain and see not simply single cells, but also neural circuits in action.

I HAD BECOME CONVINCED THAT THE KEY TO UNDERSTANDING the molecular mechanisms of spatial memory was understanding how space is represented in the hippocampus. As one might expect because of its importance in explicit memory, the spatial memory of environments has a prominent internal representation in the hippocampus.

This is evident even anatomically. Birds in which spatial memory is particularly important—those that store food at a large number of sites, for example—have a larger hippocampus than other birds.

London taxi drivers are another case in point. Unlike cabbies elsewhere, those in London must pass a rigorous examination to obtain their license. In this test, they must demonstrate that they know every street name in London and the most efficient routes for traveling between two points. Functional magnetic resonance imaging revealed that after two years of this rigorous orientation to the streets of the city, London taxi drivers have a larger hippocampus than other persons the same age. Indeed, the size of their hippocampus continues to increase with time on the job. Moreover, brain-imaging studies show that the hippocampus is activated during imagined travel, when a taxi driver is asked to recall how to get to a particular destination. How, then, is space represented on the cellular level within the hippocampus?

To address these questions, I brought the tools and insights of molecular biology to bear on existing studies of the internal representation of space in mice. We had begun by using genetically modified mice to study the effect of specific genes on long-term potentiation in the hippocampus and on explicit memory of space. We were now ready to ask how long-term potentiation helps stabilize the internal representation of space and how attention, a defining feature of explicit memory storage, modulates the representation of space. This compound approach—extending from molecules to mind—opened up the possibility of a molecular biology of cognition and attention and completed the outlines of a synthesis that led to a new science of mind.

23

ATTENTION MUST BE PAID!

In all living creatures, from snails to people, knowledge of space is central to behavior. As John O'Keefe notes, "Space plays a role in all our behaviour. We live in it, move through it, explore it, defend it." Space is not only a critical sense but a fascinating one because unlike other senses, space is not analyzed by a specialized sensory organ. How, then, is space represented in the brain?

Kant, one of the forefathers of cognitive psychology, argued that the ability to represent space is built into our minds. He pictured people as being born with principles of ordering space and time, so that when other sensations are elicited—be they objects, melodies, or tactile experiences—they are automatically interwoven in specific ways with space and time. O'Keefe applied this Kantian logic about space to explicit memory. He argued that many forms of explicit memory (for example, memory for people and objects) use spatial coordinates—that is, we typically remember people and events in a spatial context. This is not a new idea. In 55 B.C., Cicero, the great Roman poet and orator, described the Greek technique (used to this day by some actors) of remembering words by picturing the rooms of a house in sequence, associating words with each room, and then mentally walking through the rooms in the right order.

Because we do not have a sensory organ dedicated to space, the representation of space is a quintessentially cognitive sensibility: it is the binding problem writ large. The brain must combine inputs from several different sensory modalities and then generate a complete internal representation that does not depend exclusively on any one input. The brain commonly represents information about space in many areas and many different ways, and the properties of each representation vary according to its purpose. For example, for some representations of space the brain typically uses *egocentric* coordinates (centered on the receiver), encoding, for example, where a light is relative to the fovea or where an odor or touch comes from with respect to the body. Egocentric representation is also used by people or monkeys for orienting to a sudden noise by making an eye movement to a particular location, by *Drosophila* in avoiding of an odor with unpleasant associations, or by *Aplysia* in generating its gill-withdrawal reflex. For other behaviors, like memory for space in the mouse or in people, it is necessary to encode the organism's position relative to the outside world and the relationship of external objects to one another. For these purposes the brain uses *allocentric* coordinates (centered on the world).

Studies of the simpler sensory maps for touch and vision in the brain, which are based on egocentric coordinates, provided a springboard for studies of the more complex representation of allocentric space. But the spatial map discovered in 1971 by O'Keefe differs radically from the egocentric sensory maps for touch and vision discovered by Wade Marshall, Vernon Mountcastle, David Hubel, and Torsten Wiesel because it is not dependent on any given sensory modality. Indeed, in 1959, when Alden Spencer and I tried to decipher how sensory information comes into the hippocampus, we recorded from individual nerve cells while we stimulated different individual senses, but we failed to obtain a brisk response. We did not realize that the hippocampus is concerned with perception of the environment and therefore represents multisensory experience.

John O'Keefe was the first to realize that the hippocampus of rats contains a multisensory representation of extrapersonal space.

O'Keefe found that as an animal walks around an enclosure, some place cells fire action potentials only when that animal moves into a particular location, while others fire when the animal moves to another place. The brain breaks down its surroundings into many small, overlapping areas, similar to a mosaic, each represented by activity in specific cells in the hippocampus. This internal map of space develops within minutes of the rat's entrance into a new environment.

I BEGAN TO THINK ABOUT THE SPATIAL MAP IN 1992, WONDERING how it is formed, how it is maintained, and how attention directs its formation and maintenance. I was struck by the fact that O'Keefe and others had found that the spatial map of even a simple locale does not form instantaneously but over ten to fifteen minutes of the rat's entrance into the new environment. This suggests that the formation of the map is a learning process; practice makes perfect also for space. Under optimal circumstances this map remains stable for weeks or even months, much like a memory process.

Unlike vision, touch, or smell, which are prewired and based on Kantian *a priori* knowledge, the spatial map presents us with a new type of representation, one based on a combination of *a priori* knowledge and learning. The *general* capability for forming spatial maps is built into mind, but the *particular* map is not. Unlike neurons in a sensory system, place cells are not switched on by sensory stimulation. Their collective activity represents the location where the animal *thinks* it is.

I now wanted to know whether the same molecular pathways needed to induce long-term potentiation and spatial memory in our experiments on the hippocampus also form and maintain the spatial map. Although O'Keefe had discovered place cells in 1971, and Bliss and Lømo had discovered long-term potentiation in the hippocampus in 1973, no attempt had been made to connect the two findings. When we began studying spatial maps in 1992, nothing was known about the molecular steps whereby one is formed. This situation illustrates once again why working at the borders between disciplines—in this case, between the biology of place cells and the molecular biology of intracellular signaling—can be highly informative. What a scientist explores in an experiment is in good part determined by the intellectual context in which that scientist functions. Few things are more exhilarating than bringing a new way of thinking to another discipline. This cross-fertilization of disciplines is what Jimmy Schwartz, Alden Spencer, and I had in mind back in 1965 when we called our new division at NYU "neurobiology *and* behavior."

In collaboration with Robert Muller, one of the pioneers in studying place cells, we found that some of the same molecular actions responsible for long-term potentiation are indeed necessary for preserving a spatial map over a long period. We knew that protein kinase A turns on the genes and thus initiates the protein synthesis necessary for the late phase of long-term potentiation. Similarly, we found that although neither protein kinase A nor protein synthesis is needed for the initial formation of a map, they both are essential for the map to become "fixed" over the long term, so that the mouse recalls the same map every time it enters the same space.

Finding that protein kinase A and protein synthesis are required for the stabilization of the spatial map raised a further question: Does the spatial map we record in the hippocampus enable animals to have explicit memory—that is, to act as if they are familiar with a given environment? Are these maps the actual internal representation, the neural correlates of explicit memory of space? In his initial formulation O'Keefe considered the cognitive map an internal representation of space that the animal uses for navigation. He therefore saw the map as more a navigational representation, much like a compass, than a representation of the memory itself. We explored this question and found that indeed, when we blocked protein kinase A or inhibited protein synthesis, we interfered not only with the long-term stability of the spatial map but also with the ability to retain long-term spatial memories. Thus we had direct genetic evidence that the map correlates with spatial memory. Moreover, we found that in spatial memory, as in the simple explicit memory underlying the gill-withdrawal reflex in *Aplysia*, there is a distinction between the processes involved in acquiring the map (and holding onto it for a few hours) and maintaining the map in stable form for the long term.

DESPITE CERTAIN SIMILARITIES, THE EXPLICIT MEMORY OF SPACE in people differs from implicit memory in profound ways. In particular, explicit memory requires selective attention for encoding and for recall. Therefore, to examine the relation between neural activity and explicit memory, we now needed to address the issue of attention.

Selective attention is widely recognized as a powerful factor in perception, action, and memory—in the unity of conscious experience. At any given moment, animals are inundated with a vast number of sensory stimuli, yet they pay attention to only one or a very small number of them, ignoring or suppressing the rest. The brain's capacity for processing sensory information is more limited than its receptors' capacity for measuring the environment. Attention therefore acts as a filter, selecting some objects for further processing. It is in large part because of selective attention that internal representations do not replicate every detail of the external world and sensory stimuli alone do not predict every motor action. In our moment-to-moment experience, we focus on specific sensory information and exclude the rest (more or less). If you raise your eyes from this book to look at a person entering the room, you are no longer paying attention to the words on the page. At the same time, you are not attending to the decor of the room or other people in the room. If asked to report your experience later, you are more likely to remember that a person had entered the room than, say, that there was a small scratch on the wall. This focusing of the sensory apparatus is an essential feature of all perception, as William James noted in his seminal book, *The Principles of Psychology*, in 1890:

> Millions of items . . . are present to my senses which never properly enter into my experience. Why? Because they have no interest for me. My experience is what I agree to attend to. . . . Everyone knows what attention is. It is the taking possession by the mind, in clear and vivid form, of one out of what seem several simultaneously possible objects or trains of thought. Focalization, concentration of consciousness, are of its essence. It implies withdrawal from some things in order to deal effectively with others.

Attention also allows us to bind the various components of a spatial image into a unified whole. Cliff Kentros, a postdoctoral fellow, and I chose to address the link between attention and spatial memory by asking whether attention is required for the spatial map. If so, does attention alter the formation or the stability of the map? To test these ideas, we exposed mice to four conditions that require increasing degrees of attention. The first, basal or ambient attention is the attention that is present even in the absence of further stimulation. Here, animals walked around in an enclosure without distracting stimuli. Second, we required animals to forage for food, a task that necessitates a bit more attention; third, we asked the animals to discriminate between two environments; and finally, we demanded that the animals actually learn a spatial task. We engineered things so that as the mouse walked around in its enclosure, lights and sounds, which the mouse hates, would periodically come on. The only way the mouse could turn them off was to find a small, unmarked goal region and sit there for a moment. Mice learn this task very well.

We found that even ambient attention is sufficient to allow a spatial map to form and become stable for a few hours, but such a map becomes unstable after three to six hours. Long-term stability correlates strongly and systematically with the degree to which an animal is required to pay specific attention to its environment. Thus, when a mouse is forced to pay a lot of attention to a new environment, by having to learn a spatial task at the same time that it is exploring the new space, the spatial map remains stable for days and the animal readily remembers a task based on knowledge of that environment.

What is this attentional mechanism in the brain? How does it contribute to the strong encoding of information about space and the ready recall of that information after long periods of time? I already knew that attention was not simply a mysterious force in the brain but a modulatory process. Michael Goldberg and Robert Wurtz at NIH had found that in the visual system, attention enhances the response of neurons to stimuli. A modulatory pathway that had been strongly implicated in attention-related phenomena was the one mediated by dopamine. The cells that make dopamine are clustered in the midbrain, and their axons project to the hippocampus. Indeed, we found that blocking the action of dopamine in the hippocampus blocked the stabilization of the spatial map in an animal that was paying attention. Conversely, activating dopamine receptors in the hippocampus stabilized the spatial map of an animal that was *not* paying attention. The axons of the dopamine-producing neurons in the midbrain send signals to a number of sites, including the hippocampus and the prefrontal cortex. The prefrontal cortex, which is recruited for voluntary action, signals back to the midbrain, adjusting the firing of these neurons. Our finding that some of the same regions of the brain that are recruited for voluntary behaviors are also recruited for attentional processes reinforced the idea that selective attention is critical to the unitary nature of consciousness.

In *The Principles of Psychology* William James pointed out that there is more than one form of attention. There are at least two types: involuntary and voluntary. Involuntary attention is supported by automatic neural processes, and it is particularly evident in implicit memory. In classical conditioning, for example, animals learn to associate two stimuli if, and only if, the conditioned stimulus is salient or surprising. Involuntary attention is activated by a property of the external world—of the stimulus—and it is captured, according to James, by "big things, bright things, moving things, or blood." Voluntary attention, on the other hand, such as paying attention to the road and traffic while driving, is a specific feature of explicit memory and arises from the internal need to process stimuli that are not automatically salient.

James argued that voluntary attention is obviously a conscious process in people; therefore, it is likely to be initiated in the cerebral cortex. Viewed from a reductionist perspective, both kinds of attention recruit biological signals of salience, such as modulatory neurotransmitters, that regulate the function or configuration of a neural network.

Our molecular studies in *Aplysia* and mice support James's contention that these two forms of attention, involuntary and voluntary, exist. One of the key differences between them is not the absence or presence of salience, but whether the signal of salience is perceived consciously. Thus I will consciously pay attention when I need to learn how to find my way from my home in Riverdale to my son Paul's house in Westchester. But I will put my foot on the brakes automatically when a car suddenly cuts in front of me while I am driving on the road. The studies also suggest that, as James had argued, the determining factor in whether memory is implicit or explicit is the manner in which the attentional signal for salience is recruited.

In both types of memory, as we have seen, conversion of short-term to long-term memory requires the activation of genes, and in each case modulatory transmitters appear to carry an attentional signal marking the importance of a stimulus. In response to that signal, genes are turned on and proteins are produced and sent to all the synapses. Serotonin triggers protein kinase A in *Aplysia*, for example, and dopamine triggers protein kinase A in the mouse. But these signals of salience are called up in fundamentally different ways for the implicit memory underlying sensitization in *Aplysia* and the explicit memory required to form the spatial map in the mouse.

In implicit memory storage, the attentional signal is recruited involuntarily (reflexively), from the bottom up: the sensory neurons of the tail, activated by a shock, act directly on the cells that release serotonin. In spatial memory, dopamine appears to be recruited voluntarily, from the top down: the cerebral cortex activates the cells that release dopamine, and dopamine modulates activity in the hippocampus (figure 23-1).

Consistent with this idea that similar molecular mechanisms are used for top-down and bottom-up attentional processes, we have found a mechanism that may be involved in stabilizing the memory in both cases. The hippocampus of the mouse contains at least one prion-like protein similar to the one Kausik Si discovered in *Aplysia*. Martin Theis, a postdoctoral student from Germany, and I found that, in much the same way that serotonin modulates the amount and state of the CPEB protein in *Aplysia*, dopamine modulates the amount of the prion-like CPEB protein (CPEB-3) in the mouse hippocampus. This discovery raised the intriguing possibility—so far only that—that spatial maps may become fixed when an animal's

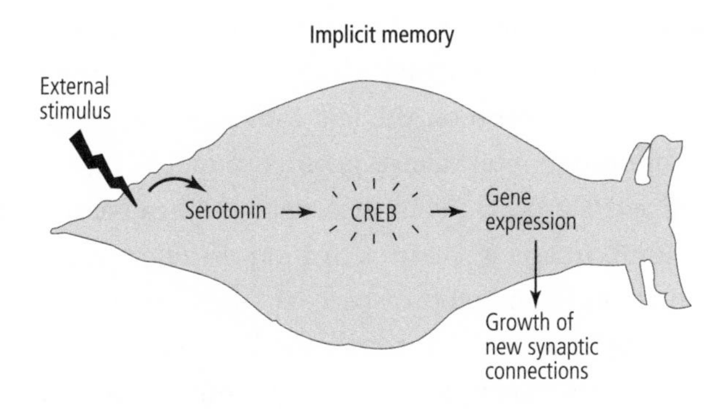

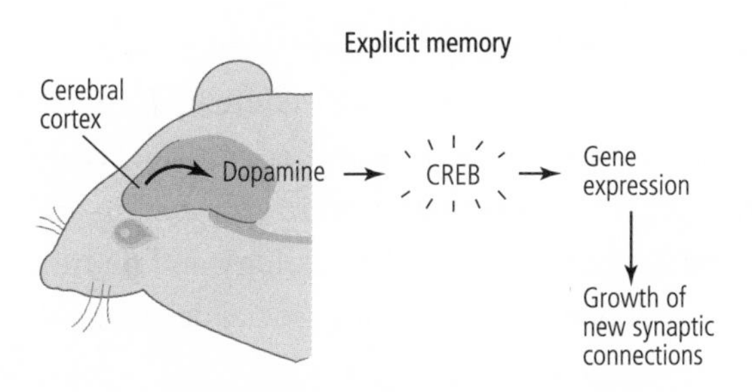

23-1 The signal of salience for long-term implicit and explicit memory. In *implicit* (unconscious) memory, an outside stimulus automatically triggers a salience signal (serotonin) in the animal. This activates genes that lead to long-term memory storage. In *explicit* (conscious) memory, the cerebral cortex voluntarily recruits a salience signal (dopamine), which causes the animal to attend. This modulates activity in the hippocampus, leading to long-term memory storage.

attention triggers the release of dopamine in the hippocampus and that dopamine initiates a self-perpetuating state also mediated by CPEB.

THE IMPORTANCE OF ATTENTION IN STABILIZING THE SPATIAL MAP raises another question: Is the spatial map, a map formed by learning, similar in all of us? Specifically, do men and women use the same strategies to find their way around an environment? This is a fascinating question and one that biologists are just beginning to explore.

O'Keefe, who first discovered place cells in the hippocampus, has extended his research on spatial orientation to gender differences. He has found clear differences in the way women and men attend to and orient themselves to the space around them. Women use nearby cues or landmarks. Thus when asked for directions, a woman is likely to say, "Turn right at the Walgreen's drugstore, and then drive until you see a white colonial house on the left with green window shutters." Men rely more on an internalized geometric map. They are likely to say, "Drive five miles north, then turn right and head east for another half mile." Brain imaging shows activation of different areas in men and women as they think about space: the left hippocampus in men and the right parietal and right prefrontal cortex in women. These studies point out the potential benefits for group effectiveness of optimizing both strategies.

Gender differences in forming the spatial map take on additional significance when considered in a broader context: To what degree do men's and women's brain structures and cognitive styles differ? Are those differences innate, or do they stem from learning and socialization? It is in questions such as these that biology and neuroscience can provide us basic guidance for far-reaching social decisions.

FIVE

There are many aspects of humanity that we still need to understand for which there are no useful models. Perhaps we should pretend that morality is known only to the gods and that if we treat humans as model organisms for the gods, then in studying ourselves we may come to understand the gods as well.

—Sydney Brenner, Nobel Lecture (2002)

24

A LITTLE RED PILL

Anyone working on memory becomes keenly aware of the crying need for drugs that can improve memories ravaged by disease or weakened by age. But before new drugs can be brought to market, they must be tested in animal models. Clearly, given the animal models of implicit and explicit memory storage we were developing, we could begin thinking about new therapeutic approaches to memory disorders. Once again, timing proved crucial. Just when genetically modified mice were being generated in the early 1990s to analyze the nature of memory and its disorders, a new industry emerged to look for novel ways to develop drugs.

Until 1976, new scientific insights could not be translated rapidly into better modes of treatment, nor were academic scientists in the United States like myself particularly interested in working with the pharmaceutical industry to develop new drugs. In that year, however, the situation changed dramatically. Robert Swanson, a twenty-eight-year-old venture capitalist who recognized the potential of genetic engineering for developing new drugs, persuaded Herbert Boyer, a professor at the University of California, San Francisco and a pioneer in the field, to join forces with him in the formation of Genentech (short for genetic engineering technologies). This was the first biotechnology company focused on commercializing genetically engineered proteins for medical purposes. Swanson and Boyer shook hands, and each contributed $500 to the venture. Swanson then raised a few hundred thousand more to get the company off the ground. It is now worth about $20 billion.

Molecular biologists had recently discovered how to sequence DNA rapidly and had developed the powerful techniques of genetic engineering: snipping specific sequences of DNA out of chromosomes, stitching the sequences together, and inserting the recombined DNA into the genome of the *E. coli* bacterium, which makes many copies of the new gene and expresses the protein that its gene encodes. Boyer was one of the first molecular biologists to appreciate that one could use bacteria to express genes that come from higher animals, even from people. Indeed, he had been instrumental in developing some of the key techniques for doing so.

Genentech planned to use recombinant DNA technology to synthesize large quantities of two human hormones of great medical importance—insulin and growth hormone. Insulin is released into the bloodstream by the pancreas to regulate sugar in the body. Growth hormone is released by the pituitary gland to regulate development and growth. To prove that it could synthesize these two fairly complex proteins, the company first focused on a simpler protein called somatostatin, a hormone released into the bloodstream by the pancreas to turn off the release of insulin.

Before 1976 the supply of medically available somatostatin, insulin, and growth hormone was limited. Insulin and somatostatin were in short supply because they had to be purified from pigs or cows. Because the amino acid sequences of the animal hormones are slightly different from those of the human hormone, they occasionally caused allergic reactions in people. Growth hormone was derived from human pituitary glands that had been removed from cadavers. In addition to being limited, this source was occasionally contaminated by prions, the infectious proteins that cause Creutzfeldt-Jacob disease, the terrible dementia that struck Irving Kupfermann. Recombinant DNA opened up the possibility of synthesizing proteins from human genes and producing them more cheaply and in unlimited quantities without having to worry about issues of safety. It was clear to Boyer and Swanson that by cloning human genes, they could manufacture these and other medically important proteins and eventually cure genetically based illnesses by substituting cloned genes for patients' defective ones.

In 1977, a year after joining Swanson, Boyer developed gene-cloning methods that allowed him to synthesize large amounts of somatostatin, thereby establishing the principle that recombinant DNA could produce medically important and commercially valuable drugs. Three years later, Genentech succeeded in cloning insulin.

Genentech was followed two years later by Biogen, a second powerhouse in biotechnology. But those two years had made a gigantic difference. Biogen was created not by a young entrepreneur acting initially on his own, but by C. Kevin Landry and Daniel Adams, two mature investors, each representing well-established venture groups. They brought to the table not $1000 and a handshake, but $750,000 and a set of contracts to build a biotechnology dream team. They approached the best and the brightest scientists in the world: first Walter Gilbert at Harvard, then Philip Sharp at MIT, Charles Weissman at the University of Zurich, Peter Hans Hofschneider at the Max Planck Institute for Biochemistry in Munich, and Kenneth Murray of the University of Edinburgh. After some discussion, all agreed to join the venture, and Gilbert agreed to chair the scientific advisory board.

Soon, an entire industry was launched. The biotechnology industry not only produced its own new products, it also transformed the pharmaceutical industry. In 1976 most of the big pharmaceutical companies were neither sufficiently bold nor sufficiently agile to carry off recombinant DNA research by themselves, but by investing in some biotechnology companies and buying others, they rapidly became competent.

BIOTECHNOLOGY COMPANIES ALSO TRANSFORMED THE ACADEMIC community, particularly its attitude toward the commercialization of science. Unlike academics in most European countries, American academics had a negative atttitude toward participating in industry.

France's great biologist, Louis Pasteur, who in the nineteenth century laid the foundation for understanding that germs cause infectious disease, had many ties to industry. He discovered the biological basis for the fermenting of wine and beer. His methods of identifying and destroying bacteria that infect silkworms, wine, and milk saved both the silk and the wine industry and led to the pasteurization of milk to prevent infection and spoilage. He developed the first vaccine against rabies, and to this day the Pasteur Institute in Paris, established in his honor while he was still alive, receives a substantial part of its income from the making of vaccines. Henry Dale, the English scientist who helped discover the chemical basis of synaptic transmission, moved freely between his academic position at Cambridge University to the Wellcome Physiological Research Laboratories, a pharmaceutical company, and then back again to an academic position at the National Institute for Medical Research in London.

In America things were different. Gilbert soon realized that three conditions would induce him and other academic biologists to change their minds about combining science and business. First, they needed evidence that a company could do something useful. Second, they needed assurance that involvement in the company would not be too great a distraction from basic scientific work. Finally, they needed to be certain that their scientific independence—so valued by university professors—could not be compromised.

By 1980, when Genentech successfully produced human insulin, the first condition—that of usefulness—had been met. A steady trickle of biologists established contact with the biotechnology industry. Once these biologists had experienced sin, they found, to their surprise, that they liked it. They liked the fact that science led to medically useful drugs, and they liked the idea that they could do well financially by doing good for the public—that they could make money by developing much-needed drugs. Whereas most academics had shunned involvement with industry and disdained colleagues who consulted for pharmaceutical companies, all that changed after 1980. Moreover, academics found that given appropriate safeguards, they could limit their commitment of time and maintain their independence. Indeed, most academics found that not only did they contribute their own knowledge, they also learned new ways of doing science by working in industry.

As a result, universities started to encourage entrepreneurial skills in their faculty. Columbia was a pioneer in this regard. In 1982 Richard Axel, together with several colleagues, developed a method of expressing any gene, including a human gene, in a cell in tissue culture. Since Axel is on the Columbia faculty, the university patented the method. It was immediately adopted by several major pharmaceutical companies, which used it to make new, therapeutically important drugs. Over the next twenty years—the lifetime of the patent—Columbia earned $500 million from this one patent alone. The funds allowed the university to recruit new faculty and strengthen its research efforts. Axel and the other inventors shared in the bounty.

At nearly the same time, Cesare Milstein at the Medical Research Council Laboratory in Cambridge, England, discovered how to make monoclonal antibodies, highly specific antibodies that target just one region of a protein. His technique, too, was immediately snapped up by the pharmaceutical industry and used to make new drugs. But the Medical Research Council and Cambridge University were still thinking in an earlier mode. They did not take out a patent on the method and lost the opportunity to receive an income that they had rightly earned and that could have supported much good science. As other universities watched these events, most that did not already have an intellectual property group began to form one.

BEFORE LONG, MOST SELF-RESPECTING MOLECULAR BIOLOGISTS had been recruited to the advisory board of one new biotechnology company or another. In this early period, companies focused mostly on hormones and antiviral agents, but by the mid-1980s, financial entrepreneurs began to wonder whether neural science could be used to produce new drugs for neurological and psychiatric disorders. In 1985, Richard Axel asked me to talk about Alzheimer's disease at a meeting in New York City of the board of directors of Biotechnology General, a company based in Israel for which Axel consulted. I gave them a brief overview of the disorder, emphasizing that Alzheimer's is emerging as a major epidemic because of the dramatic increase in the population over age sixty-five. Finding a treatment would have great public health benefits.

The facts I was relaying were fairly obvious to the neural science community, but they were not obvious to the venture capital community. After that meeting, Fred Adler, the chairman of the board of Biotechnology General, asked Richard and me to join him for lunch the next day. There, he proposed that we start a new biotechnology company focused exclusively on the brain, one that would use the insights of molecular science to focus on diseases of the nervous system.

At first, I was reluctant to become involved in biotechnology because I thought such an endeavor would be uninteresting. I shared the view held earlier by a large part of the academic community that biotechnology and pharmaceutical companies did humdrum science and that getting involved with a commercial venture would be intellectually unsatisfying. Richard encouraged me to join, however, pointing out that such work could be quite interesting. In 1987 we formed Neurogenetics, later called Synaptic Pharmaceuticals. Richard and Adler asked me to chair the scientific advisory board.

I asked Walter Gilbert to join the board. Wally, whom I had first met in 1984, is an extraordinary person, one of the most intelligent, gifted, and versatile biologists of the second half of the twentieth century. He had followed up on the Monod-Jacob theory of gene regulation and actually isolated the first gene regulator, showing that it was, as predicted, a protein that bound to DNA. With this remarkable accomplishment behind him, Wally went on to develop a method for sequencing DNA, which won him the Nobel Prize in Chemistry in 1980. As a cofounder of Biogen, Wally had also become knowledgeable about running a business. I thought this combination of scientific achievement and commercial know-how would make him a great asset.

Wally had left Biogen in 1984, gone back to Harvard, and turned his attention to neurobiology, a field in which he had recently become interested. Since he was new to the brain, I thought he might enjoy joining us and learning a bit more about the field. He agreed, and was an extremely valuable addition. Denise and I developed a habit that continues to this day—dining with Wally, usually at a wonderful restaurant, the night before scientific advisory board meetings.

Other scientists whom Richard and I invited to join the advisory board included Tom Jessell, our colleague at Columbia and a gifted developmental neurobiologist; Paul Greengard, a pioneer in second-messenger signaling in the brain who had moved from Yale to Rockefeller University; Lewis Roland, chairman of the neurology department at Columbia; and Paul Marks, former dean of the College of Physicians and Surgeons at Columbia and subsequently president of Memorial Sloan-Kettering Cancer Center. This was an extraordinarily strong group. We spent several months studying what direction the company should take.

We first considered specializing in amyotrophic lateral sclerosis, of which Alden Spencer had died, and then in multiple sclerosis, brain tumors, or stroke, but we eventually decided it would probably be best to do something related to receptors for the neurotransmitter serotonin. Many important drugs—almost all of the antidepressants, for example—act through serotonin, and Richard had just isolated and cloned the first serotonin receptor. Unlocking the molecular biology of these receptors could open up the study of a number of diseases. Moreover, the receptor cloned by Richard was only one of a large class of metabotropic receptors, so it could be used to try to clone similarly constructed receptors for other transmitters that act through second messengers.

We were strongly encouraged to work along these lines by Kathleen Mullinex, the associate provost at Columbia, whom we had recruited as chief executive officer. Although Mullinex knew no neurobiology, she had the idea that receptors could be useful to screen for new drugs. The board sharpened that idea. We would clone receptors for serotonin and dopamine, see how they function, and then design new chemical compounds to control them. Paul Greengard and I wrote the document that spelled this out, and we used as our first example Richard Axel's successful cloning of the first serotonin receptor.

The company got off to a good start. We recruited a fine scientific staff, which proved adept at cloning new receptors, and we formed effective partnerships with Eli Lilly and Merck. The company went public in 1992 and disbanded its extraordinary scientific advisory board. I remained for a while as a scientific consultant, but three years later I started a company focused on my own area of research.

THE IDEA FOR THIS NEW VENTURE ORIGINATED ONE NIGHT IN 1995, as Denise and I were having one of our dinners with Walter Gilbert. Wally and I were discussing results I had recently obtained suggesting that memory loss in aged mice can be reversed, when Denise suggested that we start a company to develop a "little red pill" for age-related memory loss. Following up on this idea, Wally and I joined forces with Jonathan Fleming, a venture capitalist from the Oxford Partners group who had supported Synaptic Pharmaceuticals. Jonathan helped us recruit Axel Unterbeck from Bayer Pharmaceuticals. In 1996 the four of us formed a new company, Memory Pharmaceuticals.

Starting a company based so directly on my work in memory was exciting, but running a company, even one derived from one's own research, is exceedingly time-consuming. Some academics leave the university in order to do it. I had no intention of leaving Columbia or the Howard Hughes Medical Institute. I wanted to help found the company and, once that was done, consult for it on a part-time basis. Both Columbia and Howard Hughes have experienced lawyers who helped me work out consulting agreements—first with Synaptic Pharmaceuticals and then with Memory Pharmaceuticals—that met both institutional guidelines and my own concerns.

Participating in these two biotechnology companies expanded my horizons. Memory Pharmaceuticals allowed me to help translate my basic research into potentially useful drugs for treating people. In addition, it exposed me to how a company works. In a typical academic department, junior faculty members are independent; in the early stage of their career they are encouraged not to collaborate with senior faculty but to develop their own research programs. In business, people must work together for the good of the company using intellectual and financial resources in a way that pushes each potential product in promising directions. Although this cooperative characteristic of industry is generally not found in universities, there are important exceptions, such as the Human Genome Project, which involved a similar merging of individual efforts for a common good.

The new company was based on the idea that the study of memory will expand into an applied science and that one day our growing understanding of the mechanisms of memory function will lead to treatments for disorders of cognition. As I had pointed out to the board of Biotechnology General, disorders of memory are more evident today than they were when I began practicing medicine fifty years ago because people are living longer now. Even in a normal, healthy population of seventy-year-olds, only about 40 percent have as good a memory as they had in their mid-thirties. The remaining 60 percent experience a modest decline in memory. In the early stages, this decline does not affect other aspects of cognitive function—it does not affect language or the ability to solve most problems, for instance. Half of the 60 percent have a slight memory impairment, sometimes called benign senescent forgetfulness, that progresses only slowly, if at all, with time and age. The remaining half, however (or 30 percent of the population over age seventy), develop Alzheimer's disease, a progressive degeneration of the brain.

In its early stages, Alzheimer's is characterized by mild cognitive impairment that is indistinguishable from benign senescent forgetfulness. But in later stages of the disease, dramatic and progressive deficits in memory and other cognitive functions develop. The vast majority of symptoms in the late, debilitating stages of the disease are attributed to the loss of synaptic connections and to the death of nerve cells. This degeneration of tissue is caused in large part by the accumulation of an abnormal material known as β-amyloid in the form of insoluble plaques in the spaces between brain cells.

I FIRST TURNED MY ATTENTION TO BENIGN SENESCENT FORGETfulness in 1993. The term is a bit euphemistic, since the disorder does not begin with senescence nor is it completely benign. It first becomes evident in some people in their forties and typically becomes slightly more pronounced with time. I hoped that the ever expanding understanding of the mechanisms of memory storage in *Aplysia* and mice might enable us to understand the underlying defect of this distressing aspect of aging and then to develop therapies for counteracting the memory loss.

As I read the literature on benign senescent forgetfulness, it became clear to me that the disorder is similar in character, if not in severity, to a memory deficit associated with damage to the hippocampus: the inability to form new long-term memories. Like H.M., people with benign senescent forgetfulness can carry on a normal conversation and retain ideas in short-term memory, but they cannot readily convert new short-term memory into long-term memory. For example, an elderly person who is introduced to someone new at a dinner party may remember the new name for a short while, but forget it completely by the next morning. This similarity gave me the first clue that age-related memory loss may involve the hippocampus. Later examination of people and of experimental animals revealed that this is in fact the case. An additional clue was provided by the finding that there is, with aging, a loss of synapses that release dopamine in the hippocampus. We had earlier found that dopamine is important for maintaining long-term facilitation and for modulating attention in spatial memory.

To obtain a better understanding of this form of memory loss, my colleagues and I developed a naturally occurring model of it in the mouse. Laboratory mice live to be two years old. Thus, mice are young when they are three to six months old. At twelve months, they are middle-aged, and at eighteen months, elderly. We used a land maze similar to the one we had used earlier to examine the role of genes in spatial memory. Placed in the center of a large circular platform surrounded by a rim of forty holes, mice learn to find the one hole that leads to an escape chamber by discovering the spatial relationship between the hole and markings on the wall. We found that most young mice rapidly go through random and serial escape strategies and learn to use the more efficient spatial strategy rather soon. Many aged mice, however, have difficulty ever learning the spatial strategy (figure 24-1).

We also found that not all older mice are impaired: the memory of some is as good as that of young animals. In addition, the memory deficit in impaired mice occurs just in explicit memory; we carried out a number of behavioral tests and found that their implicit memory for simple perceptual and motor skills was unaffected. Finally, the memory deficits are not necessarily confined to old age; some began in midlife. All of these findings suggested to us that what pertains to people also pertains to mice.

If a mouse has a defect in spatial memory, it implies that something

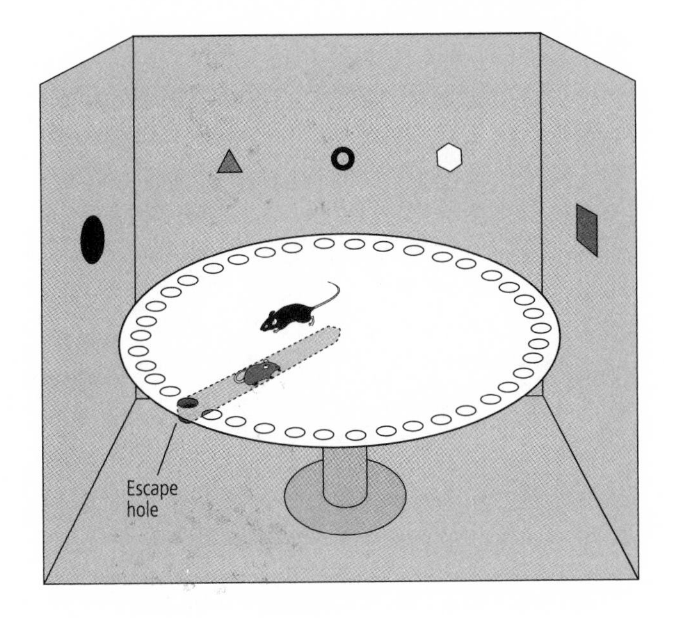

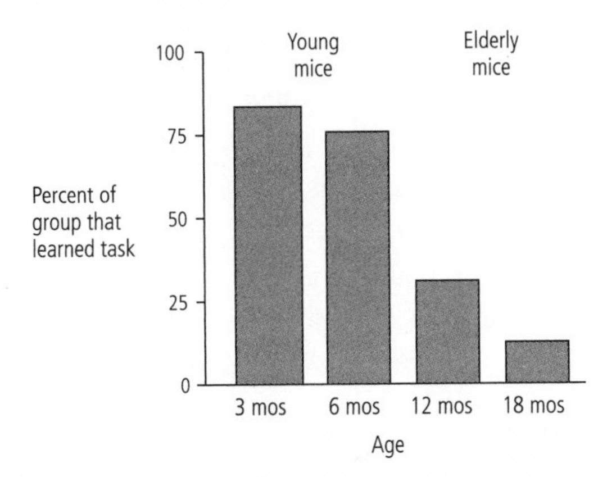

24-1 Mice show age-related memory loss in a spatial task. The Barnes Maze (above) provides an escape hole and several visual cues to orient the mouse. Aged mice have difficulty learning the spatial relationships between these cues and the escape hole (below). This correlates with defective functioning of the hippocampus.

is wrong with the hippocampus. We explored the Schaffer collateral pathway in the hippocampus of older mice with age-related memory deficits and found that the late phase of long-term potentiation, which we and others had found to be strongly correlated with long-term explicit memory, was defective. Moreover, older mice that remembered well had normal long-term potentiation, as did younger mice with normal spatial memory.

We had found earlier that the late phase of long-term potentiation is mediated by cyclic AMP and protein kinase A and that this signaling pathway is activated by dopamine. When dopamine binds to its receptor in the pyramidal cells of the hippocampus, the concentration of cyclic AMP increases. We found that drugs which activate these dopamine receptors, and thereby increase cyclic AMP, overcome the deficit in the late phase of long-term potentiation. They also reverse the hippocampus-dependent memory deficit.

Mark Barad, a postdoctoral fellow, and I wondered whether the deficit in older mice's long-term spatial memory might also be ameliorated by manipulating the cyclic AMP pathway in another way. Cyclic AMP is normally broken down by an enzyme so that signaling does not continue indefinitely. The drug Rolipram inhibits that enzyme, extending the life of cyclic AMP and increasing signaling. In old mice, Barad and I found, Rolipram significantly improves learning that involves the hippocampus; indeed, older animals given Rolipram performed as well as younger mice on memory tasks. Rolipram even increased long-term potentiation and hippocampus-dependent memory in young animals.

These results support the notion that the decline in hippocampus-dependent learning in older animals is due, at least in part, to an age-related deficit in the late phase of long-term potentiation. Perhaps more important, they suggest that benign senescent forgetfulness may be reversible. If it is, the elderly may be treated in the near future with drugs developed from such studies of the mouse.

The prospect that benign senescent memory loss is treatable led the leadership of Memory Pharmaceuticals to wonder what other forms of memory impairment might be treated if we knew more about the molecular mechanisms underlying memory formation. With this idea in mind, Memory Pharmaceuticals turned its attention to the early phase of Alzheimer's disease.

ONE OF THE INTERESTING FEATURES OF ALZHEIMER'S DISEASE is the mild deficit in memory that precedes the deposition of β-amyloid plaques in the hippocampus. Since the early cognitive deficits in Alzheimer's are so similar to age-related memory loss, Michael Shelanski at Columbia began to wonder whether the same pathways are disturbed in each. To find out, he studied the hippocampus of mice.

He exposed the mouse hippocampus to the most toxic component of β-amyloid plaques, known as the Aβ peptide, and found that long-term potentiation was impaired before any neurons had died or plaques had formed. In addition, animal models of early Alzheimer's disease displayed memory deficits before any detectable accumulation of plaque or evidence of cell death. While examining gene expression in hippocampal cells exposed to the Aβ peptide, Shelanski discovered that the peptide decreases the activity of cyclic AMP and protein kinase A. This finding suggested to him that the peptide may compromise the cyclic AMP-protein kinase A system. Indeed, he found that increasing cyclic AMP via Rolipram prevents Aβ toxicity in mouse neurons.

The same drugs that prevent age-related memory loss in mice also prevent memory deficits in mice in the early stages of Alzheimer's disease. Ottavio Arancio from Columbia University went on to show that Rolipram protects against some of the damage to neurons sustained in Alzheimer's, thereby suggesting that cyclic AMP not only strengthens the function of pathways whose efficiency has declined, but also helps protect against nerve cell damage and perhaps even leads to regeneration of lost connections in the mouse model of Alzheimer's disease.

Memory Pharmaceuticals and other companies developing drugs to combat memory loss are now tackling both of these disorders. Indeed, most companies have broadened their base since their founding and are now developing drugs not only for age-related memory loss and Alzheimer's disease, but also for a variety of memory problems that accompany other neurological and psychiatric disorders. One such disorder is depression, which in its more severe forms is associated with a dramatic loss of memory. Another is schizophrenia, which is characterized by a defect in working memory and in executive functions, such as ordering a sequence of events or attending to priorities.

MEMORY PHARMACEUTICALS IS NOW LOCATED IN MONTVALE, New Jersey. In 2004 the company went public. It has developed four new families of drugs for age-related memory loss that are substantially better than the off-the-shelf compounds my colleagues and I at Columbia had used for our experiments. Some of the compounds improve a rat's memory of a new task for months!

The era of biotechnology holds enormous promise for developing new drugs to treat people with mental diseases. In another decade we may find that our understanding of the molecular mechanisms that underlie memory formation has led to therapeutic advances that were scarcely imaginable in the 1990s. The therapeutic implications of these drugs are obvious. Less obvious are the effects the biotechnology industry will have on the new science of mind and on academic life. Not only do academics serve on advisory boards, but some of the very best scientists are leaving superb jobs at universities to take what they perceive are even better jobs in biotechnology. Richard Scheller, the extraordinary molecular biologist who was the postdoctoral fellow working with Richard Axel and with me when we began our efforts to apply molecular biology to the nervous system, left Stanford University and Howard Hughes Medical Institute to become vice president for research at Genentech. He was joined shortly thereafter by Marc Tessier-Lavigne, an outstanding developmental neurobiologist from Stanford. Corey Goodman, an acknowledged leader in the study of the development of the *Drosophila* nervous system, left the University of California, Berkeley to run his own company, Renovis. The list goes on.

The biotechnology industry now stands as a parallel career pathway for both young and mature scientists. Since the quality of science at the best companies is very high, it is likely that scientists will move freely between academic science and the biotechnology industry.

While the emergence of Memory Pharmaceuticals and other biotechnology companies has bolstered the hope of alleviating memory loss and created new career paths for scientists working on the brain, it also has raised the ethical issue of cognitive enhancement. Is it desirable to improve memory in normal people? Would it be desirable for young people who could afford them to buy memory-enhancing drugs before taking the college entrance exams? There is a range of opinions on this issue, but mine is that *healthy* young people are capable of studying and learning on their own and in school without the aid of chemical memory enhancers (students with learning disabilities might be considered differently). Studying well is, without a doubt, the best cognitive enhancer for those capable of learning.

In a larger sense, these issues raise ethical questions parallel to those raised regarding gene cloning and stem cell biology. The biological community is working in areas in which honest and well-informed people disagree about the ethical implications of the products of research.

How do we link advances in science with an adequate discussion of the ethical implications of science? Here two issues converge. The first relates to scientific research. Freedom to do research is like free speech, and we as a democratic society should, within rather broad limits, protect the freedom of scientists to carry out research wherever it takes them. If we in the United States prohibit research in a particular area of science, we can be sure it will be done elsewhere, perhaps even in a part of the world where human life is not valued as highly or thought about as extensively as it is here. The second issue relates to an evaluation of how, if at all, a scientific discovery is to be used. This evaluation should not be left to scientists, because it affects society at large. Scientists can contribute to discussions about how the products of science are to be used, but final decisions require the participation of ethicists, lawyers, patients' rights groups, and clergy, as well as scientists.

Ethics, a subfield of philosophy, historically has been concerned with the moral issues of mankind. Biotechnology gave rise to the specialized field of bioethics, which is concerned with the social and moral implications of biological and medical research. To address the particular issues raised by the new science of mind, William Safire, a columnist for *The New York Times* and president of the DANA

Foundation, a public-interest group devoted to familiarizing the general public with the importance of brain science, encouraged the foundation in 2002 to stimulate studies in the field of neuroethics. To start this off, Safire sponsored a symposium entitled *Neuroethics: Mapping the Field*. The symposium brought together scientists, philosophers, lawyers, and clergy to address how the new view of mind affects issues that range from personal responsibility and free will to the competence of a mentally ill person to stand trial and the implications for society and the individual of new pharmacological modes of treatment.

To address the issues surrounding cognitive enhancers, I joined in 2004 with Martha Farah from the University of Pennsylvania, Judy Illes from Stanford's Center for Biomedical Ethics, Robin Cook-Deegan of the Center for Genome Ethics, Law and Policy of Duke University, and several other scholars. We published our declaration in *Nature Reviews Neuroscience* as a review article entitled "Neurocognitive Enhancement: What Can We Do and What Should We Do?"

The DANA Foundation is keeping an open discussion of neuroethical issues going. As Steven Hyman, the provost of Harvard University, put it in a recent DANA publication, "Matters . . . ranging from brain privacy to enhancement of mood and memory should be matters of vigorous discussion, and ideally those discussions will mature before continued scientific advances force societies to respond."

25

MICE, MEN, AND MENTAL ILLNESS

Just as my studies of explicit memory in the 1990s had drawn me back to the issues that had attracted me to psychoanalysis in college, so the ability to study age-related memory disorders in mice at the beginning of the new millenium drew me irresistibly to the issues that had fascinated me as a resident in psychiatry. This renewed fascination with mental disorders was the result of several factors.

First, the biological research on memory that I was doing had progressed to the point that I could begin to address problems related to complex forms of memory and to the role of selective attention in memory, and this encouraged me to try to develop other animal models of mental illness. I was further attracted by the discovery that some forms of mental illnesses, such as post-traumatic stress disorders, schizophrenia, and depression, are accompanied by one or another type of memory impairment. As my understanding of the molecular biology of memory deepened and as I learned how instructive mouse models of age-related memory loss had proven to be, it became possible to think about the role of memory dysfunction in other forms of mental illness, and even in the biology of mental wellness.

Second, psychiatry had undergone a major shift toward biology in the course of my career. In the 1960s, when I was a resident at the Massachusetts Mental Health Center, most psychiatrists thought that the social determinants of behavior were completely independent of the biological determinants and that each acted on different aspects of mind. Psychiatric illnesses were classified into two major groups—organic illnesses and functional illnesses—based on presumed differences in their origin. That classification, which dated to the nineteenth century, emerged from postmortem examinations of the brains of mental patients.

The methods available for examining the brain at that time were too limited to detect subtle anatomical changes. As a result, only mental disorders that entailed significant loss of nerve cells and brain tissue, such as Alzheimer's disease, Huntington's disease, and chronic alcoholism, were classified as organic, or based in biology. Schizophrenia, the various forms of depression, and the anxiety states produced no loss of nerve cells or other obvious changes in brain anatomy and therefore were classified as functional, or not based in biology. Often, a special social stigma was attached to the so-called functional mental illnesses because they were said to be "all in a patient's mind." This notion was accompanied by the suggestion that the illness may have been put into the patient's mind by his or her parents.

We no longer think that only certain diseases affect mental states through biological changes in the brain. Indeed, the underlying precept of the new science of mind is that *all* mental processes are biological—they all depend on organic molecules and cellular processes that occur literally "in our heads." Therefore, any disorder or alteration of those processes must also have a biological basis.

Finally, I was asked in 2001 to write a paper for the *Journal of the American Medical Association* about molecular biological contributions to neurology and psychiatry with Max Cowan, a longtime friend who was vice president and senior scientific officer of the Howard Hughes Medical Institute. In writing the review, I was struck by the radical way in which molecular genetics and animal models of disease had transformed neurology, but had not psychiatry. This led me to wonder why molecular biology has not had a similar transformative effect on psychiatry.

The fundamental reason is that neurological diseases and psychiatric diseases differ in several important ways. Neurology has long been based on the knowledge of where in the brain specific diseases are located. The diseases that form the central concern of neurology—strokes, brain tumors, and the degenerative diseases of the brain—produce clearly discernible structural damage. Studies of those disorders taught us that, in neurology, location is key. We have known for almost a century that Huntington's disease is a disorder of the caudate nucleus of the brain, Parkinson's disease is a disorder of the substantia nigra, and amyotrophic lateral sclerosis (ALS) is a disorder of motor neurons. We know that each of these diseases produces its distinctive disturbances of movement because each involves a different component of the motor system.

In addition, a number of common neurological illnesses, such as Huntington's, the fragile X form of mental retardation, some forms of ALS, and the early onset form of Alzheimer's, were found to be inherited in a relatively straightforward way, implying that each of these diseases is caused by a single defective gene. Pinpointing the genes that produce these diseases has been relatively easy. Once a mutation is identified, it becomes possible to express the mutant gene in mice and flies and thus to discover how the gene gives rise to disease.

As a result of knowing the anatomical location, the identity, and the mechanism of action of specific genes, physicians no longer diagnose neurological disorders solely on the basis of behavioral symptoms. Since the 1990s, in addition to examining patients in the office, physicians can order tests for the dysfunction of specific genes, proteins, and nerve cell components, and they can examine brain scans to see how specific regions have been affected by a disorder.

TRACING THE CAUSES OF MENTAL ILLNESS IS A MUCH MORE difficult task than locating structural damage in the brain. A century of postmortem studies of the brains of mentally ill persons failed to reveal the clear, localized lesions seen in neurological illness. Moreover, psychiatric illnesses are disturbances of higher mental function. The anxiety states and the various forms of depression are disorders of emotion, whereas schizophrenia is a disorder of thought.

Emotion and thinking are complex mental processes mediated by complex neural circuitry. Until quite recently, little was known about the neural circuits involved in normal thought and emotion.

Furthermore, although most mental illnesses have an important genetic component, they do not have straightforward inheritance patterns, because they are not caused by mutations of a single gene. Thus, there is no single gene for schizophrenia, just as there is no single gene for anxiety disorders, depression, or most other mental illnesses. Instead, the genetic components of these diseases arise from the interaction of several genes with the environment. Each gene exerts a relatively small effect, but together they create a genetic predisposition—a potential—for a disorder. Most psychiatric disorders are caused by a combination of these genetic predispositions and some additional, environmental factors. For example, identical twins have identical genes. If one twin has Huntington's disease, so will the other. But if one twin has schizophrenia, the other has only a 50 percent chance of developing the disease. To trigger schizophrenia, some other, nongenetic factors in early life—such as intrauterine infection, malnutrition, stress, or the sperm of an elderly father—are required. Because of this complexity in the pattern of inheritance, we have not yet identified most of the genes involved in the major mental illnesses.

In moving from implicit memory in *Aplysia* to explicit memory and the internal representation of space in the mouse, I had moved from a relatively simple realm to a far more complex one, a realm that held many questions of broad significance for human behavior but few solid insights. In trying to explore animal models of mental disorders I was taking a further step into uncertainty. Moreover, whereas I had been early in the study of implicit memory in *Aplysia* and had entered at an interesting midpoint in the study of explicit memory in the mouse, I was a late entry into the biology of mental disorders. Many other people had worked on animal models of mental disorders before me.

Lack of knowledge about the anatomy, genetics, and neural circuitry involved in mental disorders made it difficult to model them in animals. The one clear exception, and the one I focused on initially, was anxiety states. It is difficult to know whether a mouse ever suffers from schizophrenia, whether it is deluded or hallucinating. It is similarly difficult to recognize a mouse that is psychotically depressed. But every animal with a well-developed central nervous system—from snails to mice to monkeys to people—can become afraid, or anxious. In addition, fear has distinctive, easily recognizable features in each of these animals. Thus, not only do animals experience fear, but we can tell when they are anxious. We can, so to speak, read their thoughts. This insight was first set out by Charles Darwin in his classic 1872 study *The Expression of the Emotions in Man and Animals*.

The key biological fact that Darwin appreciated, and that has facilitated the development of animal models of anxiety states, is that anxiety—fear itself—is a universal, instinctive response to a threat to one's body or social status and is therefore critical for survival. Anxiety signals a potential threat, which requires an adaptive response. As Freud pointed out, normal anxiety contributes to the mastery of difficult situations and thus to personal growth. Normal anxiety exists in two major forms: instinctive anxiety (instinctive or innate fear), which is built into the organism and is under more rigid genetic control, and learned anxiety (learned fear), to which an organism may be genetically predisposed but which is basicaly acquired through experience. As we have seen, instinctive anxiety can easily become associated through learning with a neutral stimulus. Since any capability that enhances survival tends to be conserved through evolution, both instinctive and learned fear are conserved throughout the animal kingdom (figure 25-1).

Both forms of fear can be deranged. Instinctive anxiety is pathological when it is excessive and persistent enough to paralyze action.

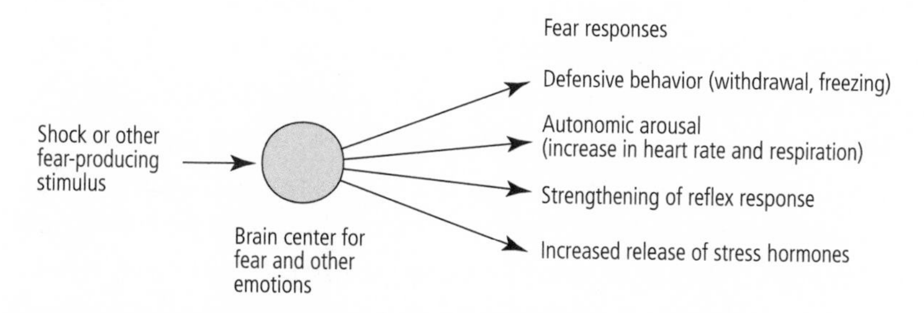

25-1 Defensive responses to fear that have been conserved through evolution.

Learned anxiety is pathological when it is provoked by events that present no real threat, as when a neutral stimulus comes to be associated in the brain with instinctive anxiety. Anxiety states were of particular interest to me because they are by far the most common mental illnesses: at some point in their lives, 10 to 30 percent of people in the general population suffer from these anxiety disorders!

By studying instinctive and learned fear in people and in experimental animals, we have gained much insight into both the behavioral and the biological mechanisms of instinctive and learned fear in people. One of the first behavioral insights was stimulated by the theories of Freud and the American philosopher William James, who realized that fear has both conscious and unconscious components. What was not clear was how the two components interact.

Traditionally, fear in people was thought to begin with conscious perception of an important event, such as seeing one's house on fire. This recognition produces in the cerebral cortex an emotional experience—fear—that triggers signals to the heart, blood vessels, adrenal glands, and sweat glands to mobilize the body in preparation for defense or escape. Thus, according to this view, a conscious, emotional event initiates the later unconscious, reflexive, and autonomic defensive responses in the body.

James rejected this view. In a highly influential article published in 1884 and entitled "What Is Emotion?" he proposed that the cognitive experience of emotion is secondary to the physiological expression of emotion. He suggested that when we encounter a potentially dangerous situation—for example, a bear sitting in the middle of our path—our evaluation of the bear's ferocity does not generate a consciously experienced emotional state. We do not experience fear until after we have run away from the bear. We first act instinctively and then invoke cognition to explain the changes in the body associated with that action.

Based on this idea, James and the Danish psychologist Carl Lange proposed that the conscious experience of emotion occurs only *after* the cortex has received signals about changes in one's physiological state. In other words, conscious feelings are preceded by certain unconscious physiological changes—an increase or decrease in blood pressure, heart rate, and muscular tension. Thus, when you see a fire, you feel afraid because your cortex has just received signals about your racing heart, knocking knees, and sweaty palms. James wrote: "We feel sorry because we cry, angry because we strike, afraid because we tremble, and not that we cry, strike or tremble because we are sorry, angry or fearful, as the case may be." According to this view, emotions are cognitive responses to information from bodily states mediated in good part by the autonomic nervous system. Our everyday experience confirms that information from the body contributes to emotional experience.

Experimental evidence soon supported some aspects of the James-Lange theory. For example, objectively distinguishable emotions are correlated with specific patterns of autonomic, endocrine, and voluntary responses. Furthermore, people whose spinal cord has been accidentally severed, cutting off feedback from the autonomic nervous system in regions of the body below the injury, appear to experience less intense emotions.

With time it became clear, however, that the James-Lange theory explains only one aspect of emotional behavior. If physiological feedback were the only controlling factor, emotions should not outlast physiological changes. Yet feelings—the thoughts and actions in response to emotion—can be sustained long after a threat has subsided. Conversely, some feelings arise much more rapidly than changes in the body. Thus there may be more to emotions than the interpretation of feedback from physiological changes in the body.

An important modification of the James-Lange view has come from the neurologist Antonio Damasio, who argues that the experience of emotion is essentially a higher order representation of the bodily reactions and that this representation can be stable and persistent. As a result of Damasio's work, a consensus is emerging on how emotions are generated. The first step is thought to be the unconscious, implicit evaluation of a stimulus, followed by physiological responses, and finally by conscious experience that may or may not persist.

To determine directly the degree to which the initial experience of emotion is dependent upon conscious or unconscious processes, scientists had to study the internal representation of emotion with the same cellular and molecular biological tools used to study conscious and unconscious cognitive processes. They did this by combining the study of animal models with the study of people. As a result, the neural pathways of emotion have been identified with some precision in the last two decades. The unconscious component of emotion, which was identified primarily by means of animal models, involves the operation of the autonomic nervous system and the hypothalamus, which regulates it. The conscious component of emotion, studied in people, involves the evaluative functions of the cerebral cortex, which are carried out by the cingulate cortex. Central to both components is the amygdala, a group of nuclei clustered together and lying deep in the cerebral hemispheres. The amygdala is thought to coordinate the conscious experience of feeling and the bodily expression of emotion, particularly fear.

Studies of people and of rodents have found that the neural systems that store unconscious, implicit, emotionally charged memories are different from those that generate the memory of conscious, explicit feeling states. Damage to the amygdala, which is concerned with the memory of fear, disrupts the ability of an emotionally charged stimulus to elicit an emotional response. In contrast, damage to the hippocampus, which is concerned with conscious memory, interferes with the ability to remember the context in which the stimulus occurred. Thus the conscious cognitive systems give us a choice of actions, but the unconscious emotional appraisal mechanisms limit those options to a few that are appropriate to the situation. An attractive feature of this view is that it brings the study of emotion in line with studies of memory storage. The unconscious recall of emotional memory has now been shown to involve implicit memory storage, whereas conscious remembrance of the feeling state has been shown to involve explicit memory storage and therefore to require the hippocampus.

ONE STRIKING FEATURE ABOUT FEAR IS THAT IT CAN READILY become associated with neutral stimuli through learning. Once this happens, the neutral stimuli can be powerful triggers of long-term emotional memories in people. Such learned fear is a key component of post-traumatic stress disorder, as well as social phobias, agoraphobia (fear of open spaces), and stage fright. In stage fright and other forms of anticipatory anxiety, a future event (being on stage, for example) is associated with the prospect of something going wrong (forgetting one's lines). Post-traumatic stress disorder occurs following an extremely stressful event, such as life-threatening combat, physical torture, rape, abuse, or natural disasters. It is manifested as recurrent episodes of fear, often triggered by reminders of the initial trauma. One of the striking features of this disorder, and of learned fear in general, is that the memory of the traumatic experience remains powerful for decades and is readily reactivated by a variety of stressful circumstances. Indeed, after a single exposure to a threat, the amygdala can retain the memory of that threat throughout an organism's entire life. How does this come about?

My entry into studies of learned fear in the mouse was in a way a natural extension of my work in *Aplysia*. In *Aplysia*, classical conditioning of fear teaches an animal to associate two stimuli: one that is neutral (a light touch on the siphon), and one that is strong enough to produce instinctive fear (a shock to the tail). Like the shock to *Aplysia*'s tail, an electric shock to the feet of a mouse elicits an instinctive fear response—withdrawal, crouching, and freezing. The neutral stimulus for mice, a simple tone, does not elicit this response. When the tone and the shock are paired repeatedly, however, the animal learns to associate the two. It learns that the tone predicts the shock. As a result, the tone by itself comes to elicit a fear response (figure 25-2).

Although the neural circuitry of learned fear in the mouse is much more complicated than that in *Aplysia*, a fair amount is known about it from the studies of Joseph LeDoux at NYU and Michael Davis, now at Emory. They found that in rodents, as in people, both innate and learned fear recruit a neural circuit focused on the amygdala. In addition, they delineated how information from the conditioned and unconditioned stimuli reaches the amygdala and how the amygdala initiates a fear response.

When a tone is paired with a shock to the feet, information about the tone and the shock are initially carried by different pathways. The tone, the conditioned stimulus, activates sensory neurons in the cochlea, the organ in the ear that receives sound. These sensory neu-

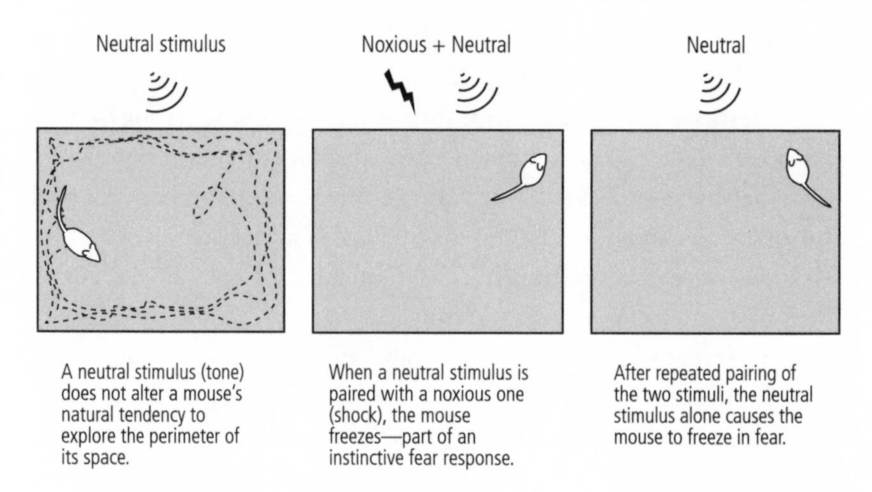

25-2 Creating learned fear in mice.

rons send their axons to a cluster of neurons in the thalamus that is concerned with hearing. The neurons in the thalamus form two pathways: a direct pathway that goes straight to the lateral nucleus of the amygdala without ever contacting the cortex, and an indirect pathway that goes first to the auditory cortex and then to the lateral nucleus (figure 25-3). Both pathways that carry information about the tone terminate on and form synaptic connections with pyramidal neurons, the main type of nerve cell in the lateral nucleus.

Information about pain from the unconditioned stimulus, the shock to the feet, activates pathways that terminate in a different cluster of neurons in the thalamus, one that processes painful stimuli. These neurons in the thalamus also form direct and indirect pathways to the pyramidal cells of the lateral nucleus. In this case, the indirect pathway goes through the somatosensory cortex.

The existence of separate pathways—one that goes through the cortex and one that bypasses it completely—provided direct evidence that the unconscious evaluation of a frightening stimulus precedes conscious, cortical evaluation of fear, as the James-Lange theory had predicted. By activating the fast, direct pathway that bypasses the cortex, a frightening stimulus can cause our hearts to race and our palms to sweat before we consciously realize through the slow pathway that a gun has gone off in our vicinity.

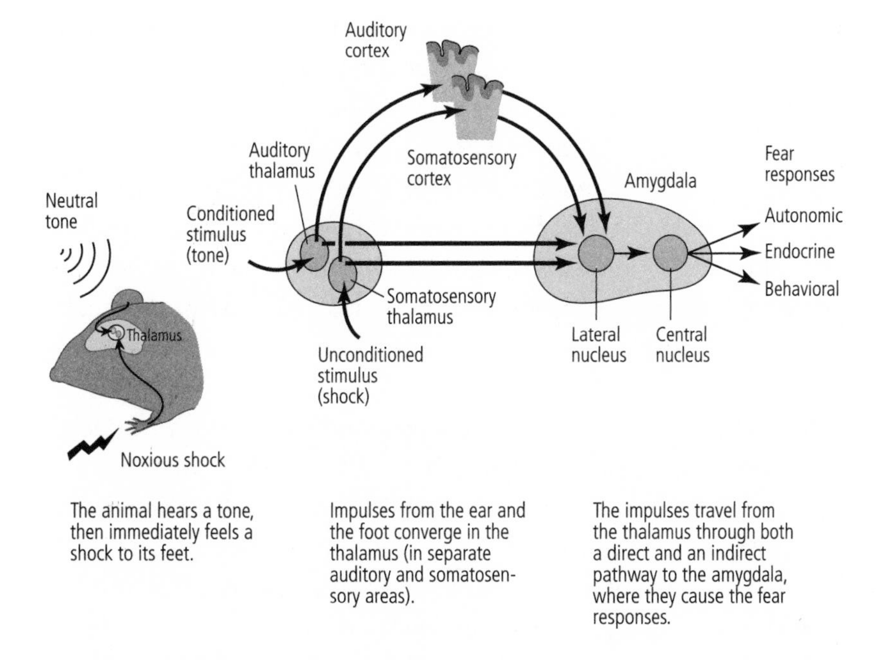

25-3 The neural pathways of learned fear.

In addition to serving as a convergence point for information about the conditioned stimulus (tone) and the unconditioned stimulus (shock), the lateral nucleus of the amygdala mobilizes adaptive responses through the connections it forms with the hypothalamus and the cingulate cortex. The hypothalamus is critical for the body's expression of fear, triggering the fight-or-flight response (an increase in heart rate, sweating, dry mouth, and muscle tension). The cingulate cortex is concerned with conscious evaluation of fear.

HOW, THEN, DOES LEARNED FEAR IN THE MOUSE WORK? DOES IT lead to changes in synaptic strength in the pathways affected by the conditioned stimulus, as is the case in *Aplysia*? To address this question, a number of scientists, including my colleagues and myself, studied slices of the mouse amygdala. Earlier studies had shown that both the direct and indirect pathways, when stimulated electrically at a rate similar to that used by Bliss and Lømo in the hippocampus, are strengthened through a variant of long-term potentiation. We studied this variant of long-term potentiation biochemically and found that although it differs a bit from its counterpart in the hippocampus, it is almost identical to the long-term facilitation that contributes to sensitization and classical conditioning (two forms of learned fear) in *Aplysia*. Both have a molecular signaling pathway that includes cyclic AMP, protein kinase A, and the regulatory gene CREB. These findings illustrate again that long-term facilitation and the various forms of long-term potentiation are part of a family of molecular processes capable of strengthening synaptic connections for long periods of time.

In 2002 Michael Rogan, who had previously worked with LeDoux, joined me, and we turned from studying slices of the mouse brain to studying intact animals. We examined the response of neurons in the amygdala to a tone and found, much as Rogan and LeDoux had found earlier in the rat, that learned fear increases that response (figure 25-4). This phenomenon resembled the long-term potentiation we had seen in slices of the amygdala. Our collaborator Vadim Bolshakov, at Harvard, then reasoned that if learned fear strengthens synapses in the amygdala of an intact mouse, electrical stimulation of slices of that same mouse's amygdala should fail to produce much further synaptic strengthening. That is exactly what we found. Thus, learning acts on the same sites and in a similar manner in the amygdala of the living animal as electrical stimuli do in slices of the amygdala.

We then used a well-established behavioral test for learned fear. We placed a mouse in a large, brightly lit box. The mouse is a nocturnal animal and fears bright light, so it normally skitters along the sides of the box, making only occasional forays toward the center. This protective behavior is a compromise between the animal's need to avoid predators and its need to explore the environment. When we sounded a tone, the mouse continued to move along the sides of the open box as if nothing had happened. But when we repeatedly followed the tone with an electric shock, the animal learned to associate the tone with the shock. When it now heard a tone, it no longer moved along the sides or entered into the center of the box; instead, it remained crouched in one corner, usually in a freezing position (figure 25-2).

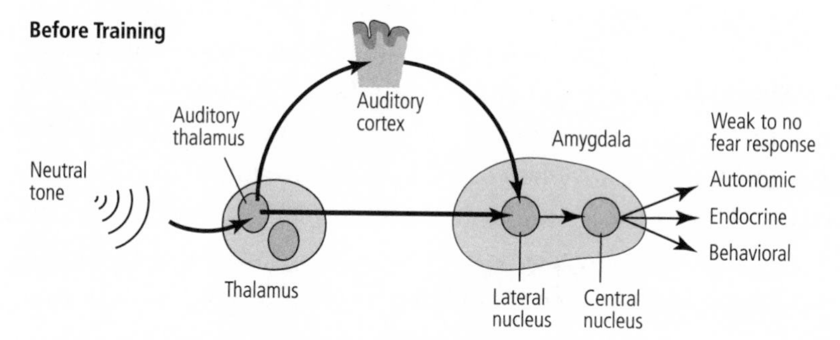

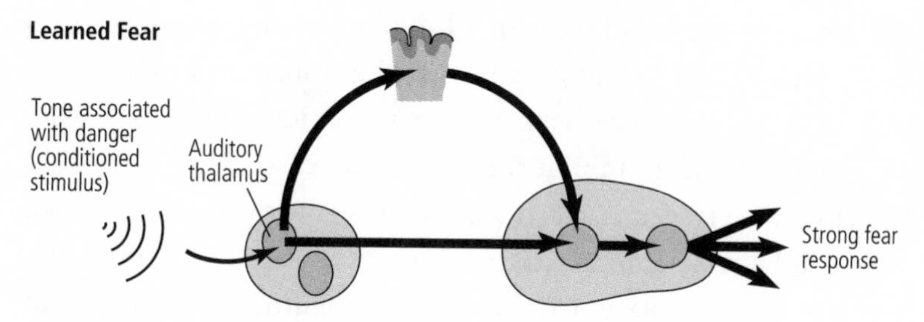

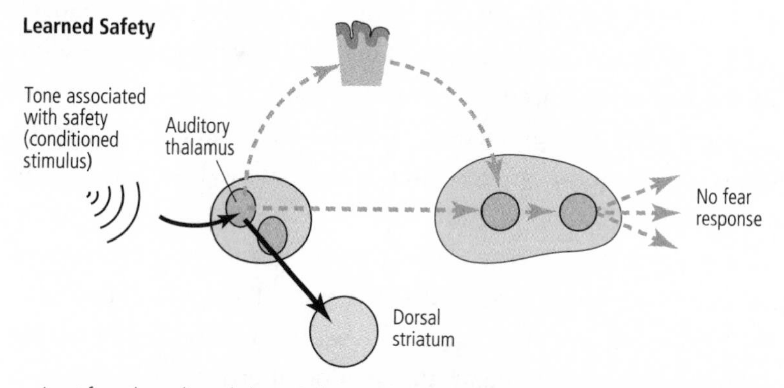

Input from the auditory thalamus is depressed, and the dorsal striatum, which is associated with a sense of well-being, is activated.

25-4 Modifying fear pathways through learning.

WITH THIS UNDERSTANDING OF THE ANATOMY AND PHYSIOLOGY of learned fear, we felt encouraged to explore its molecular basis. Gleb Shumyatsky, a postdoctoral fellow, and I set out to search for genes that might be expressed only in the lateral nucleus of the amygdala, the region we had been studying. We found that the pyramidal cells express a gene that encodes a peptide neurotransmitter called gastrin-releasing peptide. Pyramidal cells use this peptide as an excitatory transmitter in addition to, and in conjunction with, glutamate, releasing it from their presynaptic terminals onto target cells in the lateral nucleus. We next found that the target cells are a special population of inhibitory interneurons that contain receptors for gastrin-releasing peptide. Like all inhibitory interneurons in the lateral nucleus, these target cells release the transmitter GABA. The target cells then connect back to the pyramidal cells and, when active, release GABA to inhibit the pyramidal cells.

The circuit we traced is called a negative feedback circuit: a neuron excites an inhibitory interneuron that then inhibits the neuron that excited it in the first place. Could such an inhibitory feedback circuit be designed to hold an organism's fear in check? To find out, we tested a genetically modified mouse whose receptors for gastrin-releasing peptide had been deleted, thus interrupting the inhibitory feedback circuit. We speculated that the resultant shift toward greater excitation might lead to increased, uncontrolled fear.

Consistent with our prediction, we found dramatically enhanced long-term potentiation in the lateral nucleus and a significantly enhanced and persistent memory of fear. The effect proved to be remarkably specific to learned fear: the same mutant mice showed normal innate fear on a variety of other tests. This finding is consistent with the fundamental distinction between learned and innate fear. Thus, a combined cellular and genetic approach allowed us to identify a neural circuit that is important for holding learned fear in check. The discovery could lead to the development of drugs that counteract learned fear associated with such psychiatric syndromes as post-traumatic stress disorders and phobias.

WHAT ABOUT THE OPPOSITE OF FEAR? WHAT ABOUT FEELING SAFE, confident, and happy? In this context I cannot help being reminded of the first sentence of *Anna Karenina*, Leo Tolstoy's novel about the tragic consequences of a socially unacceptable love affair: "Happy families are all alike; every unhappy family is unhappy in its own way." Tolstoy here suggests, in a statement that has more literary than scientific power, that anxiety and depression can take many forms but that positive emotions—security, safety, and happiness—have common features.

With this idea in mind, Rogan and I explored the neurobiological characteristics of learned safety, presumably a form of happiness. We argued as follows. When a tone is paired with a shock, the animal learns that the tone predicts the shock. Thus if a tone and a shock are always given separately, the animal will learn that the tone never predicts the shock; instead, the tone predicts safety. When we carried out this experiment, we found exactly what we had predicted: when a mouse that was given shocks and tones separately heard the tone in a novel environment, it stopped acting defensively. It walked into the center of an open field as if it owned the place, showing no signs of fear (figure 25-5). When we looked in the lateral nucleus of mice that had undergone safety training, we found the opposite of long-term potentiation: namely, a long-term depression in the neural response to the tone, suggesting that the signal to the amygdala had been dramatically curtailed (figure 25-4).

We next asked whether safety training gives rise to a true sense of safety, an actual sense of self-confidence, or whether it simply lowers the baseline of fear that is always present in all of us. To distinguish between the two possibilities, we recorded from the striatum, an area of the brain normally involved in positive reinforcement and in feeling good. (This is the area activated by cocaine and other addictive drugs, which hijack the positive reinforcing neural system and entice a person to use the drug more often.) We found that neural activity in the striatum following a tone is not altered when the animal learns fear—that is, when it learns to associate the tone with a shock. But when an animal learns to associate the tone with safety, the response in the striatum is dramatically enhanced, consistent with the positive sensation of feeling safe.

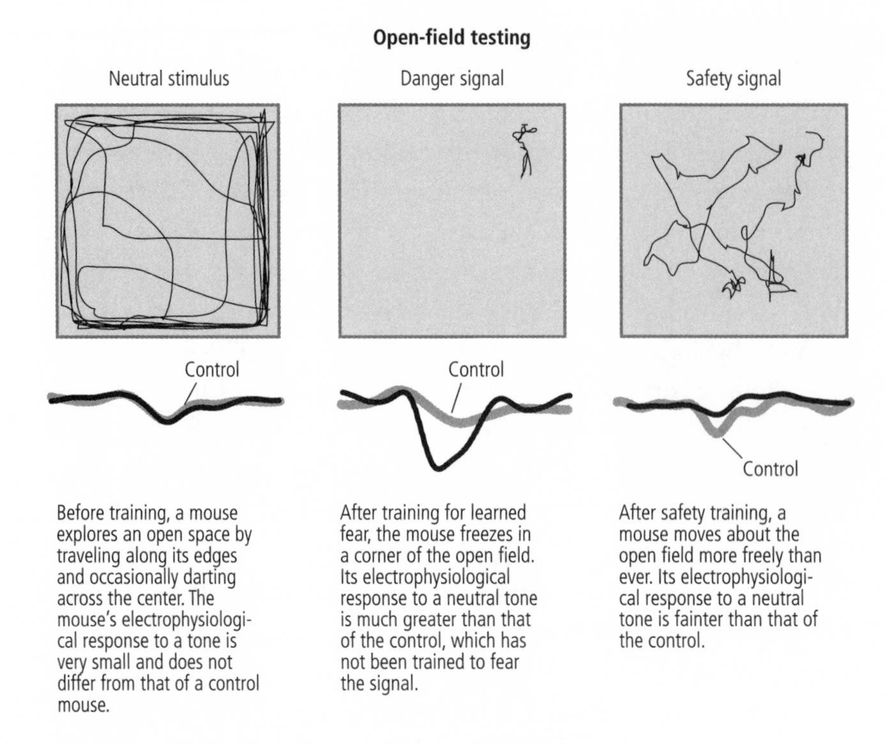

25-5 Effects of signals for learned fear and learned safety.

Our studies of learned safety have opened a new view of both positive feelings of happiness and security as well as negative feelings of anxiety and fear. They point to a second system deep in the brain that is concerned with positive emotions. Indeed, both the neurons in the thalamus that respond to the tone and the neurons in the lateral nucleus of the amygdala send connections to the striatum to convey information about contentment and safety. The striatum connects to many areas, including the prefrontal cortex, which inhibits the amygdala. So it is conceivable that by enhancing the signal in the striatum, learned safety not only enhances feelings of safety and security but also reduces fear by inhibiting the amygdala.

As these studies imply, we may be entering an era in which the molecular biology of cognition and emotion can open up ways of enhancing a person's sense of security or self-worth. Might certain anxiety states, for example, represent a defect in the neural signals that normally convey a sense of security? Since the 1960s we have had medications that alleviate certain anxiety states, but these drugs are not useful for all anxiety disorders, and some of them, such as Librium and Valium, are addictive and therefore need to be monitored extremely carefully. Therapies that enhance the activity of the neural circuitry for safety and well-being might well provide a more effective approach to treating anxiety disorders.

26

A NEW WAY TO TREAT MENTAL ILLNESS

Can mouse models be used to investigate disorders that are even more complex, more serious and disabling than anxiety states? Can they be used to study schizophrenia, the most persistent and devastating mental disorder of humankind and the one most in need of new treatments?

Schizophrenia is, surprisingly, a fairly common disorder. It strikes about 1 percent of the population worldwide and seems to affect men slightly more frequently and more severely than women. An additional 2 to 3 percent of the general population has schizotypal personality disorder, often considered to be a milder form of the disease because patients do not manifest psychotic behavior.

Schizophrenia is characterized by three types of symptoms: positive, negative, and cognitive. The positive symptoms, which last at least six months, are odd or even bizarre behaviors and disturbances in mental functioning. They are most prominent during psychotic episodes, the phases of the illness in which patients are not able to interpret reality correctly. Patients then are unable to examine their beliefs and perceptions realistically or to compare them to what is actually occurring in the world around them. The hallmarks of this inability to interpret reality are delusions (aberrant beliefs that fly in the face of facts and that are not changed by evidence that the beliefs are unreasonable), hallucinations (perceptions occurring without an external stimulus, such as hearing voices commenting on one's actions), and illogical thinking (loss of normal connections or associations between ideas, known as loosening of associations or derailment, which, when severe, results in incoherent thoughts and speech).

The negative symptoms of schizophrenia are an absence of certain normal social and interpersonal behaviors, accompanied by social withdrawal, poverty of speech, and a loss of the ability to feel and express emotions, called flattening of affect. The cognitive symptoms include poor attention and deficits in a form of explicit short-term memory known as working memory, which is critical for executive functions such as organizing one's day, and planning and carrying out a sequence of events. The cognitive symptoms are chronic, persisting even during nonpsychotic periods, and are the most difficult aspects of the disease to manage.

Between psychotic episodes, patients exhibit primarily negative and cognitive symptoms: they behave eccentrically, are socially isolated, and have a low level of emotional arousal, impoverished social drive, poverty of speech, poor attention span, and lack of motivation.

Most people working on schizophrenia have recognized for some time that the entire spectrum of symptoms cannot possibly be modeled in mice. Positive symptoms cannot be modeled readily, because we do not know how to identify delusions or hallucinations in mice. It is equally difficult to model the negative symptoms. However, following the pioneering work of Patricia Goldman-Rakic in monkeys, carried out at Yale University, my colleagues Eleanor Simpson, Christoph Kellendonk, and Jonathan Polan wanted to know if it were possible to use mouse models to investigate the molecular basis of some aspects of the cognitive symptoms of schizophrenia. We thought we could model a key component of the cognitive symptoms—notably, the defect in working memory. Working memory has been well described and is known to be critically dependent on the prefrontal cortex, a part of the frontal lobe that mediates our most complex mental processes. We also believed that understanding the cognitive deficits would improve our understanding of how the prefrontal cortex functions during normal mental states.

STUDY OF THE PREFRONTAL CORTEX DATES TO 1848, WHEN JOHN Harlow described the now famous case of railroad foreman Phineas Gage. An accidental explosion drove a tamping iron through Gage's prefrontal cortex. He survived the incident with his general intelligence, perception, and long-term memory intact, but his personality was changed. Before the accident, he was conscientious and hardworking; afterward, he drank a great deal and eventually became an unreliable drifter. Subsequent studies of people with injuries to the prefrontal cortex confirm that this region of the brain plays a critical role in judgment and long-term planning.

In the 1930s Carlyle Jacobsen, a psychologist at Yale, began to study the function of the prefrontal cortex in monkeys and provided the earliest evidence that it is involved in short-term memory. Four decades later, the British cognitive psychologist Alan Baddeley described a form of short-term memory that he called working memory because it integrates moment-to-moment perceptions over a relatively short period and relates those perceptions to established memories of past experiences, an essential feature in planning and executing complex behavior. Shortly thereafter, Joaquin Fuster at the University of California, Los Angeles and Goldman-Rakic linked Jacobsen's work on the prefrontal cortex to Baddeley's studies of working memory. They found that removing the prefrontal cortex of monkeys does not result in a generalized deficit in short-term memory but rather in a deficit in the functions that Baddeley described as working memory.

The finding that the prefrontal cortex is involved in the planning and execution of complex behaviors—functions that are disturbed in schizophrenia—led investigators to explore the prefrontal cortex of schizophrenic patients. Brain images revealed that metabolic activity in the prefrontal cortex is subnormal in these patients, even when they are not engaged in any specific mental activity. When normal individuals are challenged by a task that requires working memory, metabolic function in their prefrontal areas increases dramatically. The increase is much smaller in schizophrenic individuals.

Given that schizophrenia has a genetic component, it is perhaps not surprising that working memory is also moderately impaired in 40 to 50 percent of first-degree relatives (parents, children, and siblings) of patients with schizophrenia, even though these relatives lack clinical symptoms of the disease. Furthermore, the same relatives exhibit abnormal functioning of the prefrontal cortex, emphasizing the importance of this region in the genetic expression of schizophrenia.

The fact that the cognitive symptoms of schizophrenia resemble the behavioral defects seen when the frontal lobes are surgically disconnected from the rest of the brain in experimental animals caused us to ask: What are the molecular underpinnings of the defect in working memory in the prefrontal cortex?

MUCH OF WHAT WE KNOW ABOUT THE BIOLOGY OF SCHIZOPHRENIA comes from the study of drugs that ameliorate the disorder. In the 1950s Henri Laborit, a French neurosurgeon, had the idea that the anxiety many patients experience before surgery might be caused by the body's release of massive amounts of histamine. Histamine is a hormone-like substance produced in response to stress; it causes dilation of the blood vessels and a decrease in blood pressure. Laborit argued that the excess histamine might contribute to some of the undesirable side effects of anesthesia, such as agitation, shock, and sudden death. In his search for a drug that would block the action of histamine and calm patients, he came across chlorpromazine, which had just been developed by the French pharmaceutical firm Rhône-Poulence. Laborit was so impressed with the tranquilizing action of chlorpromazine that he began to wonder whether it might also calm agitated patients with psychiatric disorders. Two French psychiatrists, Jean Delay and Pierre Deniker, followed up on the idea and found that a high dosage of chlorpromazine indeed calmed agitated and aggressive patients with symptoms of schizophrenia.

In time, chlorpromazine and related drugs were found to be not only tranquilizers, calming patients without sedating them unduly, but also antipsychotic agents, dramatically reducing the psychotic symptoms of schizophrenia. These drugs, the first to be effective against a major mental disorder, revolutionized psychiatry. They also focused the interest of the psychiatric community on the question of how an antipsychotic agent produces its effects.

The first clue to chlorpromazine's mechanism of action came from an analysis of one of its side effects, a syndrome resembling Parkinson's disease. In 1960 Arvid Carlsson, a professor of pharmacology at the University of Göteborg in Sweden with whom I would later share the Nobel Prize, made three remarkable discoveries that provided critical insights into both Parkinson's disease and schizophrenia. First, he discovered dopamine and showed it to be a neurotransmitter in the brain. Next, he found that when he lowered the concentration of dopamine in the brain of experimental animals by a critical amount, he produced a model of Parkinson's disease. From this finding he argued that parkinsonism may result from a lowered concentration of dopamine in regions of the brain that are involved in motor control. He and others tested this idea and found that they could reverse the symptoms of Parkinson's disease by giving patients additional dopamine.

In the course of these investigations, Carlsson noticed that when patients were given an overly large dose of dopamine, they developed psychotic symptoms resembling those seen in schizophrenia. This observation caused him to suggest that the underlying cause of schizophrenia is excessive dopamine transmission. Antipsychotic agents produce their therapeutic effect, he reasoned, by blocking dopamine receptors. This action reduces dopamine transmission along several critical neural pathways and thus lessens the consequences of excess dopamine production. Carlsson's suggestion was later confirmed experimentally. Further support for his idea came from the finding that in treating patients, the antipsychotic drugs often produced Parkinsonian symptoms as a side effect of treatment, suggesting in still another way that these drugs block the action of dopamine in the brain.

In Carlsson's view, the overactivity of dopamine-producing neurons was responsible for all of the symptoms of schizophrenia—positive, negative, and cognitive. He suggested that excess dopamine in the pathway to the hippocampus, to the amygdala, and to related structures might give rise to the positive symptoms, while excess dopamine in the pathway to the cortex, especially with that pathway's abundant synaptic connections to the prefrontal cortex, might give rise to the negative and cognitive symptoms. In time, it became clear that all of the medications that alleviate the symptoms of schizophrenia target primarily a particular type of dopamine receptor, the D2 receptor. Solomon Snyder at Johns Hopkins and Philip Seeman at the University of Toronto both found a strong correlation between the effectiveness of antipsychotic drugs and their ability to block the D2 receptor. At the same time, however, it became clear that antipsychotic drugs help only with the positive symptoms of schizophrenia. They mitigate and even abolish delusions, hallucinations, and some types of disordered thinking without significantly affecting the negative or cognitive symptoms of the disease. This discrepancy was difficult to explain.

IN 2004 A NUMBER OF INVESTIGATORS DISCOVERED THAT ONE genetic predisposition or susceptibility to schizophrenia is an abnormally large number of D2 receptors in the striatum, an area of the brain that, as we have seen, is usually involved in feeling good. Having an unusually large number of D2 receptors available to bind dopamine results in increased dopamine transmission. Simpson, Kellendonk, Polan, and I wanted to explore the role of this genetic susceptibility in producing the cognitive deficits of schizophrenia, so we engineered mice with a gene that expresses a superabundance of D2 receptors in the striatum. We found that such mice do indeed have deficits in working memory, consistent with Carlsson's hypothesis.

We wanted to know why drugs that block D2 receptors fail to ameliorate the cognitive symptoms of schizophrenia, so we carried out another experiment, using genetic tools we had developed ten years earlier. Once a mouse reached adulthood, we turned off the transgene responsible for the production of excessive dopamine receptors and found that the defect in working memory was unabated. In other words, correcting the molecular defect in adult brains did not correct the cognitive deficit.

This result suggested that an overabundance of D2 receptors during development causes changes in the mouse brain that persist into adulthood. These changes might be the reason antipsychotic drugs do not work on the cognitive symptoms of schizophrenia. The overproduction of D2 receptors in the striatum exerts its impact early in development, well before the disease manifests itself, perhaps by producing fixed and irreversible changes in the dopamine system of some other part of the brain. Once this happens, the deficits in function of the prefrontal cortex, the structure involved in cognitive symptoms in the striatum, may no longer be reversible by reducing to normal the number of D2 receptors.

We have now tracked down at least one change that occurs in the prefrontal cortex as a result of the overproduction of D2 receptors: a decrease in the activation of another dopamine receptor, the D1 receptor. Earlier experiments by Goldman-Rakic had suggested that decreasing D1 receptor activation also decreases cyclic AMP, causing a deficiency in working memory.

These experiments demonstrate that genetically engineered mice may serve as valuable models in the study of complex psychiatric diseases by permitting us to break down the disease into simpler, more easily analyzed molecular components. Not only can we explore the genetic contributions to schizophrenia in mutant mice, we can also manipulate the environment of the mice, in utero and during early development, to examine what gene-environment interactions may trigger the onset of the disease.

DEPRESSION, ANOTHER COMMON ILLNESS THAT DESTROYS PSYCHIC well-being, was first described in the fifth century B.C. by the Greek physician Hippocrates, who thought that moods depend on the balance of the four humors: blood, phlegm, yellow bile, and black bile. An excess of black bile was believed to cause depression. In fact, melancholia, the ancient Greek term for depression, means "black bile." Although Hippocrates' explanation of depression seems fanciful today, the underlying view that psychological disorders reflect physiological processes is becoming generally recognized.

The clinical features of depression are easily summarized. In Hamlet's words, "How weary, stale, flat, and unprofitable seem to me all the uses of this world!" Untreated, an episode of depression typically lasts four months to a year. It is characterized by an unpleasant mood that is present day in and day out for a majority of the time, as well as intense mental anguish, inability to experience pleasure, and a generalized loss of interest in the world. Depression is often associated with a disturbance of sleep, diminished appetite, loss of weight, loss of energy, decreased sex drive, and slowing down of thoughts.

Depression affects about 5 percent of the world's population at some time in their lives. In the United States, 8 million people are affected at any given time. Severe depression can be profoundly debilitating: in extreme cases, patients stop eating or maintaining basic personal hygeine. Although some people have only a single episode, the illness is usually recurrent. About 70 percent of people who have one major depressive episode will have at least one more. The average age of onset is about twenty-eight years, but the first episode can occur at almost any age. Indeed, depression can affect young children, although it often goes unrecognized in them. Depression also occurs in the elderly; often, older people who become depressed have not had an earlier episode, and their depression is more resistant to treatment. Women are affected two to three times as often as men.

SEVERAL EFFECTIVE DRUGS HAVE BEEN DEVELOPED TO COMBAT depression. The first one—a monoamine oxidase inhibitor (MAOI)—was initially developed to fight a very different disorder, tuberculosis. MAOIs act by decreasing the breakdown of serotonin and norepinephrine, thereby making more of these neurotransmitters available for release at synapses. Physicians soon noticed that patients receiving these MAOIs were amazingly upbeat, considering the continued seriousness of their illness. Before long, physicians realized the MAOIs are more effective against depression than against tuberculosis. This insight led to the development of a group of drugs that are now effective in 70 percent of patients with major depression.

Following in the wake of the discovery of antipsychotic agents, the discovery of antidepressants moved psychiatry into a new era. Far from being a field without effective treatments for the seriously ill, psychiatry now had an effective therapeutic armamentarium comparable to that of other areas of medicine.

Drugs that are effective against depression act primarily on two modulatory transmitter systems in the brain, one of which is serotonin; the other is norepinephrine. The evidence is particularly clear regarding serotonin, which is strongly correlated with mood states in humans: high concentrations of serotonin are associated with feelings of well-being, whereas low concentrations are associated with symptoms of depression. Indeed, people who commit suicide have extremely low concentrations of serotonin.

The most effective antidepressant drugs are known as selective serotonin reuptake inhibitors. These drugs increase the concentration of serotonin in the brain by inhibiting the molecular transport system that removes serotonin from the synaptic cleft, where it is released by presynaptic neurons. Based on this finding, a hypothesis was developed which holds that depression represents decreased availability in the brain of serotonin, norepinephrine, or both.

Although this hypothesis explains some aspects of a patient's response to antidepressant drugs, it fails to account for a number of important phenomena. In particular, it fails to explain why antidepressant medications take only hours to inhibit the reuptake of serotonin in neurons, yet take at least three weeks to alleviate the symptoms of depression in people. If antidepressant drugs actually produce all of their actions by inhibiting the uptake, and thereby promoting the accumulation, of serotonin at the synapses, what accounts for the delay in the response? Perhaps it takes at least three weeks for the increased serotonin to affect key neural circuits throughout the brain—for the brain to "learn" how to become happy again. In addition, we now know that antidepressants affect other processes besides the reuptake and accumulation of serotonin.

One important clue to depression has come from the work of Ronald Duman at Yale and Rene Hen at Columbia. They have found that antidepressant drugs also increase the ability of a small region of the hippocampus, the dentate gyrus, to generate new nerve cells. Although the vast majority of nerve cells do not divide, this small nest of stem cells does divide and gives rise to differentiated nerve cells. Over a period of two to three weeks, the time it takes antidepressant drugs to work, a few of the cells are incorporated into the neural networks of the dentate gyrus. The function of these stem cells is unclear. To explore it, Hen used radiation to destroy the dentate gyrus in a mouse model of depression caused by stress. He found that antidepressants could no longer reverse depressionlike behavior in mice lacking the stem cells.

These remarkable new findings raise the possibility that antidepressants exert their effects on behavior in part by stimulating the production of neurons in the hippocampus. This idea is consistent with the finding that depression often compromises memory severely. Perhaps the damage done to the brain by depression can be overcome by restoring the ability of the hippocampus to produce new nerve cells. A remarkable idea! And one that will be challenging the imagination and skill of a new generation of psychiatric researchers in the decades ahead.

CLEARLY, MOLECULAR BIOLOGY IS POISED TO ACCOMPLISH FOR psychiatry what it has already begun to do for neurology. Genetic models of major mental illnesses in mice might therefore be useful in at least two ways. First, as studies of human patients lead to the discovery of variant genes that may predispose people to mental disease (such as the variant of the D2 receptor gene that is a risk factor for schizophrenia), these genes can be inserted into mice and used to test specific hypotheses about the origins and development of particular illnesses. Second, genetic studies in mice will enable us to explore the complex molecular pathways underlying disease at a level of detail and precision impossible in human patients. Such basic neurobiological studies will enhance our ability to diagnose and classify mental disorders and will provide a rational foundation for the development of new molecular therapies.

In a larger sense, we are moving from a decade concerned with probing the mysteries of brain function to a decade of exploring treatments for brain dysfunction. In the fifty years since I entered medicine, basic science and clinical science have ceased to be worlds apart. Some of the most interesting questions in neural science today are directly related to pressing issues in neurology and psychiatry. As a result, translational research is no longer a limited endeavor carried out by a few people in white coats. Rather, the potential for therapeutic use guides much of the research done in neuroscience.

During the 1990s, known as the Decade of the Brain, we all became translational researchers. In the first decade of the twenty-first century, our progress is being transformed into the Decade of Brain Therapeutics. As a consequence, the disciplines of psychiatry and neurology are being brought intellectually closer to each other. One can foresee the day in the not-too-distant future when resident physicians in both disciplines will share a common year of training, comparable to the year of residency training in internal medicine for physicians who go on to specialize in widely different areas, such as heart disease or gastrointestinal disorders.

27

BIOLOGY AND THE RENAISSANCE OF PSYCHOANALYTIC THOUGHT

When psychoanalysis emerged from Vienna in the first decades of the twentieth century, it represented a revolutionary way of thinking about mind and its disorders. The excitement surrounding the theory of unconscious mental processes increased as the century reached its midpoint and psychoanalysis was brought to the United States by émigrés from Germany and Austria.

As an undergraduate at Harvard, I shared this enthusiasm, not only because psychoanalysis presented a view of mind that appeared to have great explanatory power, but also because it conjured up the intellectual environment of Vienna in the early twentieth century, an environment that I admired and had missed out on. Indeed, what I so enjoyed in the intellectual life that surrounded Anna Kris and her parents were the insights and perspectives it gave me on life in Vienna in the 1930s. There was talk about *Die Neue Frei Presse* (*The New Free Press*), Vienna's most important newspaper, which, according to the Krises, was neither terribly new nor terribly free. The Krises also recalled the dramatic, even histrionic, lectures of Karl Kraus, the cultural critic and student of language whom I admired very much. Kraus lashed out against Viennese hypocrisy, and his great play *The*

Last Days of Mankind predicted what was to come: World War II and the Holocaust.

But by 1960, when I began clinical training in psychiatry, my enthusiasm had stalled. My marriage to Denise, an empirical sociologist, and my research experiences—first in Harry Grundfest's laboratory at Columbia and then in Wade Marshall's laboratory at the National Institute of Mental Health—tempered my enthusiasm for psychoanalysis. While I still admired the rich, nuanced view of mind that psychoanalysis had introduced, I was disappointed during my clinical training to see how little progress psychoanalysis had made toward becoming empirical, toward testing its ideas. I also was disappointed in many of my teachers at Harvard, physicians who were motivated to enter psychoanalytic psychiatry out of humanistic concerns, as I was, but who had little interest in science. I sensed that psychoanalysis was moving backward into an unscientific phase and, in the process, was taking psychiatry with it.

UNDER THE INFLUENCE OF PSYCHOANALYSIS, PSYCHIATRY WAS transformed in the decades following World War II from an experimental medical discipline closely related to neurology into a nonempirical specialty focused on the art of psychotherapy. In the 1950s academic psychiatry abandoned some of its roots in biology and experimental medicine and gradually became a therapeutic discipline based on psychoanalytic theories. As such, it was strangely unconcerned with empirical evidence or with the brain as the organ of mental activity. In contrast, medicine evolved during this period from a therapeutic art into a therapeutic science, based on a reductionist approach derived first from biochemistry and later from molecular biology. During medical school, I had witnessed and been influenced by this evolution. I therefore could not help but note the peculiar position of psychiatry within medicine.

Psychoanalysis had introduced a new method of examining the mental life of patients, a method based on free association and interpretation. Freud taught psychiatrists to listen carefully to patients and to do so in new ways. He emphasized a sensitivity to both the latent and the manifest meaning of the patient's communications. He also created a provisional schema for interpreting what might otherwise appear as unrelated and incoherent reports.

So novel and powerful was this approach that for many years not only Freud but other intelligent and creative psychoanalysts as well could argue that psychotherapeutic encounters between patient and analyst provided the best context for scientific inquiry into mind, particularly into unconscious mental processes. Indeed, in the early years psychoanalysts made many useful and original observations that contributed to our understanding of mind simply by listening carefully to their patients and by testing the ideas that arose from psychoanalysis—such as childhood sexuality—in observational studies of normal child development. Other original contributions included the discovery of different types of unconscious and preconscious mental processes, the complexities of motivation, transference (the displacing of past relationships onto the patient's current life), and resistance (the unconscious tendency to oppose a therapist's efforts to effect change in the patient's behavior).

Sixty years after its introduction, however, psychoanalysis had exhausted much of its novel investigative power. By 1960 it was clear, even to me, that little in the way of new knowledge or insights remained to be learned by observing individual patients and listening carefully to them. Although psychoanalysis had historically been scientific in its ambitions—it had always wanted to develop an empirical, testable science of mind—it was rarely scientific in its methods. It had failed over the years to submit its assumptions to replicable experimentation. Indeed, it was traditionally far better at generating ideas than at testing them. As a result, psychoanalysis had not made the same progress as some other areas of psychology and medicine. Indeed, it seemed to me that psychoanalysis was losing its way. Rather than focusing in on areas that could be tested empirically, psychoanalysis expanded its scope, taking on mental and physical disorders that it was not optimally suited to treat.

Initially, psychoanalysis was used to treat what were called neurotic illnesses: phobias, obsessional disorders, and hysterical and anxiety states. However, psychoanalytic therapy gradually extended its reach to almost all mental illnesses, including schizophrenia and depression.

By the late 1940s, many psychiatrists, influenced in part by their successful treatment of soldiers who had developed psychiatric problems in battle, had come to believe that psychoanalytic insights might be useful in treating medical illnesses that did not respond readily to drugs. Diseases such as hypertension, asthma, gastric ulcers, and ulcerative colitis were thought to be psychosomatic—that is, induced by unconscious conflicts. Thus by 1960 psychoanalytical theory had become for many psychiatrists, particularly those on the East and West coasts of the United States, the prevailing model for understanding all mental and some physical illnesses.

This expanded therapeutic scope appeared on the surface to strengthen psychoanalysis's explanatory power and clinical insight, but in reality it weakened psychiatry's effectiveness and hindered its attempt to become an empirical discipline aligned with biology. When Freud first explored the role of unconscious mental processes in behavior in 1894, he was also engaged in an effort to develop an empirical psychology. He tried to work out a neural model of behavior, but because of the immaturity of brain science at the time, he abandoned the biological model for one based on verbal reports of subjective experiences. By the time I arrived at Harvard to train in psychiatry, biology had begun to make important inroads in understanding higher mental processes. Despite these advances, a number of psychoanalysts took a far more radical stance—biology, they argued, is irrelevant to psychoanalysis.

This indifference to, if not disdain for, biology was one of the two problems I encountered during my residency training. An even more serious problem was the lack of concern among psychoanalysts for conducting objective studies, or even for controlling investigator bias. Other branches of medicine controlled bias by means of blind experiments, in which the investigator does not know which patients are receiving the treatment being tested and which ones are not. However, the data gathered in psychoanalytic sessions are almost always private. The patient's comments, associations, silences, postures, movements, and other behaviors are privileged. Of course, privacy is central to the trust that must be earned by the analyst—and therein lies the rub. In almost every case, the only record is the analyst's subjective accounts of what he or she believes happened. As research psychoanalyst Hartvig Dahl has long argued, such interpretation is not accepted as evidence in most scientific contexts. Psychoanalysts, however, are rarely concerned about the fact that accounts of therapy sessions are necessarily subjective.

As I began my residency in psychiatry, I sensed that psychoanalysis could be immeasurably enriched by joining forces with biology. I also thought that if the biology of the twentieth century were to answer some of the enduring questions about the human mind, those answers would be richer and more meaningful if they were arrived at in collaboration with psychoanalysis. Such a collaboration would also provide a firmer scientific foundation for psychoanalysis. I believed then, and I believe more strongly now, that biology may be able to delineate the physical basis of several mental processes that lie at the heart of psychoanalysis—namely, unconscious mental processes, psychic determinism (the fact that no action or behavior, no slip of the tongue is entirely random or arbitrary), the role of the unconscious in psychopathology (that is, the linking of psychological events, even disparate ones, in the unconscious), and the therapeutic effect of psychoanalysis itself. What particularly fascinated me, because of my interest in the biology of memory, was the possibility that psychotherapy, which presumably works in part by creating an environment in which people learn to change, produces structural changes in the brain and that one might now be in a position to evaluate those changes directly.

FORTUNATELY, NOT EVERYONE IN THE PSYCHOANALYTIC community thought that empirical research was irrelevant for the future of the discipline. Two trends have gained momentum in the forty years since I completed my clinical training, and they are beginning to exert a significant impact on psychoanalytic thought. One trend is the insistence on an evidence-based psychotherapy. The second, more difficult trend is the attempt to align psychoanalysis with the emerging biology of mind.

Perhaps the most important force driving the first trend has been Aaron Beck, a psychoanalyst at the University of Pennsylvania. Influenced by modern cognitive psychology, Beck found that a patient's major cognitive style—that is, the person's way of perceiving, representing, and thinking about the world—is a key element in a number of disorders, such as depression, anxiety disorders, and obsessive-compulsive states. By emphasizing cognitive style and ego functioning, Beck was continuing a line of thought initiated by Heinz Hartmann, Ernst Kris, and Rudolph Lowenstein.

Beck's emphasis on the role of conscious thought processes in mental disorders was novel. Traditionally, psychoanalysis had taught that mental problems arise from unconscious conflicts. For example, in the late 1950s, when Beck began his investigations, depressive illness was commonly viewed as "introjected anger." Freud had argued that depressed patients feel hostile and angry toward someone they love. Because patients cannot deal with negative feelings about someone who is important, needed, and valued, they handle those feelings by repressing them and unconsciously directing them against themselves. It is this self-directed anger and hatred that leads to low self-esteem and feelings of worthlessness.

Beck tested Freud's idea by comparing the dreams of depressed patients with those of patients who were not depressed. He found that depressed patients exhibited not more, but less hostility than other patients. In the course of carrying out this study and listening carefully to his patients, Beck found that rather than expressing hostility, depressed people express a systematic negative bias in the way they think about life. They almost invariably have unrealistically high expectations of themselves, overreact dramatically to any disappointment, put themselves down whenever possible, and are pessimistic about their future. This distorted pattern of thinking, Beck realized, is not simply a symptom, a reflection of a conflict lying deep within the psyche, but a key agent in the actual development and continuation of the depressive disorder. Beck made the radical suggestion that by identifying and addressing the negative beliefs, thought processes, and behaviors, one might be able to help patients replace them with healthy, positive beliefs. Moreover, one could do so independent of personality factors and the unconscious conflicts that may underlie them.

To test this idea clinically, Beck presented patients with evidence from their own experiences, actions, and accomplishments that countered, challenged, and corrected their negative views. He found that they often improved with remarkable speed, feeling and functioning better after a very few sessions. This positive result led Beck to develop a systematic, short-term psychological treatment for depression that focuses not on a patient's unconscious conflict, but on his or her conscious cognitive style and distorted way of thinking.

Beck and his associates initiated controlled clinical trials to evaluate the effectiveness of this mode of therapy, compared with placebo and with antidepressant medication. They found that cognitive behavioral therapy is usually as effective as antidepressant medication in treating people with mild and moderate depression; in some studies, it appeared superior at preventing relapses. In later controlled clinical trials, cognitive behavioral therapy was successfully extended to anxiety disorders, especially panic attacks, post-traumatic stress disorders, social phobias, eating disorders, and obsessive-compulsive disorders.

Beck went beyond introducing a novel form of psychotherapy and testing it empirically. He also developed scales and inventories for assessing the symptoms and extent of depression and other psychiatric disorders, measures that have introduced new scientific rigor into psychotherapy-based research. In addition, he and his colleagues wrote manuals on how the treatments were to be carried out. Beck has thus brought to the psychoanalytic therapy of mind a critical attitude, a quest for empirical data, and a desire to find out whether a given therapy works.

Influenced by Beck's approach, Gerald Klerman and Myrna Weissman created a second scientifically valid form of short-term psychotherapy, known as interpersonal psychotherapy. This treatment focuses on correcting patients' mistaken beliefs and on changing the nature of their communications in various interactions with others. Like cognitive behavioral therapy, it has proven efficacious in controlled trials for mild and moderate depression and has been codified in teaching manuals. Interpersonal therapy seems to be particularly effective in situational crises, such as the loss of a partner or a child, whereas cognitive therapy appears to be particularly effective in treating chronic disorders. Similarly, although not yet as extensively studied, Peter Sifneous and Habib Davanloo have formalized a third short-term treatment, brief dynamic therapy, which focuses on the patient's defenses and resistance, and Otto Kernberg has introduced a psychotherapy focused on transference.

Unlike traditional psychoanalysis, all four short-term modes of psychotherapy attempt to gather empirical data and to use it in determining the effectiveness of treatment. As a result, they have brought about a major change in how short-term (and even long-term) therapy is conducted, and they have begun to move the discipline toward evidence-based process and outcome studies.

The long-term effects of the new psychotherapies are still uncertain, however. Although they often achieve results, both therapeutically and in terms of basic understanding, within five to fifteen sessions, the improvement is not always long-lasting. Indeed, it would appear that for some patients to achieve sustained improvement, therapy must continue for one or two years, perhaps because treating symptoms of their disorder without addressing the underlying conflicts is not always efficacious. Even more important from a scientific viewpoint is the fact that Beck and most other proponents of evidence-based therapeutics come from a psychoanalytic tradition of observation, not from a biological tradition of experimentation. With rare exceptions, the leaders of this trend in psychotherapy have not yet turned to biology to try to understand the underlying basis of observed behavior.

WHAT IS NEEDED IS A BIOLOGICAL APPROACH TO PSYCHOTHERAPY. Until quite recently, there have been few biologically compelling ways to test psychodynamic ideas or to evaluate the efficacy of one therapeutic approach over another. A combination of effective short-term psychotherapy and brain imaging may now give us just that—a way of revealing both mental dynamics and the workings of the living brain. In fact, if psychotherapeutic changes are maintained over time, it is reasonable to conclude that different forms of psychotherapy lead to different structural changes in the brain, just as other forms of learning do.

The idea of using brain imaging to evaluate the outcome of different forms of psychotherapy is not an impossible dream, as studies of obsessive-compulsive disorder have shown. This disorder has long been thought to reflect a disturbance of the basal ganglia, a group of structures that lies deep in the brain and plays a key role in modulating behavior. One of the structures of the basal ganglia, the caudate nucleus, is the primary recipient of information coming from the cerebral cortex and other regions of the brain. Brain imaging has found that obsessive-compulsive disorder is associated with increased metabolism in the caudate nucleus. Lewis R. Baxter, Jr., and his colleagues at the University of California, Los Angeles have found that obsessive-compulsive disorder can be reversed by cognitive behavioral psychotherapy. It can also be reversed pharmacologically by inhibiting the reuptake of serotonin. Both the drugs and psychotherapy reverse the increased metabolism of the caudate nucleus.

Brain-imaging studies of patients with depression commonly reveal a decrease in activity in the dorsal side of the prefrontal cortex but an increase in activity in the ventral side. Again, both psychotherapy and drugs reverse these abnormalities. Had imaging been available in 1895, when Freud wrote "On a Scientific Psychology," he might well have directed psychoanalysis along very different lines, keeping it in close relationship with biology, as he outlined in this essay. In this sense, combining brain imaging with psychotherapy represents top-down investigation of mind and continues the scientific program Freud had originally envisioned.

As we have seen, short-term psychotherapy now comes in at least four different forms, and brain imaging may provide a scientific means of distinguishing among them. If so, it may reveal that all effective psychotherapies work through the same anatomical and molecular mechanisms. Alternatively, and more likely, imaging may show that psychotherapies achieve their goals through distinctly different mechanisms in the brain. Psychotherapies are also likely to have adverse side effects, as drugs do. Empirical testing of psychotherapies could help us maximize the safety and effectiveness of these important treatments, much as it does for drugs. It could also help predict the outcome of particular types of psychotherapy and would direct patients to the ones most appropriate for them.

THE COMBINATION OF SHORT-TERM PSYCHOTHERAPY AND BRAIN imaging may at last allow psychoanalysis to make its own distinctive contribution to the new science of mind. And not a moment too soon. There is an enormous public health need for effective therapies in a variety of mild and moderately serious mental illnesses. Studies by Ronald Kessler at Harvard suggest that almost 50 percent of the general population have had a psychiatric problem at one point in their lives. In the past, many of these people were treated with drugs. Drugs have been an enormous boon to psychiatry, but they can have side effects. Moreover, drugs alone are often not effective. Many patients do better when some form of psychotherapy is combined with drugs, while a surprising number of patients do reasonably well with psychotherapy alone.

In her book *An Unquiet Mind*, Kay Jamison describes the benefits of both modes of treatment for even a serious illness—in her case, bipolar disorder. Lithium treatment for the disorder prevented her disastrous highs, kept her out of the hospital, saved her life by preventing her from committing suicide, and made long-term psychotherapy possible. "But, ineffably," she writes, "psychotherapy *heals*. It makes some sense of the confusion, reins in the terrifying thoughts and feelings, returns some control and hope and possibility of learning from it all. Pills cannot, do not, ease one back into reality."

What I find so fascinating about Jamison's insight is her view of psychotherapy as a learning experience that allows her to pull together the strands of her experiences—her life story. It is, of course, memory that weaves one's life into a coherent whole. As psychotherapy is subjected to more rigorous tests of effectiveness and more biological studies of its effects, we will be able to examine the workings of memory and mind. We will be able to explore, for example, various styles of thinking to see how they affect the way we feel about the world and how we behave in it.

A REDUCTIONIST APPROACH TO PSYCHOANALYSIS WILL ALSO ENABLE us to reach a deeper understanding of human behavior. The most important steps in this direction have been those taken in studies of child development, an area that excited the imagination of Ernst Kris. Freud's gifted daughter, Anna, studied the traumatic effects of family disruption during World War II and found the first compelling evidence of the importance of the bonding relationship between parent and offspring during times of stress. The effects of family disruption were studied further by the New York psychoanalyst René Spitz, who compared two groups of infants separated from their mothers. One group was raised in a foundling home and cared for by nurses, each of whom was responsible for seven infants; the other group was in a nursing home attached to a women's prison, where the infants were cared for daily for brief periods of time by their mothers. By the end of the first year, the motor and intellectual performance of the children in the orphanage had fallen far below that of the children in the prison nursing home: children in the orphanage were withdrawn and showed little curiosity or gaiety. These classic studies were published in *The Psychoanalytic Study of the Child*, a set of volumes edited by three of the originators of observational studies of children: Anna Freud, Heinz Hartmann, and Ernst Kris.

In a paradigm of how reductionism can enhance our understanding of psychological processes, Harry Harlow at the University of Wisconsin extended this work by developing an animal model of maternal deprivation. He found that when newborn monkeys were isolated for six months to one year and then returned to the company of other monkeys, they were physically healthy but behaviorally devastated. They crouched in a corner of their cages and rocked back and forth, like severely disturbed or autistic children. They did not interact with other monkeys, nor did they fight, play, or show any sexual interest. Isolation of an older animal for a comparable period was innocuous. Thus, in monkeys, as in humans, there is a critical period for social development.

Harlow next found that the syndrome could be partially reversed by giving the isolated monkey a surrogate mother, a cloth-covered wooden dummy. This surrogate elicited clinging behavior in the isolated monkey but was insufficient for the development of fully normal social behavior. Normal social development could only be rescued if, in addition to a surrogate mother, the isolated animal had contact for a few hours each day with a normal infant monkey that spent the rest of the day in the monkey colony.

The work of Anna Freud, Spitz, and Harlow was expanded by John Bowlby, who formulated the idea that the defenseless infant maintains a closeness to its caretaker by means of a system of emotive and behavioral response patterns that he called the "attachment system." Bowlby conceived of the attachment system as an inborn instinctual or motivational system, much like hunger or thirst, that organizes the memory processes of the infant and directs it to seek proximity to and communication with its mother. From an evolutionary point of view, the attachment system clearly enhances the infant's chances of survival by allowing its immature brain to use the parent's mature functions to organize its own life processes. The infant's attachment mechanism is mirrored in the parent's emotionally sensitive responses to the infant's signals. Parental responses serve both to amplify and reinforce an infant's positive emotional states and to attenuate the infant's negative emotional states. These repeated experiences become encoded in procedural memory as expectations that help the infant feel secure.

These several approaches to studies of child development are now being explored in genetically modified mice to gain an even deeper insight into the nature of parent-offspring interaction.

Other experimental means of exploring psychoanalytic ideas about the functions of mind are available today. There are, for example, ways of distinguishing procedural (implicit) mental processes that are reflected in our memory for perceptual and motor skills from two other types of unconscious mental processes: the dynamic unconscious, which represents our conflicts, sexual strivings, and repressed thoughts and actions, and the preconscious unconscious, which is concerned with organization and planning and has ready access to consciousness.

Biological approaches to psychoanalytic theory could, in principle, explore all three types of unconscious processes. One way of doing so—which I will explain in the next chapter—is to compare images of activity generated by unconscious and conscious perceptual states and to identify the regions of the brain recruited by each. Most aspects of our cognitive processes are based on unconscious inferences, on processes that occur without our awareness. We see the world effortlessly and as a unified whole—the foreground of a landscape and the horizon beyond it—because visual perception, the binding of the various elements of the visual image with one another, occurs without our being aware of it. As a result, most students of the brain believe, as Freud did, that we are not conscious of most cognitive processes, only of the end result of those processes. A similar principle seems to apply to our conscious sense of free will.

Bringing biology to bear on psychoanalytic ideas is likely to invigorate the role of psychiatry in modern medicine and to encourage empirically based psychoanalytic thought to join the forces that are shaping the modern science of mind. The goal of this merger is to join radical reductionism, which drives basic biology, with the humanistic effort to understand the human mind, which drives psychiatry and psychoanalysis. This, after all, is the ultimate goal of brain science: to link the physical and biological studies of the natural world and its living inhabitants with an understanding of the intimate textures of the human mind and human experience.

28

CONSCIOUSNESS

Psychoanalysis introduced us to the unconscious in its several forms. Like many scientists now working on the brain, I have long been intrigued by the biggest question about the brain: the nature of consciousness and how various unconscious psychological processes relate to conscious thought. When I first talked with Harry Grundfest about Freud's structural theory of mind—the ego, the id, and the superego—the central focus of my thinking was: How do conscious and unconscious processes differ in their representation in the brain? But only recently has the new science of mind developed the tools for exploring this question experimentally.

To develop productive insights into consciousness, the new science of mind first had to settle on a working definition of consciousness as a state of perceptual awareness, or selective attention writ large. At its core, consciousness in people, is an awareness of self, an awareness of being aware. Consciousness thus refers to our ability not simply to experience pleasure and pain but to attend to and reflect upon those experiences, and to do so in the context of our immediate lives and our life history. Conscious attention allows us to shut out extraneous experiences and focus on the critical event that confronts us, be it pleasure or pain, the blue of the sky, the cool northern light of a Vermeer painting, or the beauty and calm we experience at the seashore.

UNDERSTANDING CONSCIOUSNESS IS BY FAR THE MOST CHALLENGING task confronting science. The truth of this assertion can best be seen in the career of Francis Crick, perhaps the most creative and influential biologist of the second half of the twentieth century. When Crick first entered biology, after World War II, two great questions were thought to be beyond the capacities of science to answer: What distinguishes the living from the nonliving world? And what is the biological nature of consciousness? Crick turned first to the easier problem, distinguishing animate from inanimate matter, and explored the nature of the gene. By 1953, after just two years of collaboration, he and Jim Watson had helped solve that mystery. As Watson later described in *The Double Helix*, "at lunch Francis winged into the Eagle [Pub] to tell everyone within hearing distance that we had found the secret of life." In the next two decades, Crick helped crack the genetic code: how DNA makes RNA and RNA makes protein.

In 1976, at age sixty, Crick turned to the remaining scientific mystery: the biological nature of consciousness. This he studied for the rest of his life in partnership with Christof Koch, a young computational neuroscientist. Crick brought his characteristic intelligence and optimism to bear on the question; moreover, he made consciousness a focus of the scientific community, which had previously ignored it. But, despite almost thirty years of continuous effort, Crick was able to budge the problem only a modest distance. Indeed, some scientists and philosophers of mind continue to find consciousness so inscrutable that they fear it can never be explained in physical terms. How can a biological system, a biological machine, they ask, feel anything? Even more doubtful, how can it think about itself?

These questions are not new. They were first posed in Western thought during the fifth century B.C. by Hippocrates and by the philosopher Plato, the founder of the Academy in Athens. Hippocrates was the first physician to cast superstition aside, basing his thinking on clinical observations and arguing that all mental processes emanate from the brain. Plato, who rejected observations and experiments, believed that the only reason we can think about ourselves and our mortal body is that we have a soul that is immaterial and immortal. The idea of an immortal soul was subsequently incorporated into Christian thought and elaborated upon by St. Thomas Aquinas in the thirteenth century. Aquinas and later religious thinkers held that the soul—the generator of consciousness—is not only distinct from the body, it is also of divine origin.

In the seventeenth century, René Descartes developed the idea that human beings have a dual nature: they have a body, which is made up of material substance, and a mind, which derives from the spiritual nature of the soul. The soul receives signals from the body and can influence its actions but is itself made up of an immaterial substance that is unique to human beings. Descartes' thinking gave rise to the view that actions like eating and walking, as well as sensory perception, appetites, passions, and even simple forms of learning, are all mediated by the brain and can be studied scientifically. Mind, however, is sacred and as such is not a proper subject of science.

It is remarkable to reflect that these seventeenth-century ideas were still current in the 1980s. Karl Popper, the Vienna-born philosopher of science, and John Eccles, the Nobel laureate neurobiologist, espoused dualism all of their lives. They agreed with Aquinas that the soul is immortal and independent of the brain. Gilbert Ryle, the British philosopher of science, referred to the notion of the soul as "the ghost in the machine."

TODAY, MOST PHILOSOPHERS OF MIND AGREE THAT WHAT WE call consciousness derives from the physical brain, but some disagree with Crick as to whether it can ever be approached scientifically. A few, such as Colin McGinn, believe that consciousness simply cannot be studied, because the architecture of the brain poses limitations on human cognitive capacities. In McGinn's view, the human mind may simply be incapable of solving certain problems. At the other extreme, philosophers such as Daniel Dennett deny that there is any problem at all. Dennett argues, much as neurologist John Hughlings Jackson did a century earlier, that consciousness is not a distinct operation of the brain; rather, it is the combined result of the computational workings of higher-order areas of the brain concerned with later stages of information processing.

Finally, philosophers such as John Searle and Thomas Nagel take a middle position, holding that consciousness is a discrete set of biological processes. The processes are accessible to analysis, but we have made little headway in understanding them because they are very complex and represent more than the sum of their parts. Consciousness is therefore much more complicated than any property of the brain that we understand.

Searle and Nagel ascribe two characteristics to the conscious state: unity and subjectivity. The unitary nature of consciousness refers to the fact that our experiences come to us as a unified whole. All of the various sensory modalities are melded into a single, coherent, conscious experience. Thus when I approach a rosebush in the botanical garden at Wave Hill near my house in Riverdale, I sniff the exquisite fragrance of the blossoms at the same time that I see their beautiful red color—and I perceive this rosebush against the background of the Hudson River and the cliffs of the Palisade mountain ridge behind it. My perception is not only whole during the moment I experience it, it is also whole two weeks later, when I engage in mental time travel to recapture the moment. Despite the fact that there are different organs for smell and vision, and that each uses its own individual pathways, they converge in the brain in such a way that my perceptions are unified.

The unitary nature of consciousness poses a difficult problem, but perhaps not an insurmountable one. This unitary nature can break down. In a surgical patient whose brain is severed between the two hemispheres, there are two conscious minds, each with its own unified percept.

Subjectivity, the second characteristic of conscious awareness, poses the more formidable scientific challenge. Each of us experiences a world of private and unique sensations that is much more real to us than the experiences of others. We experience our own ideas, moods, and sensations directly, whereas we can only appreciate another person's experience indirectly, by observing or hearing about it. We therefore can ask, Is your response to the blue you see and the jasmine you smell—the meaning it has for you—identical to my response to the blue I see and the jasmine I smell and the meaning these have for me?

The issue here is not one of perception per se. It is not whether we each see a very similar shade of the same blue. That is relatively easy to establish by recording from single nerve cells in the visual system of different individuals. The brain does reconstruct our perception of an object, but the object perceived—the color blue or middle C on the piano—appears to correspond to the physical properties of the wavelength of the reflected light or the frequency of the emitted sound. Instead, the issue is the significance of that blue and that note for each of us. What we do not understand is how electrical activity in neurons gives rise to the meaning we ascribe to that color or that wavelength of sound. The fact that conscious experience is unique to each person raises the question of whether it is possible to determine objectively any characteristics of consciousness that are common to everyone. If the senses ultimately produce experiences that are completely and personally subjective, we cannot, the argument goes, arrive at a general definition of consciousness based on personal experience.

Nagel and Searle illustrate the difficulty of explaining the subjective nature of consciousness in physical terms as follows: Assume we succeed in recording the electrical activity of neurons in a region known to be important for consciousness while the person being studied carries out some task that requires conscious attention. For example, suppose we identified the cells that fire when I look at and become aware of a red image of the blossoms on a rosebush at Wave Hill. We have now taken a first step in studying consciousness—namely, we have found what Crick and Koch have called the neural correlate of consciousness for this one percept. For most of us, this would be a great advance because it pinpoints a material concomitant of conscious perception. From there we could go on to carry out experiments to determine whether these correlates also meld into a coherent whole, that is, the background of the Hudson River and the Palisades. But for Nagel and Searle, this is the easy problem of consciousness. The hard problem of consciousness is the second mystery, that of subjective experience.

How is it that I respond to the red image of a rose with a feeling that is distinctive to me? To use another example, what grounds do we have for believing that when a mother looks at her child, the firing of cells in the region of the cortex concerned with face recognition accounts for the emotions she feels and for her ability to summon the memory of those emotions and that image of her child?

As yet, we do not know how the firing of specific neurons leads to the subjective component of conscious perception, even in the simplest case. In fact, according to Searle and Nagel, we lack an adequate theory of how an objective phenomenon, such as electrical signals in the brain, can cause a subjective experience, such as pain. And because science as we currently practice it is a reductionist, analytical view of complicated events, while consciousness is irreducibly subjective, such a theory lies beyond our reach for now.

According to Nagel, science cannot take on consciousness without a significant change in methodology, a change that would enable scientists to identify and analyze the elements of subjective experience. Those elements are likely to be basic components of brain function, much as atoms and molecules are basic components of matter, but to exist in a form we cannot yet imagine. The reductions performed routinely in science are not problematic, Nagel holds. Biological science can readily explain how the properties of a particular type of matter arise from the objective properties of the molecules of which it is made. What science lacks are rules for explaining how subjective properties (consciousness) arise from the properties of objects (interconnected nerve cells).

Nagel argues that our complete lack of insight into the elements of subjective experience should not prevent us from discovering the neural correlates of consciousness and the rules that relate conscious phenomena to cellular processes in the brain. In fact, it is only by accumulating such information that we will be in a position to think about the reduction of something subjective to something physical and objective. But to arrive at a theory that supports this reduction, we will first have to discover the elements of subjective consciousness. This discovery, says Nagel, will be enormous in its magnitude and its implications, requiring a revolution in biology and most likely a complete transformation of scientific thought.

The aim of most neural scientists working on consciousness is much more modest than this grand perspective would imply. They are not deliberately working toward or anticipating a revolution in scientific thought. Although they must struggle with the difficulties of defining conscious phenomena experimentally, they do not see those difficulties as precluding all experimental study under existing paradigms. Neural scientists believe, and Searle for one agrees with them, that they have been able to make considerable progress in understanding the neurobiology of perception and memory without having to account for individual experience. For example, cognitive neural scientists have made advances in understanding the neural basis of the perception of the color blue without addressing the question of how each of us responds to the same blue.

WHAT WE DO NOT UNDERSTAND IS THE HARD PROBLEM OF consciousness—the mystery of how neural activity gives rise to subjective experience. Crick and Koch have argued that once we solve the easy problem of consciousness, the unity of consciousness, we will be able to manipulate those neural systems experimentally to solve the hard problem.

The unity of consciousness is a variant of the binding problem first identified in the study of visual perception. An intimate part of my experiencing the subjective pleasure of the moment at Wave Hill is how the look and the smell of roses in the gardens are bound together and unified with my view of the Hudson, the Palisades, and all the other component images of my perception. Each of these components of my subjective experience is mediated by different brain regions within my visual and olfactory and emotional systems. The unity of my conscious experience implies that the binding process must somehow connect and integrate all of these separate areas in the brain.

As a first step toward solving the easy problem of consciousness, we need to ask whether the unity of consciousness—a unity thought to be achieved by neural systems that mediate selective attention—is localized in one or just a few sites, which would enable us to manipulate them biologically. The answer to this question is by no means clear. Gerald Edelman, a leading theoretician on the brain and consciousness, has argued effectively that the neural machinery for the unity of consciousness is likely to be widely distributed throughout the cortex and thalamus. As a result, Edelman asserts, it is unlikely that we will be able to find consciousness through a simple set of neural correlates. Crick and Koch, on the other hand, believe that the unity of consciousness will have direct neural correlates because they most likely involve a specific set of neurons with specific molecular or neuroanatomical signatures. The neural correlates, they argue, probably require only a small set of neurons acting as a searchlight: the spotlight of attention. The initial task, they argue, is to locate within the brain that small set of neurons whose activity correlates best with the unity of conscious experience and then to determine the neural circuits to which they belong.

How are we to find this small population of nerve cells that could mediate the unity of consciousness? What criteria must they meet? In Crick and Koch's last paper (which Crick was still correcting on his way to the hospital a few hours before he died, on July 28, 2004), they focused on the claustrum, a sheet of brain tissue that is located below the cerebral cortex, as the site that mediates unity of experience. Little is known about the claustrum except that it connects to and exchanges information with almost all of the sensory and motor regions of the cortex as well as the amygdala, which plays an important role in emotion. Crick and Koch compare the claustrum to the conductor of an orchestra. Indeed, the neuroanatomical connections of the claustrum meet the requirements of a conductor; it can bind together and coordinate the various brain regions necessary for the unity of conscious awareness.

The idea that obsessed Crick at the end of his life—that the claustrum is the spotlight of attention, the site that binds the various components of any percept together—is the last in a series of important ideas he advanced. Crick's enormous contributions to biology (the double helical structure of DNA, the nature of the genetic code, the discovery of messenger RNA, the mechanisms of translating messenger RNA into the amino acid sequence of a protein, and the legitimizing of the biology of consciousness) put him in a class with Copernicus, Newton, Darwin, and Einstein. Yet his intense, lifelong focus on science, on the life of mind, is something he shares with many in the scientific community, and that obsession is symbolic of science at its best. The cognitive psychologist Vilayanur Ramachandran, a friend and colleague of Crick's, described Crick's focus on the claustrum during his last weeks:

> Three weeks prior to his death I visited him in his home in La Jolla. He was eighty-eight, had terminal cancer, was in pain, and was on chemotherapy; yet he had obviously been working away nonstop on his latest project. His very large desk—occupying half the room—was covered by articles, correspondence, envelopes, recent issues of *Nature*, a laptop (despite his dislike of computers), and recent books on neuroanatomy. During the whole two hours that I was there, there was no mention of his illness—only a flight of ideas on the neural basis of consciousness. He was especially interested in a tiny structure called the claustrum which, he felt, had been largely ignored by mainstream pundits. As I was leaving he said: "Rama, I think the secret of consciousness lies in the claustrum—don't you? Why else would this tiny structure be connected to so many areas in the brain?"—And gave me a sly, conspiratorial wink. It was the last time I saw him.

Since so little is known about the claustrum, Crick continued, he wanted to start an institute to focus on its function. In particular, he wanted to determine whether the claustrum is switched on when unconscious, subliminal perception of a given stimulus by a person's sensory organs turns into a conscious percept.

ONE EXAMPLE OF SUCH SWITCHING THAT INTRIGUED CRICK AND Koch is binocular rivalry. Here, two different images—say, vertical stripes and horizontal stripes—are presented to a person simultaneously in such a way that each eye sees only one set of stripes. The person may combine the two images and report seeing a plaid, but more commonly the person will see first one image, then the next, with horizontal and vertical stripes alternating back and forth spontaneously.

Using MRI, Eric Lumer and his colleagues at University College, London have identified the frontal and parietal areas of the cortex as the regions of the brain that become active when a person's conscious attention switches from one image to another. These two regions have a special role in focusing conscious attention on objects in space. In turn, the prefrontal and posterior parietal regions of the cortex seem to relay the decision regarding which image is to be enhanced to the visual system, which then brings the image into consciousness. Indeed, people with damage to the prefrontal cortex have difficulty switching from one image to the other in situations of binocular rivalry. Crick and Koch might argue that the frontal and parietal areas of the cortex are recruited by the claustrum, which switches attention from one eye to the other and unifies the image presented to conscious awareness by each eye.

As these arguments make clear, consciousness remains an enormous problem. But through the efforts of Edelman on the one hand, and Crick and Koch on the other, we now have two specific and testable theories worthy of exploration.

AS SOMEONE INTERESTED IN PSYCHOANALYSIS, I WANTED TO TAKE the Crick-Koch paradigm of comparing unconscious and conscious perception of the same stimulus to the next step: determining how visual perception becomes endowed with emotion. Unlike simple visual perception, emotionally charged visual perception is likely to differ between individuals. Therefore, a further question is, How and where are unconscious emotional perceptions processed?

Amit Etkin, a bold and creative M.D.-Ph.D. student, and I undertook a study in collaboration with Joy Hirsch, a brain imager at Columbia, in which we induced conscious and unconscious perceptions of emotional stimuli. Our approach paralleled in the emotional sphere that of Crick and Koch in the cognitive sphere. We explored how normal people respond consciously and unconsciously to pictures of people with a clearly neutral expression or an expression of fear on their faces. The pictures were provided by Peter Ekman at the University of California, San Francisco.

Ekman, who has cataloged more than 100,000 human expressions, was able to show, as did Charles Darwin before him, that irrespective of sex or culture, conscious perceptions of seven facial expressions—happiness, fear, disgust, contempt, anger, surprise, and sadness—have virtually the same meaning to everyone (figure 28-1). We therefore argued that fearful faces should elicit a similar response from the healthy young medical and graduate student volunteers in our study, regardless of whether they perceived the stimulus consciously or unconsciously. We produced a conscious perception of fear by presenting the fearful faces for a long period, so people had time to reflect on them. We produced unconscious perception of fear by presenting the same faces so rapidly that the volunteers were unable to report which type of expression they had seen. Indeed, they were not even sure they had seen a face!

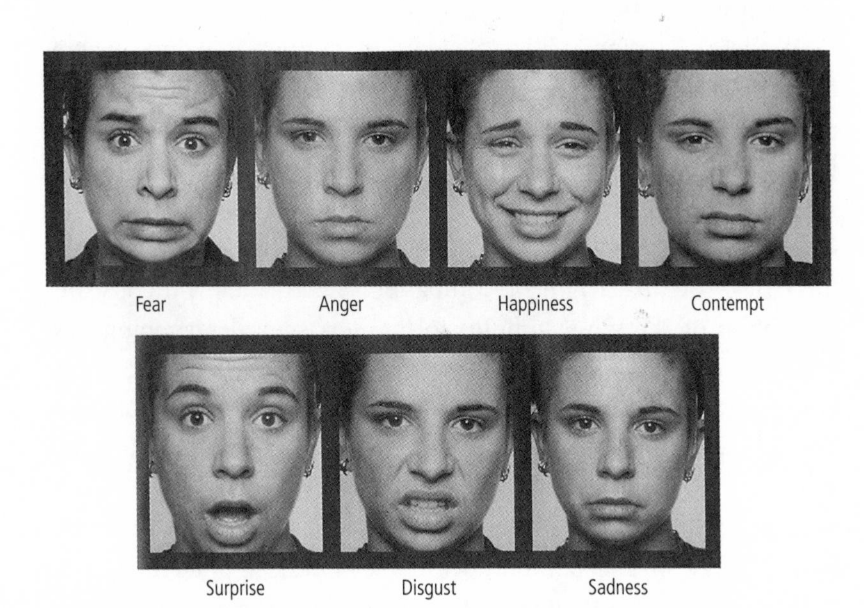

28-1 Ekman's seven universal facial expressions. (Courtesy of Paul Ekman.)

Since even normal people differ in their sensitivity to a threat, we gave all of the volunteers a questionnaire designed to measure background anxiety. In contrast to the momentary anxiety most people feel in a new situation, background anxiety reflects an enduring baseline trait.

Not surprisingly, when we showed the volunteers pictures of faces with fearful expressions, we found prominent activity in the amygdala, the structure deep in the brain that mediates fear. What was surprising was that conscious and unconscious stimuli affected different regions of the amygdala, and they did so to differing degrees in different people, depending on their baseline anxiety.

Unconscious perception of fearful faces activated the basolateral nucleus. In people, as in mice, this area of the amygdala receives most of the incoming sensory information and is the primary means by which the amygdala communicates with the cortex. Activation of the basolateral nucleus by unconscious perception of fearful faces occurred in direct proportion to a person's background anxiety: the higher the measure of background anxiety, the greater the person's response. People with low background anxiety had no response at all. Conscious perception of fearful faces, in contrast, activated the dorsal region of the amygdala, which contains the central nucleus, and it did so regardless of a person's background anxiety. The central nucleus of the amygdala sends information to regions of the brain that are part of the autonomic nervous system—concerned with arousal and defensive responses. In sum, unconsciously perceived threats disproportionately affect people with high background anxiety, whereas consciously perceived threats activate the fight-or-flight response in all volunteers.

We also found that unconscious and conscious perception of fearful faces activates different neural networks outside the amygdala. Here again, the networks activated by unconsciously perceived threats were recruited only by the anxious volunteers. Surprisingly, even unconscious perception recruits participation of regions within the cerebral cortex.

Thus viewing frightening stimuli activates two different brain systems, one that involves conscious, presumably top-down attention and one that involves unconscious, bottom-up attention, or vigilance, much as a signal of salience does in explicit and implicit memory in *Aplysia* and in the mouse.

These are fascinating results. First, they show that in the realm of emotion, as in the realm of perception, a stimulus can be perceived both unconsciously and consciously. They also support Crick and Koch's idea that in perception, distinct areas of the brain are correlated with conscious and unconscious awareness of a stimulus. Second, these studies confirm biologically the importance of the psychoanalytic idea of unconscious emotion. They suggest that the effects of anxiety are exerted most dramatically in the brain when the stimulus is left to the imagination rather than when it is perceived consciously. Once the image of a frightened face is confronted consciously, even anxious people can accurately appraise whether it truly poses a threat.

A century after Freud suggested that psychopathology arises from conflict occurring on an unconscious level and that it can be regulated if the source of the conflict is confronted consciously, our imaging studies suggest ways in which such conflicting processes may be mediated in the brain. Moreover, the discovery of a correlation between volunteers' background anxiety and their unconscious neural processes validates biologically the Freudian idea that unconscious mental processes are part of the brain's system of information processing. While Freud's ideas have existed for more than one hundred years, no previous brain-imaging study had tried to account for how differences in people's behavior and interpretations of the world arise from differences in how they unconsciously process emotion. The finding that unconscious perception of fear lights up the basolateral nucleus of the amygdala in direct proportion to a person's baseline anxiety provides a biological marker for diagnosing an anxiety state and for evaluating the efficacy of various drugs and forms of psychotherapy.

In discerning a correlation between the activity of a neural circuit and the unconscious and conscious perception of a threat, we are beginning to delineate the neural correlate of an emotion—fear. That description might well lead us to a scientific explanation of consciously perceived fear. It might give us an approximation of how neural events give rise to a mental event that enters our awareness. Thus, a half century after I left psychoanalysis for the biology of mind, the new biology of mind is getting ready to tackle some of the issues central to psychoanalysis and consciousness.

One such issue is the nature of free will. Given Freud's discovery of psychic determinism—the fact that much of our cognitive and affective life is unconscious—what is left for personal choice, for freedom of action?

A critical set of experiments on this question was carried out in 1983 by Benjamin Libet at the University of California, San Francisco. Libet used as his starting point a discovery made by the German neuroscientist Hans Kornhuber. In his study, Kornhuber asked volunteers to move their right index finger. He then measured this voluntary movement with a strain gauge while at the same time recording the electrical activity of the brain by means of an electrode on the skull. After hundreds of trials, Kornhuber found that, invariably, each movement was preceded by a little blip in the electrical record from the brain, a spark of free will! He called this potential in the brain the "readiness potential" and found that it occurred 1 second before the voluntary movement.

Libet followed up on Kornhuber's finding with an experiment in which he asked volunteers to lift a finger whenever they felt the urge to do so. He placed an electrode on a volunteer's skull and confirmed a readiness potential about 1 second before the person lifted his or her finger. He then compared the time it took for the person to will the movement with the time of the readiness potential. Amazingly, Libet found that the readiness potential appeared not after, but 200 milliseconds before a person felt the urge to move his or her finger! Thus by merely observing the electrical activity of the brain, Libet could predict what a person would do before the person was actually aware of having decided to do it.

This finding has caused philosophers of mind to ask: If the choice is determined in the brain before we decide to act, where is free will? Is our sense of willing our movements only an illusion, a rationalization after the fact for what has happened? Or is the choice made freely, but not consciously? If so, choice in action, as in perception, may reflect the importance of unconscious inference. Libet proposes that the process of initiating a voluntary action occurs in an unconscious part of the brain, but that just before the action is initiated, consciousness is recruited to approve or veto the action. In the 200 milliseconds before a finger is lifted, consciousness determines whether it moves or not.

Whatever the reasons for the delay between decision and awareness, Libet's findings also raise the moral question: How can one be held responsible for decisions that are made without conscious awareness? The psychologists Richard Gregory and Vilayanur Ramachandran have drawn strict limits on that argument. They point out that "our conscious mind may not have free will, but it does have free won't." Michael Gazzaniga, one of the pioneers in the development of cognitive neuroscience and a member of the American Council of Bioethics, has added, "Brains are automatic, but people are free." One cannot infer the sum total of neural activity simply by looking at a few neural circuits in the brain.

SIX

The true Vienna lover lives on borrowed memories. With a bitter-sweet pang of nostalgia, he remembers things he never knew. . . . the Vienna that never was is the grandest city ever.

—Orson Welles, "Vienna 1968"

29

REDISCOVERING VIENNA VIA STOCKHOLM

On the day of Yom Kippur, October 9, 2000, I was awakened by the ringing of the telephone at 5:15 in the morning. The phone is on Denise's side of the bed, so she answered it and gave me a shove in the ribs.

"Eric, this call is from Stockholm. It must be for you. It's not for me!"

On the telephone was Hans Jörnvall, the secretary general of the Nobel Foundation. I listened quietly as he told me that I had won the Nobel Prize in Physiology or Medicine for signal transduction in the nervous system and that I would share it with Arvid Carlsson and my longtime friend Paul Greengard. The conversation felt unreal.

The Stockholm deliberations must be among the best-kept secrets in the world. There are practically never any leaks. As a result, it is almost impossible to know who will get the prize in any given October. Yet very few people who receive the Nobel Prize are absolutely astonished by the very idea of winning it. Most people who are eligible sense that they are being considered because their colleagues speak of the possibility. Moreover, the Karolinska Institute runs periodic symposia designed to bring the world's leading biologists to Stockholm, and I had just attended such a symposium a few weeks earlier. Nonetheless I had not expected this call. Many eminently prizeworthy candidates who are talked about as being eligible are never selected as laureates, and I thought it unlikely that I would be recognized.

In my state of disbelief, I didn't know what to say except to acknowledge my gratitude. Jörnvall told me not to make any telephone calls until 6:00 A.M., when the press would be informed. After that, he said, I could make any calls I wanted.

Denise had begun to worry. I had been lying quietly with the phone to my ear for what seemed an unending period of time. She did not associate this uncommunicative state with me and feared that I was overwhelmed emotionally by the news. When I got off the phone and told her what I had just learned, she was doubly thrilled, pleased to learn that I had won the Nobel Prize and relieved that I was still alive and well. She then said, "Look, it's so early. Why don't you go back to sleep?"

"Are you kidding?" I replied. "How can I possibly sleep?"

I waited out the half hour patiently and then proceeded to call everybody. I telephoned our children, Paul and Minouche, waking Minouche on the West Coast in the middle of the night. I then called Paul Greengard to congratulate him on our shared good fortune. I called my friends at Columbia, not only to share the news but also to prepare them for the press conference that was likely to be called for the afternoon. It became clear to me that even though this call came on Yom Kippur, the Day of Atonement and the most solemn Jewish holiday of the year, the press conference would go ahead.

Before I had gotten through my initial calls, the doorbell rang, and to my astonishment and delight, our neighbors in Riverdale, Tom Jessell, his wife, Jane Dodd, and their three daughters appeared on the doorstep with a bottle of wine in hand. Although it was too early to uncork the wine, they were most welcome visitors, a bit of reality in the dizzying wonderland of the Nobel. Denise proposed that we all sit down and have breakfast, which we did, despite the telephone's ringing off the hook.

Everyone was calling—radio, television, newspapers, our friends. I found the telephone calls from Vienna most interesting because they were calling to tell me how pleased Austria was that there was yet another Austrian Nobel Prize. I had to remind them that this was an American Nobel Prize. I then received a telephone call from the Columbia press office asking me to participate in a press conference at the Alumni Auditorium at 1:30 P.M.

On the way to the press conference, I stopped off briefly at our synagogue—both to atone and to celebrate—and then went on to the lab, where I was received with jubilation. I was simply overwhelmed! I told everyone how grateful I was for their efforts and that I felt the Nobel was very much a shared prize.

The press conference was attended by many members of the faculty, who graciously gave me a standing ovation. Also present were the academic leaders of the university. David Hirsh, acting dean of the medical school, introduced me briefly to the press, and I made some comments expressing my gratitude to the university and to my family. I then explained very briefly the nature of my work. Over the next several days, more than a thousand e-mail messages, letters, and telephone calls poured in. I heard from people I had not seen in decades; girls I had dated in high school suddenly found me interesting again. Amid all the hustle and bustle, a commitment I had made earlier proved unexpectedly fortunate. I had agreed months before to give a lecture on October 17 in Italy in honor of Massimiliano Aloisi, a revered professor at the University of Padua. This seemed a wonderful opportunity for Denise and me to get away from the hurly-burly. Padua proved delightful, giving us the opportunity to visit the Scrovegni Chapel, which houses magnificent Giotto frescoes. I had also undertaken to combine the visit to Padua with a plenary lecture at the University of Turin, where I was to receive an honorary degree.

In Padua and then in Venice, which we visited briefly, we looked for gowns that Denise might wear to the Nobel ceremonies in Stockholm. We finally struck gold in Turin, where Denise was referred to the dressmaker Adrianne Pastrone. Denise loved her designs and bought several gowns. I feel enormous gratitude toward Denise, in addition to my deep love for her, because of her support of me and my work in our life together. She has had a wonderful career in epidemiology at Columbia, but there is no question in my mind that she compromised her work and, even more, her leisure by taking up the slack created by my obsession with science.

On November 29, just before we were to leave for Stockholm, the Swedish ambassador to the United States invited the seven American laureates to Washington so that the recipients and their spouses could get to know one another. The visit featured a reception in the Oval Office hosted by President Clinton, who filled the room with his presence, discussed macroeconomics with the laureates in that field, and graciously posed Denise and me, and each of the other laureates and his partner, for pictures with him. Clinton was about to leave the presidency and spoke fondly of his job, emphasizing that he had become so good at positioning people for photo opportunities that he and the White House photographer might go into business together. The visit in the Oval Office was followed by dinner at the Swedish embassy, where Denise and I chatted with the laureates in other fields.

THE NOBEL PRIZE OWES ITS EXISTENCE TO THE REMARKABLE vision of one person, Alfred Nobel. Born in Stockholm in 1833, he left Sweden when he was nine and returned only for very brief stays. He spoke Swedish, German, English, French, Russian, and Italian fluently but had no real homeland. A brilliant inventor, Nobel developed more than three hundred patents and maintained throughout his life a deep interest in science.

The invention that made his fortune was dynamite. In 1866 he discovered that liquid nitroglycerin, when absorbed by a kind of silicified earth called *kieselguhr,* was no longer unstable. In this form it could be shaped into sticks and used safely, since it now needed a detonator to explode. Dynamite sticks paved the way for the mining of minerals and for the unprecedented expansion of public works in the nineteenth century. Railways, canals (including the Suez Canal), harbors, roads, and bridges were constructed with relative ease, in large part because of dynamite's power to move huge amounts of earth.

Nobel never married, and when he died on December 10, 1896, he left an estate of 31 million Swedish kronen, then equivalent to $9 million, a huge sum in its day. His will states: "The whole of my remaining realizable estate shall . . . constitute a fund, the interest on which shall be annually distributed in the form of prizes to those who, during the preceding year, shall have conferred the greatest benefit on mankind." Nobel then went on to list the five fields in which these prizes would be given: physics, chemistry, physiology or medicine, literature, and, for "the person who shall have done the most or the best work for fraternity between nations," the Nobel Peace Prize.

Despite its extraordinary clarity and vision, the will raised problems that were not settled for several years. To begin with, several parties were interested in acquiring the inheritance: Nobel's relatives, some Swedish academies, the Swedish government, and, most important, the French government. The French claimed that France was Nobel's legal residence. He rarely visited his native Sweden after he was nine years old, he never paid taxes there (paying taxes in a country usually serves as proof of citizenship), and he lived in France for almost thirty years. However, Nobel had never applied for French citizenship.

As a first step, Ragnar Sohlman, Nobel's administrative assistant and executor (who later proved to be an effective, farsighted executive director of the Nobel Foundation), joined forces with the Swedish government to prove that Nobel was Swedish. They argued that since Nobel wrote his will in Swedish, appointed a Swede as executor, and designated various Swedish Academies to implement its provisions, he must legally be considered Swedish. In 1897 the Swedish government formally directed the country's attorney general to maintain the will under Swedish jurisdiction.

This solved only one part of the problem: there remained the hesitations of the Swedish academies. They warned that they would have to find knowledgeable nominators, translators, consultants, and evaluators in order to award the prizes, yet Nobel's will did not provide for these expenses. In the end, Sohlman encouraged the passage of a law giving each committee a part of the value of the prize for honoraria and expenses of its members and consultants. Compensation for members amounted to about one-third of a professor's annual salary.

The first Nobel Prizes were awarded on December 10, 1901, the fifth anniversary of Nobel's death. Sohlman had invested the holdings of Nobel's estate wisely, and the endowment had already grown to 3.9 billion Swedish kronen, or slightly more than $1 billion. The award for each prize was 9 million Swedish kronen. The prizes in science and literature were awarded in a ceremony in Stockholm that has been repeated on that date every year since, except during World Wars I and II.

WHEN DENISE AND I ARRIVED AT THE CHECK-IN DESK OF Scandinavian Airlines on December 2, we were given the red-carpet treatment. It continued when we arrived in Stockholm. We were met by Professor Jörnvall and assigned a driver and a limousine for our use during our stay. Irene Katzman, desk officer at the Swedish foreign service, served as an administrative coordinator for us. At the Grand Hotel, Stockholm's premier hotel, we were given a beautiful suite overlooking the harbor. That first evening we had dinner with Irene, her husband, and their children. The next day, at our request, Irene arranged a private tour of the Jewish Museum, which described how the Jewish community of Sweden had helped save a significant portion of the Jewish community of Denmark during the Hitler period.

There followed a series of activities, each with its own power and charm. On December 7, Arvid Carlsson, Paul Greengard, and I gave a press conference. That evening, we had dinner with the Nobel Committee for Physiology or Medicine, the people who had selected us. The committee members told us that they probably knew us as well as our spouses did because they had studied us in detail for more than a decade.

Denise and I were joined in Stockholm by our children—Minouche and her husband, Rick Sheinfield, and Paul and his wife, Emily—and by our older grandchildren, Paul and Emily's daughters, Allison, then age eight, and Libby, age five. (Minouche was pregnant with Maya when she joined us in Stockholm; her son, Izzy, who was then two, stayed with Rick's parents.)

Denise and I had also invited our senior colleagues from Columbia—Jimmy and Cathy Schwartz, Steve Siegelbaum and Amy Bedik, Richard Axel, Tom Jessell and Jane Dodd, and John Koester and Kathy Hilten. All of them were longtime friends to whom I owed a great deal. Bridging the two groups were Ruth and Gerry Fischbach. Ruth is Denise's second cousin and director of the Center for Bioethics at Columbia. Gerry is an outstanding neural scientist and a leader of the

29-1 My family in Stockholm. Standing, from left: Alex and Annie Bystryn (my nephew and niece), Jean-Claude Bystryn (their father, Denise's brother), Ruth and Gerry Fischbach (Ruth is Denise's cousin), Marcia Bystryn (Jean-Claude's wife). Sitting, from left: Libby, Emily, and Paul Kandel, Denise, me, Minouche and her husband, Rick, Allison. (From Eric Kandel's personal collection.)

scientific community in the United States. Shortly before our trip to Stockholm, he was offered the post of dean of the College of Physicians and Surgeons and vice president for health sciences at Columbia University. By the time he arrived, he had accepted the offer and was my new boss.

This was too good an occasion to pass up. On the one free evening of our stay in Stockholm, Denise and I threw a dinner party in a beautiful private dining room of the Grand Hotel for all the guests and relatives we had invited to Stockholm. We wanted to thank everyone for coming and for celebrating this great occasion with us. Beyond that, we wanted to celebrate Gerry's becoming dean and vice president at Columbia. It was a joyous evening (figure 29-1).

ON THE AFTERNOON OF DECEMBER 8, ARVID, PAUL, AND I GAVE our Nobel lectures in the Karolinska Institute before the faculty and students of the institute and our guests and friends. I spoke about my work, and as I introduced *Aplysia*, I could not help but comment that this was not only a very beautiful animal but a very accomplished one. I then flashed on the screen a wonderful image that Jack Byrne, one of my first graduate students, had sent me showing a proud *Aplysia* with a Nobel Prize medal draped around its neck (figure 29-2). The audience broke out in laughter.

Each year on the Saturday closest to the award dinner, the Jewish community of Stockholm, about seven thousand strong, invites the Jewish Nobel laureates to the Great Synagogue of Stockholm to personally receive the rabbi's blessing and a token gift. On December 9, I took a sizable entourage of colleagues and family with me to the synagogue. At the service I was asked to make a brief comment and was given a beautiful, small glass replica of the synagogue; Denise was given a red rose by a woman in the congregation who had also been in hiding in France during the war.

The next day, December 10, we received the Nobel Prize from King Carl XVI Gustaf. The ceremony at the Stockholm Concert Hall was the most remarkable and memorable event of all. Every detail is honed to perfection from a century of experience. To commemorate Alfred Nobel's death, the Concert Hall was decorated with flowers

29-2 ***Aplysia*** **with a Nobel Prize.** (Courtesy of Jack Byrne.)

flown in from San Remo, Italy, where Nobel spent the last years of his life. Everyone was dressed formally, the men in white tie and tails, and a marvelously festive mood was in the air. The Stockholm Philharmonic, seated on a balcony behind the stage, played at various times during the ceremony.

The ceremony began at 4:00 P.M. Once the laureates and the Nobel assembly were onstage, the king appeared, together with Queen Silvia, their three children, and the king's aunt, Princess Lilian. With the royal family in place, the standing audience of two thousand dignitaries joined in singing the royal anthem. Presiding over it all was a large painting of Alfred Nobel.

The investiture began with comments in Swedish by Bengt Samuelsson, chairman of the board of the Nobel Foundation. They were followed by representatives of the five awards committees, who described the discoveries and achievements being recognized. Our award in Physiology or Medicine was introduced by Urban Ungerstadt, a senior neurophysiologist and member of the Nobel Committee at the Karolinska Institute. After outlining in Swedish our respective contributions, he turned and addressed us in English:

> Dear Arvid Carlsson, Paul Greengard, and Eric Kandel. Your discoveries concerning "signal transduction in the nervous system" have truly changed our understanding of brain function.
>
> From Arvid Carlsson's research we now know that Parkinson's disease is due to failure in synaptic release of dopamine. We know that we can substitute the lost function by a simple molecule, L-DOPA, which replenishes the emptied stores of dopamine and, in this way, give millions of humans a better life.
>
> We know from Paul Greengard's work how this is brought about. How second messengers activate protein kinases leading to changes in cellular reactions. We begin to see how phosphorylation plays a central part in the very orchestration of the different transmitter inputs to the nerve cells.
>
> Finally, Eric Kandel's work has shown us how these transmitters, through second transmitters and protein phosphorylation, create short- and long-term memory, forming the very basis for our ability to exist and interact meaningfully in our world.

> On behalf of the Nobel Assembly at Karolinska Institute, I wish to convey our warmest congratulations, and I ask you to step forward to receive the Nobel Prize from the hands of His Majesty the King.

One by one, Arvid, Paul, and I rose and came forward. Each of us shook hands with the king and received from him a decorated certificate bound in a leather box containing the gold medal. On one side of the medal is an image of Alfred Nobel (figure 29-3), while on the other are two women, one representing the genius of medicine and the other a sick girl. The genius of medicine, holding an open book in her lap, is collecting water that is pouring out of a rock in order to quench the thirst of the sick girl. To blaring trumpets, I bowed three times, as prescribed: once to the king, once to the Nobel Assembly, and finally to Denise, Paul, Emily, Minouche, Rick, and the rest of the distinguished audience. When I sat down, the Stockholm Philharmonic played the third movement of Mozart's unsurpassed clarinet con-

29-3 My granddaughters Libby and Allison on the stage with me after the adjournment of the Nobel Prize ceremony. We are holding the Nobel Medal. (From Eric Kandel's personal collection.)

certo. On this occasion the melodic solos, written for a Viennese temperament like mine, sounded even lovelier than usual.

From the awards ceremony we went directly to a banquet in the City Hall. Completed in 1923, this magnificent building was designed by the great Swedish architect Ragnar Ostberg on the lines of a northern Italian piazza. A table set for eighty in the center of the large hall accommodated the laureates, the royal family, the prime minister, and various other dignitaries. Guests of the laureates, members of the award-granting institutions, representatives of the major universities, and high-level representatives from the government and from industry were seated at twenty-six tables surrounding the center table. A few students from each Swedish university and some of the colleges were seated around the walls.

After dinner, each laureate or a representative from each group of laureates went to the podium to say a few words. I spoke for our group:

> Engraved above the entrance to the Temple of Apollo at Delphi was the maxim, Know thyself. Since Socrates and Plato first speculated on the nature of the human mind, serious thinkers through the ages—from Aristotle to Descartes, from Aeschylus to Strindberg and Ingmar Bergman—have thought it wise to understand oneself and one's behavior. . . .
>
> Arvid Carlsson, Paul Greengard, and I, whom you honor here tonight, and our generation of scientists, have attempted to translate abstract philosophical questions about mind into the empirical language of biology. The key principle that guides our work is that the mind is a set of operations carried out by the brain, an astonishingly complex computational device that constructs our perception of the external world, fixes our attention, and controls our actions.
>
> We three have taken the first steps in linking mind to molecules by determining how the biochemistry of signaling within and between nerve cells is related to mental processes and to mental disorders. We have found that the neural networks of the brain are not fixed, but that communication between nerve cells can be regulated by neurotransmitter molecules discovered here in Sweden by your great school of molecular pharmacology.

> In looking toward the future, our generation of scientists has come to believe that the biology of the mind will be as scientifically important to this century as the biology of the gene has been to the twentieth century. In a larger sense, the biological study of mind is more than a scientific inquiry of great promise; it is also an important humanistic endeavor. The biology of mind bridges the sciences—concerned with the natural world—and the humanities—concerned with the meaning of human experience. Insights that come from this new synthesis will not only improve our understanding of psychiatric and neurological disorders, but will also lead to a deeper understanding of ourselves.
>
> Indeed, even in our generation, we already have gained initial biological insights toward a deeper understanding of the self. We know that even though the words of the maxim are no longer encoded in stone at Delphi, they are encoded in our brains. For centuries the maxim has been preserved in human memory by those very molecular processes in the brain that you graciously recognize today, and that we are just beginning to understand.

The banquet was followed by dancing. Denise and I had taken lessons to brush up on our limited and rarely practiced waltzing skills, but sadly, and to Denise's unending disappointment, we didn't get much of a chance to dance. As soon as dinner was over, we were approached by our friends, and I so enjoyed chatting with them that I found it hard to break away.

On December 11 we were invited to dinner at the Royal Palace by the king and queen. The morning of December 13, Santa Lucia's Day and the first day of Sweden's month-long celebration of Christmas, Paul, Arvid, and I were awakened by young college students—mostly women—carrying candles and singing carols in our honor. We then left the capital to give a series of lectures at the University of Uppsala. We returned to a brash and highly entertaining Santa Lucia dinner organized by the medical students in Stockholm. The next day, we left for New York.

Four years later, on October 4, 2004, Denise and I were on a Lufthansa flight from Vienna to New York when the stewardess gave me a message saying that my colleague and friend Richard Axel and Linda Buck, his former postdoctoral student, had received the Nobel Prize in Physiology or Medicine for their groundbreaking studies on the sense of smell carried out at Columbia. In December 2004 we all went back to Stockholm to celebrate Richard and Linda. Life is a circle, indeed!

A FEW WEEKS AFTER I FIRST HEARD FROM STOCKHOLM THAT I had received the Nobel Prize, the president of Austria, Thomas Klestil, wrote to congratulate me. He expressed a desire to honor me as a Nobel laureate of Viennese origins. I took the opportunity to suggest that we organize a symposium entitled "Austria's Response to National Socialism: Implications for Scientific and Humanistic Scholarship." My purpose was to compare Austria's response to the Hitler period, which was one of denial of any wrongdoing, with Germany's response, which was to try to deal honestly with the past.

President Klestil agreed enthusiastically and sent me copies of several speeches he had delivered about the awkward situation of present-day Jews in Vienna. He then put me in touch with Elisabeth Gehrer, the minister of education, to help me organize the symposium. I told her that I hoped the symposium would serve three functions: first, to help acknowledge Austria's role in the Nazi effort to destroy the Jews during World War II; second, to try to come to grips with Austria's implicit denial of its role during the Nazi period; and third, to evaluate the significance for scholarship of the disappearance of the Jewish community of Vienna.

Austria's record on these first two issues is quite clear. For a decade before Austria joined with Germany, a significant fraction of the Austrian population belonged to the Nazi party. Following annexation, Austrians made up about 8 percent of the population of the greater German Reich, yet they accounted for more than 30 percent of the officials working to eliminate the Jews. Austrians commanded four Polish death camps and held other leadership positions in the Reich: in addition to Hitler, Ernst Kaltenbrunner, who was head of the Gestapo, and Adolf Eichmann, who was in charge of the extermination program, were Austrians. It is estimated that of the 6 million Jews who perished during the Holocaust, approximately half were killed by Austrian functionaries led by Eichmann.

Yet despite their active participation in the Holocaust, the Austrians claimed to be victims of Hitler's aggression—Otto von Hapsburg, the pretender to the Austrian throne, managed to convince the Allies that Austria was the first free nation to fall victim to Hitler's war. Both the United States and the Soviet Union were willing to accept this argument in 1943, before the war ended, because von Hapsburg thought it would stimulate Austria's public resistance to the Nazis as the war ground to a halt. In later years both allies maintained this myth to ensure that Austria would remain neutral in the Cold War. Because it was not held accountable for its actions between 1938 and 1945, Austria never underwent the soul-searching and cleansing that Germany did after the war.

Austria readily accepted the mantle of injured innocence, and this attitude characterized many of Austria's actions after the war, including its treatment of Jewish financial claims. The country's initial uncompromising stand against paying reparations to the Jews was based on the premise that Austria had itself been a victim of aggression. In this way, the survivors of one of Europe's oldest, largest, and most distinguished Jewish communities were essentially disenfranchised, both financially and morally, for a second time after the war.

The Allies initially validated this alleged innocence by exempting Austria from the payment of reparations. The Allied occupation forces pressured the Austrian parliament to enact a war criminals law in 1945, but it was not until 1963 that a prosecuting agency was established to put the measures into effect. In the end, few people were tried, and most of those were acquitted.

Austria's intellectual loss is equally clear and dramatic. Within days of Hitler's arrival, the intellectual life of Vienna was in shambles. About 50 percent of the university's medical faculty—one of the largest and most distinguished in Europe—was dismissed for being Jewish. Viennese medicine has never recovered from this "cleansing." Particularly distressing is how little was done after the collapse of the Third Reich to redress the injustices committed against Jewish academics or to rebuild the academic faculty. Few Jewish academics were invited back to Vienna, and even fewer were given restitution for the property or income they had lost. Of those who did return, some were not reinstated in their university positions, and almost all had

29-4 Eduard Pernkopf, the dean of the medical school of the University of Vienna, meets his faculty in April 1938, several weeks after Hitler's entry into Vienna. The dean and the organized faculty greet one another with "Heil Hitler!" (Courtesy of Österreichische Gesellschaft für Zeitgeschichte, Wien.)

great difficulty regaining their homes or even their citizenship, of which they had been stripped.

Equally disturbing was the fact that many of the non-Jewish members of the faculty of medicine who remained in Vienna during the war were Nazis, yet they retained their academic appointments afterward. Furthermore, some who were initially forced to leave the faculty because they had committed crimes against humanity were later reinstated.

To give but one example, Eduard Pernkopf, dean of the faculty of medicine from 1938 to 1943 and rector of the University of Vienna from 1943 to 1945, was a Nazi even before Hitler entered Austria. Pernkopf had been a "supporting" member of the National Socialist party since 1932 and an official member since 1933. Three weeks after Austria joined with Germany, he was appointed dean; he appeared in Nazi uniform before the medical faculty, from which he had dismissed all Jewish physicians, and gave the "Heil Hitler" salute (figure 29-4).

After the war, Pernkopf was imprisoned in Salzburg by Allied forces, but he was released a few years later, his status having been changed from that of war criminal to a lesser category. Perhaps most shocking, he was allowed to finish his book *Atlas of Anatomy*, a work thought to be based on dissection of the bodies of people who had been killed in Austrian concentration camps.

Pernkopf was only one of many Austrians who were "rehabilitated" in the postwar period. Their rehabilitation underscores the tendency of Austria to forget, suppress, and deny the events of the Nazi period. Austrian history books gloss over the country's involvement in crimes against humanity, and blatant Nazis continued to teach a new generation of Austrians after the war ended. Anton Pelinka, one of Austria's leading political historians, has called this phenomenon the "great Austrian taboo." It is precisely this moral vacuum that induced Simon Wiesenthal to establish his documentation center for Nazi war crimes in Austria, not in Germany.

IN SOME WAYS, THE TIMIDITY OF AUSTRIAN JEWS—MYSELF included—contributed to the taboo. On my first return visit to Vienna, in 1960, when a man came up to me and recognized me as Hermann Kandel's son, neither of us even mentioned the intervening years. Twenty years later, when Stephen Kuffler and I were inducted as honorary members of the Austrian Physiological Society, neither of us protested when the academic dignitary introducing us glossed over our escape from Vienna as if it had not happened.

But by 1989 I had reached the limit of my silence. That spring, Max Birnstiel, a wonderful Swiss molecular biologist, invited me to Vienna to participate in an inaugural symposium for the Institute of Molecular Pathology. It was clear that Max was going to energize science in Vienna. The symposium took place in April, almost fifty years to the day after I had left, and I was enthusiastic about the timing.

I began my lecture with some comments about why I had left Vienna and how ambivalent I felt about the city upon my return. I described the fondness I felt for Vienna, where I first learned about the music and art that I enjoy, as well as the enormous anger, disappointment, and pain caused by the humiliation I suffered there. I added how fortunate I was to have been able to go to the United States.

After I finished my comments there was no applause, no recognition. No one said a word. Later, a little old lady walked up to me and said in typical Viennese fashion, "You know, not all Viennese were bad!"

THE SYMPOSIUM THAT I HAD PROPOSED TO PRESIDENT KLESTIL took place in June 2003. Fritz Stern, a colleague at Columbia and my good friend, helped me organize it, and he and many other outstanding historians with special knowledge of the areas covered by the symposium participated. The talks described the differences between Germany, Switzerland, and Austria in dealing with their past and the devastating consequences for Vienna's intellectual life of losing so many great scholars. That list includes Popper, Wittgenstein, and the key philosophers of the Vienna Circle; Freud, the world leader of psychoanalysis; and leaders of the great Vienna schools of medicine and mathematics. On the final day, three Viennese émigrés talked about the liberating influence of American academic life, and Walter Kohn, himself a Viennese émigré and a Nobel laureate in chemistry from the University of California, Santa Barbara, and I spoke about our experiences in Vienna.

The symposium also gave me the opportunity to establish contact with the Jewish community and to think about what made the Jewish experience there so special. I gave a lecture at the Jewish Museum and then invited several members of the audience for dinner at a nearby restaurant, where we talked about the past and the future.

The members of the Jewish community in Vienna with whom I had dinner reminded me of what had been lost. The history of Austrian culture and scholarship in the modern era largely paralleled the history of Austrian Jewry. Only in fifteenth-century Spain had the European Jewish community achieved a more productive period of creativity than it had in Vienna during the late Hapsburg period, from 1860 to 1916, and the decade thereafter. Writing in 1937 Hans Tietze stated, "Without the Jews, Vienna would not be what it is, and the Jews without Vienna would lose the brightest era of their existence during recent centuries."

In speaking of the importance of Jews for Viennese culture, Robert Wistrich wrote:

> Can one conceive of twentieth-century culture without the contributions of Freud, Wittgenstein, Mahler, Schönberg, Karl Kraus, Theodore Herzl? . . . This secularized Jewish intelligentsia changed the face of Vienna, and indeed of the modern world. They helped transform a city that which had not been in the forefront of European intellectual or artistic creativity (except in music) into an experimental laboratory for the creative triumphs and traumas of the modern world.

After the symposium, I met again some of the Viennese Jews with whom I had dined earlier and discussed with them what they thought the symposium had accomplished. They agreed that it had helped young academics in Vienna to recognize that Austria had collaborated enthusiastically with the German Nazis in the Holocaust. It had also called attention—through the newspapers, television, radio, and magazines—to the fact that a segment of the international community had begun focusing its attention on Austria's role in the Hitler era. This made me hopeful that gradually change might come.

But one incident points up Austria's continued difficulty in dealing with its heavy debt to and responsibility for the Jewish community. While we were in Vienna in June 2003, Walter Kohn and I learned that the Viennese Kultusgemeinde, the Jewish social service agency that is responsible for the synagogues, the Jewish schools and hospitals, and the Jewish cemetery in Vienna, was going bankrupt trying to protect those entities against continuing vandalism. European governments typically compensate Jewish agencies for such expenses, but the Austrian government's compensation was not adequate. As a result, the Kultusgemeinde had to empty its own coffers and spend its entire endowment. The government refused requests from Ariel Muzicant, the president of the agency, to increase its subsidy.

Back in the United States, Walter Kohn and I joined forces to see whether we could help ameliorate the situation. Walter had gotten to know Peter Launsky-Tieffenthal, the consul general of Austria to Los

Angeles, and Launsky-Tieffenthal arranged for a conference call that would include himself, Muzicant, Wolfgang Schüssel (the chancellor of Austria), Walter, and me.

We thought the conference call was all set, but at the last moment Schüssel canceled. He did so for two reasons. First, he was concerned that his participation might be taken as an indication that the Austrian government was not doing enough for the Jewish community, which he denied. Second, he was willing to speak to Walter Kohn but not to me, because I had been critical of Austria.

Fortunately, when Walter and I were in Vienna for the symposium, we had also met Michael Häupl, the mayor of the city of Vienna and governor of the state of Vienna. We were very impressed with Häupl, a former biologist, and greatly enjoyed our evening with him. He acknowledged that the Jewish agency was being shortchanged. After Schüssel refused to talk to us, Walter wrote Häupl, who swung into action below the federal level. To Walter's and my delight, he succeeded in persuading the governors of the Austrian states to help out financially. In June 2004 the states rescued the Kultusgemeinde from insolvency, at least for the time being.

In these negotiations, I felt that the Kultusgemeinde needed our support in principle—on moral grounds. As far as I knew, I had no personal involvement with the agency. A few weeks later I learned that I was wrong. In addition to principle, I had a personal obligation to support the Kultusgemeinde.

In July 2004 I received through the Holocaust Museum in Washington, D.C., my father's file from the Kultusgemeinde. In it were requests from my father for funds to pay first for my transportation and that of my brother to the United States and then to pay for the transportation of my parents. Simply stated: I owe my existence in the United States to the generosity of the Viennese Kultusgemeinde.

Despite Mayor Häupl's success, some Viennese Jews see no future for themselves or their children in Austria. The number of Jews in Vienna is small. At present, only about 9,000 Viennese have officially registered themselves as Jews with the Kultusgemeinde, and another 8,000 may be unregistered. This small number is a function of the tiny fraction of the original community who survived the war and the few who returned after the war or who immigrated to Vienna from Eastern Europe. It also speaks to the government's failure to reverse the emigration of Jews and to encourage, as Germany has done, the immigration of Eastern European Jews to Austria.

The situation in Vienna today reminds me of Hugo Bettauer's satirical novel *The City Without Jews: A Novel About the Day After Tomorrow*, written in 1922. Bettauer described the Vienna of tomorrow as a city in which the anti-Semitic government has expelled all of its Jewish citizens, including Jews who had converted to Christianity, because even they could not be trusted. Without the Jews, the intellectual and social life of Vienna deteriorated, as did its economy. A character commenting about the city, now without Jews, says:

> I always keep my eyes and ears open—in the morning when I do my shopping and at concerts, at the opera, and on the tramway. And, I also hear people recalling the past more and more wistfully and speaking of it as if it had been very beautiful. . . . "In the old days when the Jews were still here"—they say that in every imaginable tone of voice but never with hatred[;] you know, I think people actually are becoming lonesome for the Jews.

The city fathers in Bettauer's book had no choice but to plead with the Jews to return to Vienna. Sadly, that ending is as unrealistic today as it was eighty years ago.

I RETURNED TO VIENNA IN SEPTEMBER 2004 TO CELEBRATE THE publication of the symposium volume and to attend the fall meeting of the Orden pour le Mérite. The Orden, originally established in 1748 by Frederick the Great of Prussia, consists of leading scholars, scientists, and artists, half of whom are native Germans and half foreigners who speak German. In addition, Denise and I decided, at the urging of our children, to observe Yom Kippur in the main synagogue of Vienna.

When we arrived at the synagogue, it was surrounded by security guards who were worried about both Austrian and Arab anti-Semitic violence. Once we were allowed to enter, we discovered that the congregation had reserved a seat for each of us in the first row of the men's and women's sections, respectively. At one point in the service, the rabbi, Paul Chaim Eisenberg, wanted to honor me and asked me to come up on the stage and open the curtains of the Ark that contains the Torah scrolls. My eyes filled with tears; I froze and could not bring myself to do it.

The next day I joined the meeting of the Orden. We met together with an Austrian honorary society, the Ehrenzeichen für Wissenschaft und Kunst, and heard the vigorous and well-known eighty-year-old urban geographer Elisabeth Lichtenberger, a student of the social and economic structure of Vienna's Ringstrasse, present an anti-American lecture on the future of Europe. As we broke for lunch, Lichtenberger sought me out to obtain my thoughts about the differences between life in Austria and the United States. I told her that I was a poor person to ask: for me there was no comparison. I barely escaped with my life from Vienna in 1939, whereas I have had a privileged life in the United States.

Lichtenberger then leaned over to me and said, "Let me explain what happened in 1938 and 1939. There was massive unemployment in Vienna until 1938. I felt that in my family, people were poor and oppressed. The Jews controlled everything—the banks, the newspapers. Most physicians were Jewish, and they were simply squeezing every penny out of these impoverished people. It was terrible. That's why it all happened."

At first I thought she was joking, but as I realized she was not, I turned to her and literally screamed, *"Ich glaube nicht was Sie mir sagen!"* "I can't believe you are talking to me this way! You, an academic, are blindly mouthing anti-Semitic Nazi propaganda!"

Within a few minutes everyone around our table turned with astonishment as I continued to berate her. Finally, seeing that I had no effect on her, I turned my back on her and engaged the person on the other side of me in conversation.

My confrontation with Lichtenberger was the first of three revealing conversations I had with Austrians of different ages during that visit in September 2004. The second occurred when a woman of about fifty, a Vienna-born secretary for an Austrian colleague in the Orden, quantum physicist Anton Zeilinger, turned to me and said, "I am so glad I read your comments at the symposium last year. Until then I knew nothing about the Kristallnacht!" Finally, a young Austrian businessman in the hotel lobby recognized me and said, "It is so wonderful for you to come to Vienna again. It must be so difficult for you to do so!"

These opinions probably reflect accurately the spectrum of Austrian attitudes toward the Jews, a spectrum dependent largely on age. My hope is that the difference in the three generations' attitudes may signal a lessening of anti-Semitism in Austria. Even some of the Jews in Vienna see this.

Two other events were even more encouraging. The first was at the book conference, when Georg Winkler, dean of the faculty at the University of Vienna, introduced me. Winkler went out of his way to acknowledge the collaboration of the university with the Nazis and to apologize for it. "The University of Vienna has waited too long to do its own analysis and make transparent its involvement with National Socialism," he stated.

The second took place at a social event I attended with the Orden at the Hofburg, the royal palace formerly occupied by the Hapsburgs. While in Vienna, I had learned that President Klestil, who had invited me four years earlier to organize the symposium, had recently died. At the event I met the newly elected president of Austria, Heinz Fischer. He immediately recognized my name and invited Denise and me to join him and his wife for a private dinner at the Hotel Sacher. The president told us that his wife's father had been put in a concentration camp by the Nazis in 1938 and was released only because he had obtained a visa for Sweden. Both President Fischer and his wife had made a major effort to encourage Karl Popper and other former Jewish émigrés to return and settle in Vienna.

The new president is even more involved with Jewish life in Vienna than the former one. In addition, I found it uplifting to think that sixty-five years after being forced to leave Vienna, I would be invited by the president of Austria to join with him in a private and frank conversation about Jewish life in Vienna over wine, dinner, and Sacher torte at the Hotel Sacher.

ON OCTOBER 4, OUR LAST DAY IN VIENNA, DENISE AND I stopped at Severingasse 8 on our way to the airport. We did not try to enter the apartment house or visit the small set of rooms I had left sixty-five years before. We merely stood outside and watched the sun's rays bathe the peeling wooden door. I felt amazingly at peace: so glad to have survived, and to have emerged from that building and from the Holocaust relatively unscathed.

30

LEARNING FROM MEMORY: PROSPECTS

After fifty years of teaching and research, I continue to find that doing science at a university—in my case, Columbia University—is unendingly interesting. I derive great joy from thinking about how memory works, developing specific ideas about how it persists, shaping those ideas through discussions with students and colleagues, and then seeing how they are corrected as the experiments play out. I continue to explore the science in which I work almost like a child, with a naïve joy, curiosity, and amazement. I feel particularly privileged to be working in the biology of mind, an area that—unlike my first love, psychoanalysis—has grown magnificently in the last fifty years.

In reviewing those years, I am impressed with how little there was initially to suggest that biology would become the passion of my professional life. Had I not been exposed in Harry Grundfest's laboratory to the excitement of actually doing research, of carrying out experiments to discover something new, I would have ended up with a very different career and, I presume, a very different life. In the first two years of medical school I took the required basic science courses, but until I had actually done research, I saw my scientific education as a prerequisite for doing what I really cared about—practicing medicine, taking care of patients, understanding their illnesses, and preparing to become a psychoanalyst. I was astonished to discover that working in the laboratory—*doing* science in collaboration with interesting and creative people—is dramatically different from taking courses and reading about science.

Indeed, I find the process of doing science, of exploring biological mysteries on a day-to-day basis, deeply rewarding, not only intellectually but also emotionally and socially. Doing experiments gives me the thrill of discovering anew the wonders of the world. Moreover, science is done in an intense and endlessly engrossing social context. The life of a biological scientist in the United States is a life of discussion and debate—it is the Talmudic tradition writ large. But rather than annotate a religious text, we annotate texts written by evolutionary processes working over hundreds of millions of years. Few other human endeavors engender as great a feeling of camaraderie with colleagues young and old, students and mentors alike, as making an interesting discovery together.

The egalitarian social structure of American science encourages this camaraderie. Collaboration in a modern biology laboratory is dynamic, extending not only from the top down but also, importantly, from the bottom up. Life at an American university bridges gaps in both age and status in ways that I have always found inspiring. François Jacob, the French molecular geneticist whose work so influenced my thinking, told me that what impressed him most about the United States on his first visit was the fact that graduate students called Arthur Kornberg, a world-famous DNA biochemist, by his first name. That was no surprise to me. Grundfest and Purpura and Kuffler always treated me and all their students as equals. Yet this would not—could not—have taken place in the Austria, the Germany, the France, or perhaps even the England of 1955. In the United States, young people speak up and are listened to if they have interesting things to say. Therefore, I have learned not only from my mentors, but also from my daily interaction with an extraordinary group of graduate students and postdoctoral fellows.

In thinking about the students and postdoctoral fellows with whom I have collaborated in my laboratory, I am reminded of the painting workshop of the Renaissance artist Andrea del Verrocchio. In the period from 1470 to 1475, his workshop was filled with a succession of gifted young artists, including Leonardo da Vinci, who studied there and, while doing so, made major contributions to the canvases that Verrocchio painted. To this day, people point to Verrocchio's *Baptism of Christ*, which hangs in the Uffizi Gallery, in Florence, and say, "That beautiful kneeling angel on the left was painted in 1472 by Leonardo." Similarly, when I give talks and project giant drawings of *Aplysia* neurons and their synapses onto an auditorium screen, I tell my audience, "This new culture system was developed by Kelsey Martin, this CREB activator and repressor were found by Dusan Bartsch, and these wonderful prion-like molecules at the synapse were discovered by Kausik Si!"

AT ITS BEST, THE SCIENTIFIC COMMUNITY IS INFUSED WITH A marvelous sense of collegiality and common purpose, not only in the United States but throughout the world. As pleased as I am about what my colleagues and I have been able to contribute to the emerging picture of memory storage in the brain, I am even more proud to be part of the accomplishments of the international community of scientists that has given rise to a new science of mind.

Within the span of my career the biological community has advanced almost unerringly from understanding the molecular nature of the gene and the genetic code to reading the code of the entire human genome and unraveling the genetic basis of many human diseases. We now stand at the threshold of understanding many aspects of mental functioning, including mental disorders, and perhaps someday even the biological basis of consciousness. The total accomplishment—the synthesis that has occurred within the biological sciences in the last fifty years—is phenomenal. It has brought biology, once a descriptive science, to a level of rigor, mechanistic understanding, and scientific excitement comparable to that of physics and chemistry. At the time I entered medical school, most physicists and chemists regarded biology as a "soft science"; today, physicists and chemists are flocking into biological fields, along with computer scientists, mathematicians, and engineers.

Let me give an example of this synthesis in the biological sciences. Soon after I began to use cell biology to link neurons to brain function and behavior in *Aplysia*, Sydney Brenner and Seymour Benzer began to look for genetic approaches to link neurons to brain function and behavior in two other simple animals. Brenner studied the behavior of the tiny worm *C. elegans*, which has only 302 cells in its central nerve cord. Benzer studied the behavior of the fruit fly, *Drosophila*. Each experimental system has distinct advantages and drawbacks. *Aplysia* has large, easily accessible nerve cells, but it is not optimal for traditional genetics; *C. elegans* and *Drosophila* are highly suitable for genetic experiments, but their nerve cells are small and not well suited to studies of cell biology.

For twenty years these experimental systems developed within different traditions and along largely separate lines. The parallels inherent in them were not apparent. But the power of modern biology has drawn them progressively closer. In *Aplysia*, first with recombinant DNA techniques and now with a nearly complete map of the DNA in its genome, we have the power to transfer and manipulate genes in individual cells. In a complementary way, new advances in cell biology and the introduction of more sophisticated behavioral analyses make possible cellular approaches to the behavior of the fruit fly and the worm. As a result, the molecular conservation that has so powerfully characterized the biology of genes and proteins is now being seen in the biology of cells, neural circuits, behavior, and learning.

ALTHOUGH DEEPLY SATISFYING, A CAREER IN SCIENCE IS BY NO means easy. I have experienced many moments of intense pleasure along the way, and the day-to-day activity is wonderfully invigorating intellectually. But the fun of doing science is to explore domains of knowledge that are relatively unknown. Like anyone who ventures into the unknown, I have at times felt alone, uncertain, without a well-trodden path to follow. Every time I embarked on a new course, there were well-meaning people, both social friends and scientific colleagues, who advised against it. I had to learn early on to be comfortable with insecurity and to trust my own judgment on key issues.

My experience is hardly unique. Most scientists who have tried to pursue even slightly new directions in their research, with all the difficulty and frustration these paths entail, tell similar stories of cautionary advice urging them not to take risks. For most of us, however, cautions against going forward only kindle the spirit of adventure.

The most difficult career decision of my life was to leave the potential security of a practice in psychiatry for the uncertainty of research. Despite the fact that I was a well-trained psychiatrist and enjoyed working with patients, I decided in 1965, with Denise's encouragement to devote myself to full-time research. In an upbeat frame of mind, having put this decision behind us, Denise and I took a brief holiday. We accepted an invitation from my good friend Henry Nunberg to spend a few days at his parents' summer home in Yorktown Heights, New York. Henry was then pursuing a residency in psychiatry at my hospital, the Massachusetts Mental Health Center. Denise and I knew his parents moderately well.

Henry's father, Herman Nunberg, was an outstanding psychoanalyst and an influential teacher whose textbook I much admired for its clarity. He had a broad, albeit dogmatic interest in many aspects of psychiatry. At our first dinner together, I enthusiastically outlined my new career plans of learning in *Aplysia*. Herman Nunberg looked at me in amazement and muttered, "It sounds to me as if your psychoanalysis was not fully successful; you seem never really to have quite resolved your transference."

I found that comment both humorous and irrelevant—and typical of many American psychoanalysts of the 1960s, who simply could not understand that an interest in brain research need not imply a rejection of psychoanalysis. If Herman Nunberg were alive today, it is almost inconceivable that he would pass the same judgment on a psychoanalysis-oriented psychiatrist who moved into brain science.

This theme recurred periodically throughout the first twenty years of my career. In 1986, when Morton Reiser retired as chairman of the department of psychiatry at Yale University, he invited several colleagues, including me, to give a talk at a symposium held in his honor. One of the invitees was Reiser's close associate Marshall Edelson, a well-known professor of psychiatry and director of education and medical studies for the department of psychiatry at Yale. In his lecture, Edelson argued that efforts to connect psychoanalytic theory to a neurobiological foundation, or to try to develop ideas about how different mental processes are mediated by different systems in the brain, were an expression of a deep logical confusion. Mind and body must be dealt with separately, he continued. We cannot seek causal connections between them. Scientists will eventually conclude, he argued, that the distinction between mind and body is not a temporary methodological stumbling block stemming from the inadequacy of our current ways of thought, but rather an absolute, logical, and conceptual barrier that no future developments can ever overcome.

When my turn came, I gave a paper on learning and memory in the snail. I pointed out that all mental processes, from the most prosaic to the most sublime, emanate from the brain. Moreover, all mental illness, regardless of symptoms, must be associated with distinctive alterations in the brain. Edelson rose during the discussion and said that, while he agreed that psychotic illnesses were disorders of brain function, the disorders that Freud described and that are seen in practice by psychoanalysts, such as obsessive-compulsive neurosis and anxiety states, could not be explained on the basis of brain function.

Edelson's views and Herman Nunberg's more personal judgment are idiosyncratic extremes, but they are representative of the thinking of a surprisingly large number of psychoanalysts not so many years ago. The insularity of such views, particularly the unwillingness to think about psychoanalysis in the broader context of neural science, hindered the growth of psychoanalysis during biology's recent golden age. In retrospect, it was probably not that Nunberg, or perhaps even Edelson, really thought that mind and brain were separate; it was rather that they did not know how to join them.

Since the 1980s the way in which mind and brain should be joined has become clearer. Consequently, psychiatry has taken on a new role. It has become a stimulus to modern biological thought as well as a beneficiary of it. In the last few years I have seen significant interest in the biology of mind within the psychoanalytic community. We now understand that every mental state is a brain state and every mental disorder is a disorder of brain function. Treatments work by altering the structure and function of the brain.

I encountered a different type of negative reaction when I turned from studying the hippocampus in the mammalian brain to studying simple forms of learning in the sea snail. There was a strong sense at the time among scientists working on the mammalian brain that it was radically different from the brain of lower vertebrates like fish and frogs and incomparably more complex than that of invertebrates. The fact that Hodgkin, Huxley, and Katz had provided a basis for studying the nervous system by studying the giant axon of the squid and the nerve-muscle synapse of the frog was seen by these mammalian chauvinists as an exception. Of course all nerve cells are similar, they conceded, but neural circuitry and behavior are very different in vertebrates and invertebrates. This schism persisted until molecular biology began to reveal the amazing conservation of genes and proteins throughout evolution.

Finally, there were continued disputes about whether any of the cellular or molecular mechanisms of learning and memory revealed by studies of simple animals were likely to be generalizable to more complex animals. In particular, there were arguments about whether sensitization and habituation are useful forms of memory to study. The ethologists, who study behavior in animals in their natural environments, emphasized the importance and generality of these two simple forms of memory. But the behaviorists emphasized primarily associative forms of learning, such as classical and operant conditioning, which are clearly more complex.

The disputes were eventually resolved in two ways. First, Benzer proved that cyclic AMP, which we had found to be important for short-term sensitization in *Aplysia*, was also required for a more complex form of learning in a more complex animal—namely, classical conditioning in *Drosophila*. Second, and even more dramatic, the regulatory protein CREB, first identified in *Aplysia*, was found to be an important component in the switch from short- to long-term memory in many forms of learning in various types of organisms, from snails to flies to mice to people. It also became clear that learning and memory, as well as synaptic and neuronal plasticity, represent a family of processes that share a common logic and some key components but vary in the details of their molecular mechanisms.

In most cases, by the time the dust had settled, these disputations proved beneficial for science: they sharpened the question and moved the science along. That was the important thing for me, the sense that we were moving in the right direction.

WHERE IS THE NEW SCIENCE OF MIND HEADING IN THE YEARS ahead? In the study of memory storage, we are now at the foothills of a great mountain range. We have some understanding of the cellular and molecular mechanisms of memory storage, but we need to move from these mechanisms to the systems properties of memory: What neural circuits are important for various types of memory? How are internal representations of a face, a scene, a melody, or an experience encoded in the brain?

To cross the threshold from where we are to where we want to be, major conceptual shifts must take place in how we study the brain. One such shift will be from studying elementary processes—single proteins, single genes, and single cells—to studying systems properties—mechanisms made up of many proteins, complex systems of nerve cells, the functioning of whole organisms, and the interaction of groups of organisms. Cellular and molecular approaches will certainly continue to yield important information in the future, but they cannot by themselves unravel the secrets of internal representations in neural circuits or the interactions of circuits—the key steps linking cellular and molecular neuroscience to cognitive neuroscience.

To develop an approach that can relate neural systems to complex cognitive functions, we will have to move to the level of the neural circuit, and we will have to determine how patterns of activity in different neural circuits are brought together into a coherent representation. To study how we perceive and recall complex experiences, we will need to determine how neural networks are organized and how attention and conscious awareness regulate and reconfigure the actions of the neurons in those networks. Biology will therefore have to focus more on nonhuman primates and on human beings as the model systems of choice. For this, we will need imaging techniques that can resolve the activity of individual neurons and of neuronal networks.

THESE CONSIDERATIONS HAVE CAUSED ME TO WONDER WHAT questions I would take on were I to start anew. I have two requirements of a scientific problem. The first is that it allow me to open a new area that will occupy me for a very long time. I like long-term commitments, not brief romances. Second, I enjoy tackling problems at the border of two or more disciplines. With those predilections in mind, I have found three questions that appeal to me.

First, I would like to understand how the unconscious processing of sensory information occurs and how conscious attention guides the mechanisms in the brain that stabilize memory. Only then can we address in biologically meaningful terms the theories about conscious and unconscious conflicts and memory first proposed by Freud in 1900. I am much taken by Crick and Koch's argument that selective attention is not only essential in its own right but also one of the royal roads to consciousness. I would like to develop a reductionist approach to the problem of attention by focusing on how place cells in the hippocampus create an enduring spatial map only when an organism is paying attention to its surroundings. What is the nature of this spotlight of attention? How does it enable the initial encoding of the memory throughout the neural circuitry that is involved in spatial memory? What other modulatory systems in the brain besides dopamine are recruited when an animal pays attention, and how are they recruited? Do they use a prion-like mechanism to stabilize place cells and long-term memory? It obviously would be good to extend such studies to people. How does attention allow me to embark on my mental time travel to our little apartment in Vienna?

A second, related issue that fascinates me is the relation of unconscious to conscious mental processing in people. The idea that we are unaware of much of our mental life, first developed by Hermann Helmholtz, is central to psychoanalysis. Freud has added the interesting idea that although we are not aware of most instances of mental processing, we can gain conscious access to many of them by paying attention. From this perspective, to which most neural scientists now subscribe, most of our mental life is unconscious; it becomes conscious only as words and images. Brain imaging could be used to connect psychoanalysis to brain anatomy and to neural function by determining how these unconscious processes are altered in disease states and how they might be reconfigured by psychotherapy. Given the importance of unconscious psychic processes, it is reassuring to think that biology can now teach us a good bit about them.

Finally, I like the idea of applying molecular biology to link my area, the molecular biology of mind, to Denise's area, sociology, and thus develop a realistic molecular sociobiology. Several researchers have made a fine start here. Cori Bargmann, a geneticist now at Rockefeller University, has studied two variants of *C. elegans* that differ in their feeding patterns. One variant is solitary and seeks its food alone. The other is social and forages in groups. The only difference between the two is one amino acid in an otherwise shared receptor protein. Transferring the receptor from a social worm to a solitary worm makes the solitary worm social.

Male courtship in *Drosophila* is an instinctive behavior that requires a critical protein, called fruitless. Fruitless is expressed in two slightly different forms: one in male flies, the other in female flies. Ebru Demir and Barry Dickson have made the remarkable discovery that when the male form of the protein is expressed in females, the females will mount and direct the courtship toward other females or toward males that have been engineered to produce a characteristic female odor, or pheromone. Dickson went on to find that the gene for fruitless is required during development for hardwiring the neural circuitry for courtship behavior and sexual preference.

Giacomo Rizzolatti, an Italian neuroscientist, has discovered that when a monkey carries out a specific action with its hand, such as putting a peanut in its mouth, certain neurons in the premotor cortex become active. Remarkably, the same neurons become active when a monkey watches another monkey (or even a person) put food in its mouth. Rizzolatti calls these "mirror neurons" and suggests that they provide the first insight into imitation, identification, empathy, and possibly the ability to mime vocalization—the mental processes intrinsic to human interaction. Vilayanur Ramachandran has found evidence of comparable neurons in the premotor cortex of people.

In looking at just these three research strands, one can see a whole new area of biology opening up, one that can give us a sense of what makes us social, communicating beings. An ambitious undertaking of this sort might not only discern the factors that enable members of a cohesive group to recognize one another but also teach us something about the factors that give rise to tribalism, which is so often associated with fear, hatred, and intolerance of outsiders.

I AM OFTEN ASKED, "WHAT DID YOU GAIN FROM YOUR PSYCHIATRIC training? Was it profitable for your career as a neural scientist?"

I am always surprised by such questions, for it is clear to me that my training in psychiatry and my interest in psychoanalysis lie at the very core of my scientific thinking. They have provided me with a perspective on behavior that has influenced almost every aspect of my work. Had I skipped residency training and gone to France earlier to also spend time in a molecular biology laboratory, I might have worked on the molecular biology of gene regulation in the brain at a slightly earlier point in my career. But the overarching ideas that have influenced my work and fueled my interest in conscious and unconscious memory derive from a perspective on mind that psychiatry and psychoanalysis opened up for me. Thus, my initial career as an aspiring psychoanalyst was hardly a detour; rather, it was the educational bedrock of all I have been able to accomplish since.

Often, newly graduating medical students who want to do research ask me whether they should take more basic coursework or go into research right away. I always urge them to get into a good laboratory. Obviously, coursework is important—I continued to take courses throughout my years at the National Institute of Mental Health, and I continue to this day to learn from seminars and meetings, from my colleagues, and from students. But it is much more meaningful and enjoyable to read the scientific literature about experiments you are involved in yourself than to read about science in the abstract.

Few things are more exciting and stimulating to the imagination than making a new finding, no matter how modest. A new finding allows one to see for the first time a part of nature—a small piece of the puzzle of how something functions. Once I have gotten into a problem, I find it extremely helpful to get a complete perspective, to learn what earlier scientists thought about it. I want to see not only what lines of thought proved to be productive, but also where and why certain other directions proved to be unproductive. So I was very much influenced by the psychology of Freud and by the early workers in the field of learning and memory—James, Thorndike, Pavlov, Skinner, and Ulric Neisser. Their thinking, and even their errors, provided a wonderfully rich cultural background for my later work.

I also think it is important to be bold, to tackle difficult problems, especially those that appear initially to be messy and unstructured. One should not be afraid to try new things, such as moving from one field to another or working at the boundaries of different disciplines, for it is at the borders that some of the most interesting problems reside. Working scientists are constantly learning new things and are not inhibited from moving into a new area because it is unfamiliar. They follow their interests instinctively and teach themselves the necessary science as they go along. Nothing is more stimulating for self-education than working in a new area. I had no useful preparation for science before I began with Grundfest and Purpura; I knew very little biochemistry when I joined forces with Jimmy Schwartz; and I knew nothing about molecular genetics when Richard Axel and I began to collaborate. In each case, trying new things proved anxiety-provoking but also exhilarating. It is better to lose some years trying something new and fundamental than to carry out routine experiments that everyone else is doing and that others could do as well as (if not better than) you.

Most of all, I think it is important to define a problem or a set of interrelated problems that has a long trajectory. I was fortunate at the very beginning to stumble onto an interesting problem in my work on the hippocampus and memory and then to switch decisively to the study of learning in a simple animal. Both have an intellectual sweep and scope that have carried me through many experimental failures and disappointments.

As a result, I have not experienced the malaise that some of my colleagues have described when, in midlife, they become bored with the science they are doing and turn to other things. I have engaged in a variety of non-research-based academic activities, such as writing textbooks, serving on academic committees at Columbia and nationally, and helping to found a biotechnology company. But I never did any of those things because I was bored with doing science. Richard Axel talks about the reinforcing value of data—the playing in one's head with new and interesting findings—as addictive. Unless Richard sees new data coming along, he becomes despondent, a feeling many of us share.

MY WORK IN SCIENCE HAS ALSO BEEN GREATLY ENRICHED BY THE passion that Denise and I share for music and art. When we moved to New York from Boston in December 1964, we bought a hundred-year-old house in the Riverdale section of the Bronx with wonderful views of the Hudson River and the Palisades. Over the decades we have filled that house with etchings, drawings, and paintings—decorative art from the beginning of the twentieth century—a form with strong roots in Vienna as well as in France. We collect French art nouveau furniture, vases, and lamps by Louis Majorelle, Emile Gallé, and the brothers Daum, an interest that originated with Denise. Her mother got us started in this direction by giving us for a wedding present a beautiful tea table that Gallé had made for his first exhibit.

Once in New York, we began to focus our interest in graphic art on Austrian and German expressionists—Klimt, Kokoschka, and Schiele among the Austrians, and Max Beckmann, Emil Nolde, and Ernst Kirschner among the Germans. This interest originated with me. For almost every major birthday—and sometimes in between when we cannot wait—Denise and I buy each other something we think the other would like. Most of the time we select the pieces together. As I write this, I am beginning to suspect that our collecting may well be an attempt to recapture part of our hopelessly lost youth.

IN RETROSPECT, IT SEEMS A VERY LONG WAY FROM VIENNA TO Stockholm. My timely departure from Vienna made for a remarkably fortunate life in the United States. The freedom I have experienced in America and in its academic institutions made the Nobel Prize possible for me, as it has for many others. Having been trained in history and the humanities, where one learns early on how depressing life can be, I am delighted to have ultimately switched to biology, where a delusional optimism still abounds.

Once in a while, reflecting back upon my years in science as I look out at the Hudson River darkening outside my window at the end of another long, exhausting, and often exhilarating day, I find myself filled with wonder to be doing what I am doing. I entered Harvard to become a historian and left to become a psychoanalyst, only to abandon both of those careers to follow my intuition that the road to a real understanding of mind must pass through the cellular pathways of the brain. And by following my instincts, my unconscious thought processes, and heeding what then seemed an impossibly distant call, I was led into a life I have enjoyed immensely.

GLOSSARY

acetylcholine: A chemical neurotransmitter released by motor neurons at synapses with muscle cells as well as at synapses between neurons.

action potential: A large transient electrical signal about 1/10 of a volt in amplitude and 1 to 2 milliseconds in duration that propagates along the axon to the neuron's presynaptic terminal without failure or flagging. At the presynaptic terminal, the action potential triggers the release of neurotransmitter onto target neurons.

agnosia: Loss of knowledge; the inability to consciously recognize objects through otherwise normally functioning sensory pathways, e.g., depth agnosia, movement agnosia, color agnosia, and prosopagnosia (impairment of face recognition).

AMPA receptor (A-amino-3-hydroxy-5-methylisoxazole-4-proprionic acid): One of two types of postsynaptic receptors for glutamate. It is active in response to normal synaptic transmission. (Compare **NMDA receptor**.)

amygdala: The region of the brain most specifically concerned with emotions, such as fear. It coordinates autonomic and endocrine responses in conjunction with emotional states, and it underlies emotional memory. The amygdala is itself a collection of several nuclei that lie deep in the temporal lobes of the cerebral hemispheres.

aphasia: A category of language disorders resulting from lesions to specific structures in the brain. Such disorders can cause an inability to understand language (Wernicke's aphasia), express language (Broca's aphasia), or both.

associative learning: A process in which the subject of an experiment (a person or an experimental animal) learns about the relationship between two stimuli or between a stimulus and a behavioral response.

autonomic nervous system: One of two major subdivisions of the peripheral nervous system. It controls the viscera, smooth muscles, and exocrine glands, and it mediates involuntary control of heart rate, blood pressure, and respiration.

axon: The long output fiber of the neuron that ends as presynaptic terminals and sends signals to other cells.

basal ganglia: A group of brain structures lying deep within both cerebral hemispheres that helps to regulate motor activity and cognition. The basal ganglia include the putamen, caudate, globus pallidus, and substantia nigra. Together, the putamen and caudate are called the **striatum**.

behaviorism: A theory, first developed at the beginning of the twentieth century, which holds that the only appropriate approach to the study of behavior is through direct observation of a subject's actions. "Mental function" is regarded as unobservable. Behaviorism contrasts with cognitive approaches to the study of behavior, which have dominated psychological research in recent decades.

benzodiazepines: A class of anti-anxiety drugs and muscle relaxants that includes diazepam (Valium) and lorazepam (Ativan). Benzodiazepines dampen synaptic transmission by binding to receptors for the inhibitory neurotransmitter **GABA** and enhancing GABA's effect on neurons.

biochemistry: A field in biology that attempts to understand life processes by studying the various chemical pathways and reactions that occur in living organisms, particularly the role played by proteins.

brain: The organ that mediates all mental functions and all behavior. Conventionally subdivided into several main parts: the brain stem, hypothalamus and thalamus, cerebellum, and two cerebral hemispheres.

brain stem: A collective term for three anatomical structures—the medulla, pons, and midbrain—all located at the bottom of the brain, above the spinal cord. The brain stem processes sensation from the skin and joints in the head, neck, and face, as well as specialized senses, such as hearing, taste, and balance. In addition, it mediates certain life-support functions, such as breathing, heart rate, and digestion. The sensory input and motor output of the brain stem are carried by the cranial nerves. (See **brain**.)

Broca's area: A region in the posterior part of the left frontal cortex that is critically involved in the expression of language. (Compare **Wernicke's area**.)

calcium (Ca^{2+}): The positively charged calcium ion is essential to the release of neurotransmitter. An influx of calcium ions, which is controlled by voltage-gated calcium channels in the nerve cell membrane, triggers this release of neurotransmitter.

cell biology: A field in biology that attempts to understand life processes, such as growth, development, adaptation, and reproduction, within the context of the cell, its subcellular structures, and its physiological processes.

cell body: The metabolic center of the neuron. It contains the nucleus with its chromosomes. It gives rise to two types of processes, the axon and the dendrites, both of which conduct electrical signals.

cell culture: The growth of cells taken from an animal and then placed into a petri dish under controlled conditions in the laboratory.

cell theory: The idea, proposed in the 1830s by anatomists Jakob Schleiden and Theodore Schwann, that all living tissues and organs in the bodies of all animals share a common structural and functional unit, the cell, and that all cells come from other cells.

central nervous system: One of the two divisions of the nervous system, the other being the **peripheral nervous system**. The central nervous system includes the brain and the spinal cord. Although anatomically distinct, the central and peripheral nervous systems are functionally interconnected.

cerebellum: One of the major parts of the brain involved in motor control. It modulates the force and range of motion and is involved in motor coordination and the learning of motor skills. (See **brain**.)

cerebral cortex: The outer covering of the cerebral hemispheres. It is divided into four lobes (frontal, parietal, temporal, occipital).

cerebral hemisphere: The cerebral hemispheres lie on either side of the brain and are connected by a large collection of axons called the corpus callosum which assures the unity of conscious experience. The cerebral hemispheres comprise the cerebral cortex and three deep-lying structures: the basal ganglia, hippocampus, and amygdala. (See **brain**.)

channel: A membrane-spanning protein that mediates the flow of ions into and out of cells. In nerve cells, some channels are responsible for the resting potential and others trigger the changes in membrane potential that generate the **action potential**, whereas still others change the excitability of the nerve cells. Ion channels may be opened or closed by changes in membrane potential (voltage-gated) or by the binding of chemical messengers (transmitter-gated), or they may passively conduct ions (nongated, or resting). (Compare **nongated channel; transmitter-gated channel; voltage-gated channel**.)

chemical synapse: A site at which one neuron releases a chemical signal (neurotransmitter) that binds to receptors on an abutting neuron, thus exciting or inhibiting the cell that receives the signal. (Compare **electrical synapse**.)

chemical theory of synaptic transmission: A theory that implicates certain chemicals called neurotransmitters as the mediators of synaptic transmission between two neurons.

chloride (Cl^-)**:** The negatively charged chlorine ion mediates the inhibition of neurons by **GABA**.

chromosome: A structure that contains the genetic material of an organism, usually in the form of a tightly coiled, double-stranded DNA molecule intertwined with various proteins. Chromosomes replicate themselves, thereby enabling cells to reproduce and pass on their genetic material to ensuing generations. (See **DNA**.)

classical conditioning: A form of implicit learning discovered by Ivan Pavlov in which a subject learns to associate a previously neutral conditioned stimulus with an unconditioned stimulus that typically elicits a reflex action. In experiments with dogs, for example, presentation of food (the unconditioned stimulus) normally elicits salivation. Pavlov found that if the sound of a bell (the previously neutral conditioned stimulus) was consistently paired with food, the dog would learn to associate the sound of the bell with food and thus would salivate whenever it heard the bell, regardless of whether food was present. Conversely, if the sound is paired with a shock to the leg that causes the animal to lift its leg, the dog will soon lift its leg in response to the tone alone.

cognitive map: A representation in the brain of a particular external physical space. An example is a **spatial map** evident in the hippocampus.

cognitive neuroscience: A combination of the concept and methods of cognitive psychology designed to study mental processes with those of neuroscience that study the brain. The methods involved in this combined discipline include neural science, cognitive psychology, behavioral neurology, and computer science.

conditioned response: The response elicited by the conditioned stimulus after classical conditioning. This response is similar to the response originally elicited by the unconditioned stimulus. (See **classical conditioning**.)

conditioned stimulus: A neutral stimulus that, before training, produces no overt response; it can be associated with an unconditioned stimulus through classical conditioning. (See **classical conditioning**.)

connection specificity: The principle formulated by Cajal according to which neurons form specific functional interconnections, based on three anatomical observations: (1) neurons, like other cells, are separated from one another by a cell membrane; (2) neurons do not connect indiscriminately to one another or form random networks; and (3) each neuron communicates only with specific postsynaptic cells, and only at specialized sites (synapses).

CPEB (Cytoplasmic Polyadenlyation Element-binding Protein): A regulator of translation at the synapse. CPEB is thought to contribute to the stabilization of long-term memory.

CREB (Cyclic AMP Response Element-binding Protein): A gene regulator protein which is activated by the cyclic AMP and protein kinase A pathway. CREB activates the genes responsible for long-term memory. (See **cyclic AMP**, **protein kinase A**.)

cyclic AMP (cyclic adenosine-3',5'-monophosphate): A molecule that acts as a second messenger in the cell, triggering changes in protein structure and function. Cyclic AMP activates an enzyme called the cyclic AMP-dependent protein kinase, which acts on and modifies the function of many proteins, including ion channels and the proteins that regulate the transcription of DNA into RNA. (See **phosphorylation**; **protein kinase A**; **second messenger**; **transcription**.)

cytoplasm: All of the material inside the cell except the nucleus. The machinery for making proteins is located here.

dendrite: The branched structures on most nerve cells where the neuron receives signals from other neurons.

dentate gyrus: See **gyrus**.

depolarization: A change in the membrane potential of the cell toward more positive values and therefore toward the threshold for firing an action potential. Depolarization increases the likelihood that a neuron will generate an action potential and is therefore excitatory. (Compare **hyperpolarization**.)

DNA (deoxyribonucleic acid): The material of which genes are made. DNA which is made up of four subunits called nucleotides contains the instructions needed for the synthesis of proteins. A greater portion of the total genetic information encoded in DNA is expressed in the brain than in any other organ of the body. (See **chromosome**.)

dopamine: A neurotransmitter in the brain that plays a major role in long-term potentiation, the control of attention, voluntary movement and cognition, and in the action of many stimulants (e.g., cocaine). Dopamine deficiency results in Parkinson's disease; dopamine excess contributes to the positive symptoms of schizophrenia.

dynamic polarization: The principle that information within a neuron flows in a single predictable and consistent direction.

electrical synapse: A site at which one neuron connects with another, transmitting signals via an electrical current flowing through a junction between the two neurons. (Compare **chemical synapse**.)

electrode: A sensing instrument made out of glass or metal and shaped like a needle. Glass electrodes are inserted into a neuron in order to record electrical activity across the surface membrane. Metal electrodes are used to record from outside the cell.

endocrine: A class of glands that secretes chemicals called hormones directly into the bloodstream. The hormones then travel to target tissues and exert their effect.

ethology: the study of animal behavior in its natural environment.

excitability change: The change in the threshold of a nerve cell that follows activity.

excitation: The depolarization of a postsynaptic cell, increasing the likelihood that an action potential will be generated.

excitatory: Indicates a neuron or synapse that depolarizes its target, increasing the chance that the neuron will fire an action potential. (Compare **inhibitory**.)

explicit learning: A class of learning that requires conscious participation and is concerned with acquiring information about people, places, and things. Also known as declarative learning. (Compare **implicit learning**.)

explicit memory: The storage of information about people, places, and things that requires conscious attention for recall. Such memories can be described in words. Explicit memory is what most people refer to when they speak of memory. Also known as declarative memory. (Compare **implicit memory**.)

expression: See **gene expression**.

facilitation: The process by which the strength of the synaptic connection between two cells is strengthened.

fiber: An axon.

first messenger: The neurotransmitter or hormone that binds to a receptor on the cell surface and activates a chemical (the **second messenger**) inside the cell.

fornix: A bundle of axons that carries information into and out of the hippocampus.

forward genetics: A genetic technique which usually employs a chemical to produce random mutations in a single gene. These mutants are then selected for a specific phenotype.

frontal lobe: One of the four lobes of the cerebral cortex. The frontal lobe is primarily concerned with executive function, working memory, reasoning, planning, speech, and movement. The frontal lobes are disordered in schizophrenia. (Compare **occipital lobe**; **parietal lobe**; **temporal lobe**.)

functional magnetic resonance imaging (fMRI): A noninvasive biomedical imaging technique that employs a large magnet to detect changes in blood flow and oxygen consumption in the brain. Blood flow and oxygen utilization increases in regions where neurons are more active, such as during the performance of a cognitive task.

GABA (gamma-aminobutyric acid): The primary inhibitory neurotransmitter in the brain, capable of causing, among other effects, sleep, muscle relaxation, and decreased emotional activity.

ganglion (pl. ganglia): A cluster of functionally related neuron cell bodies in the peripheral nervous system of vertebrates and in the central nervous system of *Aplysia* and other invertebrate animals.

gated channel: An ion channel that opens and closes in response to a particular type of signal. (See **transmitter-gated channel**; **voltage-gated channel**.)

gene: A specific sequence of DNA that is located at a certain point on a chromosome and contains the instructions for synthesizing a particular protein.

gene expression: The production of proteins based upon the specific genetic information encoded in the DNA of an organism.

Gestalt psychology: A school of psychology that focused particularly on visual perception and emphasized the fact that perception occurs as a reconstruction of sensory information in the brain based on an analysis of the relationship between an object and its surroundings.

glutamate: A common amino acid that functions as the major excitatory neurotransmitter in the brain and spinal cord.

gyrus (pl. gyri): The crest of a convolution on the outside of the cerebral cortex. Many of the gyri are invariant in location and help identify regions of the cortex. The groove between two gyri is called a sulcus. The **dentate gyrus** is part of the hippocampal formation and sends information to the hippocampus.

habituation: A simple, nonassociative form of learning in which a subject learns about the properties of a single, innocuous stimulus. The subject learns to ignore the stimulus, resulting in decreased neuronal response to it.

heterosynaptic facilitation: A neural mechanism whereby sensitization occurs. In heterosynaptic facilitation, enhanced strength of synaptic connections between two nerve cells is brought about by the activity of another cell or group of cells.

heterosynaptic plasticity: Any change in the strength (be it enhancement or depression) of a synaptic connection between two cells brought about by activity in a third cell or a group of cells.

higher-order cortex: Any of several regions of the cerebral cortex that process information from a primary sensory or motor area of the brain.

higher-order mental processing: Neuronal processing that occurs beyond the primary sensory or motor area of the brain.

hippocampus: The hippocampus is required for the storage of explicit memory. A structure lying deep in the temporal lobe of the cerebral hemispheres. The hippocampus, the dentate gyrus, and the subiculum constitute the hippocampal formation.

homosynaptic depression: A neural mechanism by which habituation occurs. In homosynaptic depression, the strength of the synaptic connection between two cells is decreased as a result of activity in one, the other, or both cells. This abated response occurs within the same pathway that is repeatedly stimulated.

homosynaptic plasticity: A change in the strength of the synaptic connection between two cells (be it enhancement or depression) brought on by activity in one, or the other, or both of these cells.

hormone: A chemical produced by endocrine glands in the body that serves as a messenger; hormones are secreted most often by endocrine glands directly into the bloodstream, through which they travel to their target. (See **endocrine**.)

hyperpolarization: A change in the membrane potential of a nerve cell toward a more negative value. Hyperpolarization decreases the likelihood that a neuron will generate an action potential and is therefore inhibitory. (Compare **depolarization**.)

hypothalamus: A part of the brain that lies immediately below the thalamus and regulates autonomic, endocrine, and visceral functions. (See **brain**.)

implicit memory: The storage of information that does not require conscious attention for recall—usually in the form of habits, perceptual or motor strategies, and associative and nonassociative conditioning. Also called procedural memory. (Compare **explicit memory**.)

inhibition: A change of the membrane potential toward more negative values, preventing or reducing the likelihood of an action potential in that cell.

inhibitory: Indicates a neuron or synapse that hyperpolarizes its target, decreasing the chance that the neuron will fire an action potential. (Compare **excitatory**.)

inhibitory feedback: A circuit in which a neuron excites an inhibitory interneuron that, in turn, connects with the first neuron and inhibits its action. This type of circuit is a form of self-regulation.

instrumental conditioning: See **operant conditioning**.

integration: The process by which a neuron adds up all incoming excitatory and inhibitory signals and determines whether an action potential will be generated.

interneuron: One of the three major functional types of neurons. These connect or regulate other neurons. Many interneurons are inhibitory. (Compare **motor neuron**; **sensory neuron**.)

involuntary attention: Attention focused on a particular stimulus, whether internal or external, as a result of a reflexive response to some aspect of the stimulus, usually a powerful, noxious, or otherwise highly novel stimulus.

ion: An atom or molecule having a net positive or negative charge. The major ions found on the inside or outside of the nerve cell membrane are potassium, sodium, chloride, calcium, and magnesium, as well as organic ions such as certain amino acids.

ion channel: See **channel**.

ionic hypothesis: The theory developed by Hodgkin and Huxley that the movements of sodium and potassium ions through the nerve cell membrane are regulated independently and that they give rise to the action potential and the resting potential.

ionotropic receptor: A protein that spans the cell surface membrane and contains a transmitter-binding site and channel through which ions can pass. The binding of the appropriate transmitter directly opens or closes the channel to the movement of ions. (See **transmitter-gated channel**; compare **metabotropic receptor**.)

localization: A theory that specific functions are carried out by specialized parts of the nervous system. (Compare **mass action**.)

magnetic resonance imaging (MRI): A noninvasive technique that uses a large magnet for imaging living subjects; used to visualize structures in the brain.

MAP kinase (Mitogen Activated Protein): A kinase that often acts in conjunction with protein kinase A to initiate long-term memory. In *Aplysia*, it is thought to act on CREB-2 (the inhibitor of CREB-mediated transciption). (See **CREB**; **protein kinase A**.)

mass action: The view, championed by Jean Pierre Flourens and by Karl Lashley during the first half of the twentieth century, that brain function is holistic rather than subdivided into specialized and localizable subunits. These theorists believed that loss of functionality due to brain damage would be directly proportional to the amount of tissue damaged rather than to the location of the damage. Also known as aggregate field theory. (Compare **localization**.)

mediating circuit: The primary circuit involved in a reflex action; it comprises the motor neurons, sensory neurons, and interneurons directly involved in the reflex. (Compare **modulating circuit**.)

medulla: One of the parts of the brain stem, it lies directly on top of the spinal cord. The medulla includes several centers responsible for such vital autonomic functions as digestion, breathing, and control of heart rate.

membrane hypothesis: The concept that even in the resting state, there exists a steady voltage difference across the neuronal membrane.

membrane potential: See **resting membrane potential**.

memory: The storage of learned information. Memory exists in at least two stages, short-term (minutes to hours) and long-term (days to weeks). It also has two forms: explicit and implicit. (See **explicit memory**; **implicit memory**.)

messenger RNA: The form of ribonucleic acid (RNA) that carries the instructions for a particular protein from the DNA in the nucleus of a cell to the protein synthesis machinery in the cytoplasm. The process of messenger RNA production is called transcription. (See **translation**, **transcription**.)

metabotropic receptor: A protein on the cell surface that binds a transmitter or hormone (the first messenger) and then activates a chemical inside the cell (the second messenger) that initiates a cell-wide response. (Compare **ionotropic receptor**.)

midbrain: The uppermost part of the brain stem, it controls many sensory and motor functions, including eye movements and the coordination of visual and auditory reflexes.

modulating circuit: The circuit for regulatory (nonreflex) processing, such as sensitization and classical conditioning, it modifies the function of the primary circuit involved in the behavior. (Compare **mediating circuit**.)

molecular biology: A hybrid discipline of genetics and biochemistry that attempts to understand life processes at the level of the macromolecules of the cell and their structure and function.

motor neuron: One of the three major functional types of neurons. Motor neurons form synapses with muscle cells, conveying information from the central nervous system and converting it into movement. (Compare **interneuron**; **sensory neuron**.)

motor system: The part of the nervous system that mediates movement and other active functions, as opposed to the sensory system, which receives and processes stimuli.

nerve: A bundle of axons.

nerve cell: See **neuron**.

neural analog of learning: The attempt to simulate the sensory stimuli used in learning experiments by electrically stimulating axons that end on a target nerve cell in an isolated ganglion.

neural circuit: A group of several neurons that are interconnected to and communicate with one another.

neural correlate of consciousness: A process that occurs in neurons while a person is engaging in an activity that requires conscious attention.

neural map: The orderly topographical arrangement of neurons in the central nervous system that reflects the spatial relationships of neurons in the primary sense organ. The brain contains a similarly ordered motor map for movement.

neurology: The classic field of medicine concerned with the nervous system, both normal and diseased. Clinical neurology is concerned with the diagnosis and treatment of nervous system disorders, which usually do not primarily affect mental processes. Relevant disorders include strokes, seizures, Huntington's disease, Alzheimer's disease, and Parkinson's disease. Neurology has posed many of the critical questions that cognitive neural science has attempted to address. In contrast, psychiatry attempts to address disorders of the brain that affect mental processes.

neuron: The fundamental unit of any nervous system. The human brain contains about 100 billion neurons, each of which forms about 1000 synapses. Neurons are similar to other cells in having common molecular machinery for cellular function, but they have the unique ability to communicate rapidly with one another over great distances and with great precision.

neuron doctrine: The theory that individual neurons are the fundamental signaling elements of the nervous system.

neurotransmitter: A chemical substance that is released by one neuron and binds to receptors on another neuron, altering the flow of electrical current or internal biochemical events in the second cell. The specific action of a neurotransmitter depends on the properties of the receptor. There may be many different kinds of receptors for a single neurotransmitter.

NMDA receptor (N-methyl-D-aspartate)**:** One of two types of postsynaptic receptors for glutamate that are discussed in this book. The NMDA receptor plays a critical role in long-term potentiation. (Compare **AMPA receptor**.)

nongated channel: A channel in the membrane of nerve cells that passively conducts ions (most often potassium ions) across the cell membrane. The flow of ions through these channels is responsible for the resting membrane potential of the cell. Also known as a resting channel. (Compare **gated channel**.)

nucleotide base: The basic building block of DNA or RNA. There are typically four types that in combination code for genes. In DNA the four bases are thymine, adenine, cytosine, and guanine. In RNA uracil replaces thymine.

nucleus (pl. nuclei): (1) The processing center of a cell, where all of the genetic material is found. The nucleus is surrounded by a membrane that separates it from the cytoplasm. (2) A cluster of functionally related neuron cell bodies in the central nervous system. In the peripheral nervous system, or in the central nervous system of invertebrate animals, groups of neurons are arranged in ganglia. (See **cell body**; compare **cytoplasm**.)

occipital lobe: One of the four lobes of the cerebral cortex. Found at the rear of the cortex, the occipital lobe is important for vision. (Compare **frontal lobe**; **parietal lobe**; **temporal lobe**.)

operant conditioning: A form of implicit associative learning in which a subject learns through reward or punishment to perform or not to perform an action (one that is not a preexisting reflex) in response to a previously neutral conditioned stimulus. Also called instrumental conditioning or operant conditioning.

organic ions: Molecules containing carbon atoms and carrying an electrical charge (including some amino acids and proteins) that are involved in biological processes.

parietal lobe: One of the four lobes of the cerebral cortex, it is located between the frontal and occipital lobes. The parietal lobe processes sensations such as touch, pressure, and pain and is important in integrating multiple sensations into a single experience. (Compare **frontal lobe**; **occipital lobe**; **temporal lobe**.)

peripheral nervous system: The portion of the nervous system, including the autonomic nervous system, where motor or autonomic activities are mediated by neurons that lie outside the spinal cord and brain stem. The peripheral nervous system is functionally connected with the central nervous system. (Compare **central nervous system**.)

phosphorylation: The addition of a phosphate group to a protein, thereby changing the structure, charge, or activity of the protein. Phosphorylation is carried out by a special class of enzymes called protein kinases.

phrenology: A theory popular during the nineteenth century that posited a correlation between personality traits and the shape of the skull. It was thought that frequent use of underlying brain structures would lead to enlargement of those structures, which would be reflected in bumps on the skull.

place cells: Neurons of the hippocampus that fire only when an animal is in a particular location within its environment, together forming a cognitive map of that environment. When the animal moves to a different location, different place cells become active.

plasticity: The ability of synapses, neurons, or regions of the brain to change their properties in response to usage or different patterns of stimulation. Also known as plastic change.

positron-emission tomography (PET scan): A computerized tomography technique for imaging brain functions in living organisms. The technique, while conceptually similar to functional magnetic resonance imaging, employs radioactive molecules to probe specific brain activities, such as blood flow and metabolism. (See **functional magnetic resonance imaging**.)

postsynaptic cell; postsynaptic neuron: The neuron that receives signals (electrical or chemical) from another neuron at a synapse. The signals affect the excitability of the postsynaptic cell.

postsynaptic receptor: See **receptor**.

potassium (K^+)**:** A positively charged ion that is essential for nervous system function. Concentrations of potassium inside the resting neuron are higher than those outside the cell.

potentiation: The process by which activity in one neuron causes an enhancement of the strength of the synaptic connection with its target. Long-term potentiation is a persistent increase (lasting hours to days) in the synaptic response of a postsynaptic neuron following repeated stimulation of the presynaptic neuron.

prefrontal cortex: The forwardmost portion of the frontal cortex, it is associated with planning, decision making, higher-level cognition, attention, and aspects of motor function.

presynaptic cell: The neuron that sends signals (electrical or chemical) to another neuron at the synapse.

presynaptic terminal: The terminal area at the end of the axon of the presynaptic neuron from which synaptic vesicles with neurotransmitters are released onto the postsynaptic cell (chemical synapses) or that connects via electrical junctions to the postsynaptic cell (electrical synapses).

prion (proteinaceous infectious agent): A very small class of infectious proteins that can take on two functionally distinct shapes, the recessive form which is inactive or has a conventional, physiological role, and the dominant form which is self-perpetuating and toxic to nerve cells. In the dominant form, prions can cause degenerative diseases of the nervous system such as mad cow disease (bovine spongiform encephaly) and, in humans, Creutzfeldt-Jakob disease.

procedural memory: See **implicit memory**.

processes: In a neuron, protrusions where synapses can or will develop. (See **axon**; **dendrite**.)

promoter: A specific site for each gene located on the DNA to which regulatory proteins bind, thus turning the gene on or off.

propagation: (1) A process by which nerve impulses travel down the neuron. (2) In prions, a process by which one form of the prion perpetuates itself.

protein: A large molecule made up of one or more chains of amino acids, held together in a complex three-dimensional structure. Proteins play regulatory, structural, and catalytic roles in living systems.

protein kinase: An enzyme that catalyzes the phosphorylation of other proteins, thereby modifying their function.

protein kinase A: The target of cyclic AMP and the enzyme that phosphorylates target proteins. It is comprised of four subunits, two regulatory subunits which inhibit the two catalytic subunits. The catalytic subunit phosphorylates other enzymes.

psychiatry: The field of medicine concerned with normal and abnormal mental functions. Clinical psychiatry deals with such disorders as schizophrenia, depression, anxiety, and drug abuse.

pyramidal cells: A particular type of neuron, typically excitatory and found in the cerebral cortex, that is shaped roughly like a pyramid. Pyramidal cells are the major class of neurons in the hippocampus, where they encode place. (See **place cells**.)

quantum (pl. quanta): A small packet containing about 5000 molecules of neurotransmitter, that is released from the presynaptic terminal of the axon. Quanta are packaged in synaptic vesicles. (See **synaptic transmission**; **synaptic vesicle**.)

receptive field: The portion of the total sensory world that activates a particular sensory neuron. For example, the receptive field of one sensory neuron in the retina may respond to a spot of light shone on the upper left portion of a visual field.

receptor: A specialized protein in the postsynaptic cell that recognizes and binds the neurotransmitter released by the presynaptic cell. All receptors for chemical transmitters have two functions: they recognize transmitters and they carry out an effector function within the cell. For example, they can be involved in gating ion channels or in activating second messengers. Based on these gating or activation functions, receptors fall into two major categories: ionotropic and metabotropic. (See **ionotropic receptor**; **metabotropic receptor**.)

receptor cell: A sensory cell specialized to respond to a particular physical property, such as touch, light, or temperature.

recombinant DNA: A molecule of DNA formed by combining strands from two originally separate DNA molecules.

recruitment: The process by which various components necessary for a certain biochemical pathway are gathered together so that the requisite chemical reactions can occur in sequence.

reductionist analysis; reductionism: A scientific approach that seeks to eliminate features of the process studied that are not essential to its function, thereby isolating the most important features. This may involve creating a simple model for a more complicated process, as the more complicated process may be too complex to study effectively.

reflex: An unlearned, involuntary response to a stimulus. In the case of spinal reflexes, these responses are mediated by the spinal cord and do not require that messages be sent to the brain. (Compare **voluntary attention**.)

refractory period: The time during which the neuron has a higher threshold for generating further action potentials after its firing of one action potential.

replication: The formation of copies of double-stranded DNA.

The two DNA strands divide and each serves as a template, or parent strand, and is copied. The new strands, or daughter strands, carry the complement.

repressor: A regulatory protein that binds to the promoter and prevents a gene from being turned on.

resting membrane potential: The difference in electrical charge between the inside and outside surfaces of a nerve cell membrane, resulting from the uneven distribution of sodium, potassium, and chloride ions. The resting potential is about –60 to –70 millivolts in most mammalian nerve cells.

reverse genetics: Genetic technique by which a gene is either removed or introduced into a mouse's genome and the effect of the genetic alteration is tested in order to evaluate a specific hypothesis.

RNA (ribonucleic acid): A nucleotide related to DNA, the class of nucleic acid that includes messenger RNA.

Schaffer collateral pathway: A pathway in the hippocampus that is important for explicit memory storage and has thus served as an important experimental model of the synaptic change essential for memory.

second messenger: A chemical that is produced inside the cell when a neurotransmitter binds to a particular class of receptor on the surface. Cyclic AMP is a common second messenger in neurons. (Compare **first messenger**; see **cyclic AMP**; **metabotropic receptor**.)

sensation: Touch, pain, sight, hearing, smell, taste.

sensitization: A type of nonassociative learning in which exposure to a noxious stimulus produces a stronger reflex response to other stimuli, even innocuous ones. (See **heterosynaptic facilitation**.)

sensory neuron: One of the three major functional types of neurons. Sensory neurons transmit information about environmental stimuli from a sensory receptor to other neurons in a sensory pathway. (Compare **interneuron**; **motor neuron**; **sensory receptor**.)

serotonin: A modulatory neurotransmitter in the brain that has been implicated in the regulation of mood states, including depression, anxiety, food intake, and impulsive violence.

signal: A change in the membrane potential of a postsynaptic neuron as a result of input from a presynaptic neuron or activation of a sensory receptor. There are two types of signals. Local signals are synaptic potentials. These are spatially restricted and do not propagate actively. By contrast, propagated signals are action potentials. These propagate along the whole length of the axon to the synaptic terminals. The action potential signals are highly stereotyped throughout the nervous system; the "message" conveyed by an action potential depends entirely on the pathway in which the active neuron is located.

sodium (Na^+): A positively charged ion that is an essential element for nervous system function. Sodium concentrations inside the resting neuron are lower than those outside the cell.

somatosensory cortex: The portion of the cerebral cortex, located in the parietal lobe, that processes sensations, including, touch, vibration, pressure, and sense of limb position. (See **parietal lobe**.)

somatosensory system: The sensory system concerned with sensation from the skin at the body surface (touch, vibration, pressure, pain) and the sense of limb position. Signals are carried from the peripheral nervous system to the brain.

spatial map: An internal representation of the external environment, found in the hippocampus as a combination of many place cells. A type of cognitive map.

spatial memory: A form of explicit memory concerned with finding one's way around in space.

spinal cord: A part of the central nervous system that controls movements of the limbs and trunk, processes sensory information from the skin, joints, and muscles of the limbs and trunk, and controls autonomic function. (See **brain**.)

spinal reflex: An involuntary movement triggered by sensory input and produced by neural circuitry limited to the spinal cord.

stimulus: Any event that provokes a response. Stimuli possess four attributes: modality (pathway), intensity, duration, and location.

striatum: A part of the basal ganglion that plays a role in movement and cognition.The striatum consists of the putamen, the caudate nucleus, and the nucleus accumbens. It functions abnormally in people with Parkinson's disease. It is the mediator of pleasurable sensations and a site of abnormality in schizophrenia. (Compare **basal ganglia**.)

synapse: The specialized site of communication between two neurons. A synapse consists of three components: a presynaptic terminal, a postsynaptic cell, and a zone of opposition—the synaptic cleft in between. Depending on the nature of the zone of opposition, synapses can be categorized as chemical or electrical, each using a different mechanism of synaptic transmission.

synaptic cleft: The gap between two neurons at a chemical synapse.

synaptic marking: A process by which synapses are tagged, priming them for long-term strengthening.

synaptic plasticity: An increase or decrease in synaptic strength, for short or long periods, following specific patterns of neuronal activity. Shown to be critically involved in learning and memory.

synaptic potential: A graded change in the membrane potential of a postsynaptic neuron produced by a signal, usually chemical, from a presynaptic neuron. A synaptic potential can be either excitatory or inhibitory; if sufficiently strong, an excitatory synaptic potential will trigger an action potential in the postsynaptic cell. Thus the synaptic potential is an intermediate step linking an action potential in the presynaptic terminal with an action potential in the postsynaptic cell.

synaptic terminal: See **presynaptic terminal**.

synaptic transmission: The mechanism by which one neuron influences the excitability of another, either chemically or electrically. Chemical synaptic transmission is mediated by the release of a neurotransmitter from the presynaptic cell, which acts on receptors in the postsynaptic cell. Electrical synaptic transmission is mediated by the flow of current across a junction between two neurons.

synaptic vesicle: A membrane-bound sac containing about 5000 molecules of neurotransmitters to be released from the presynaptic terminal in an all-or-nothing way. (See **quantum**; **synaptic transmission**.)

temporal lobe: One of the four lobes of the cerebral cortex. Located below the frontal lobe and parietal lobe, the temporal lobe is primarily concerned with hearing and vision, as well as aspects of learning, memory, and emotion. (Compare **frontal lobe**; **occipital lobe**; **parietal lobe**.)

thalamus: A major relay point of the brain, it processes most of the sensory information reaching the cerebral cortex from the various sensory systems and motor information that is conveyed from the motor cortices to muscles for movement.

transcription: The manufacture of RNA from a DNA template.

transgene: A foreign gene that has been introduced into the genome of another organism.

transgenesis: The introduction of genes from one organism into the genome of another in such a way that the genes can be passed onto progeny.

translation: The production of proteins from messenger RNA, based on the genetic code.

transmitter: See **neurotransmitter**.

transmitter-gated channel: An ion channel whose opening and closing is regulated by the binding of a chemical messenger, such as a neurotransmitter. The binding of the transmitter can regulate the movement of ions directly or lead to the activation of a second messenger. Transmitter-gated channels can be excitatory or inhibitory. They are involved in neuron-to-neuron communication, whereas voltage-gated channels are involved in generating the action potential within a single neuron. (Compare **voltage-gated channel**.)

trial-and-error learning: See **operant conditioning**.

unconditioned stimulus: A rewarding or aversive stimulus that always produces an overt response.

visual system: A sensory pathway, stretching from the retina to the cortex, that detects stimuli in the environment and produces an image of the external world.

voltage-gated channel: An ion channel that opens and closes in response to changes in the membrane potential of the cell. Voltage-gated channels in neurons can be permeable to sodium, potassium, or calcium. Voltage-gated channels can, for example, generate the action potential or let in calcium to trigger neurotransmitter release, depending on the channel and its location in the cell. (Compare **transmitter-gated channel**.)

voluntary attention: Attention focused on a particular stimulus, whether internal or external, in accordance with one's own predisposition; it is determined internally, by one's brain processes. (Compare **reflex**.)

Wernicke's area: The portion of the left parietal lobe concerned with comprehension of language. (Compare **Broca's area.**)

working memory: A distinct type of short-term memory subserved in part by the prefrontal cortex, it integrates moment-to-moment perceptions over a relatively short period and combines them with memories of past experiences. Working memory is needed for many apparently simple aspects of everyday life, such as carrying on a conversation, adding a list of numbers, or driving a car. This memory is defective in individuals with schizophrenia.

NOTES AND SOURCES

These notes are designed to help the reader to sources of quotations and other points of reference that are referred to in each chapter and to lead to additional sources of information.

Preface

Two papers announced the structure of DNA and its implications for replication: J. D. Watson and F. H. C. Crick, "Molecular structure of nucleic acids; A structure for deoxyribose nucleic acid," *Nature* 171 (1953):737–38; and J. D. Watson and F. H. C. Crick, "Genetical implications of the structure of deoxyribonucleic acid," *Nature* 171 (1953):964–67.

The first edition of our textbook is E. R. Kandel and J. H. Schwartz, *Principles of Neural Science* (New York: Elsevier, 1981).

Some of the autobiographical details discussed in this book were described in a highly abbreviated form in my Nobel lecture, which was printed as E. R. Kandel, *The Molecular Biology of Memory Storage: A Dialog Between Genes and Synapses, Les Prix Nobel* (Stockholm: Almquist & Wiksell International, 2001).

1: Personal Memory and the Biology of Memory Storage

For a discussion of mental time travel, see D. Schacter, *Searching for Memory: The Brain, the Mind and the Past* (New York: Basic Books, 1996).

For two excellent histories on the emergence of genetics and molecular biology, see H. F. Judson, *The Eighth Day of Creation* (New York: Simon & Schuster, 1979); and F.

Jacob, *The Logic of Life: A History of Heredity* (New York: Pantheon, 1982).

For a discussion of the biology of memory, see L. Squire and E. R. Kandel, *Memory: From Mind to Molecules* (New York: Scientific American Books, 1999).

Especially valuable for the history of biology are these four books: C. Darwin, *On the Origin of Species* (1859; repr., Cambridge, Mass.: Harvard University Press, 1964); E. Mayr, *The Growth of Biological Thought: Diversity, Evolution and Inheritance* (Cambridge, Mass.: Belknap, 1982); R. Dawkins, *The Ancestor's Tale: A Pilgrimage to the Dawn of Evolution* (New York: Houghton Mifflin, 2004); and S. J. Gould, "Evolutionary Theory and Human Origins" in *Medicine, Science, and Society*, ed. K. J. Isselbacher (New York: Wiley, 1984).

For technical discussions of the emergence of the new science of mind, see T. D. Albright, T. M. Jessell, E. R. Kandel, and M. I. Posner, "Neural science: A century of progress and the mysteries that remain," *Neuron* (Suppl.) 25(S2) (2000):1–55; E. R. Kandel, J. H. Schwartz, and T. M. Jessell, *Principles of Neural Science*, 4th ed. (New York: McGraw-Hill, 2000).

Other information for this chapter was drawn from: Y. Dudai, *Memory from A to Z* (Oxford: Oxford University Press, 2002).

2: A Childhood in Vienna

I have been much influenced by the discussion of the history of the Jews in Vienna by G. E. Berkley, *Vienna and Its Jews: The Tragedy of Success, 1880s–1980s* (Cambridge, Mass.: Abt Books, 1988) and C. E. Schorske, *Fin de Siècle Vienna: Politics and Culture* (New York: Alfred A. Knopf, 1980). Berkley's book is the source for what "the Viennese have managed to do overnight" (p. 45), for William Johnston's comments about Vienna (p. 75), for Hans Ruzicka (p. 303), and for the *Reichspost*'s editorial (p. 307). Schorske's discussion of the cultural explosion in Vienna in 1900 is now a classic; quotation about middle-class culture from p. 298.

For Hitler's expectations before the Anschluss, see I. Kershaw, *Hitler, 1936–1945: Nemesis* (New York: W.W. Norton, 2000); and E. B. Bukey, *Hitler's Austria: Popular Sentiment in the Nazi Era, 1938–1945* (Chapel Hill: University of North Carolina Press, 2000).

Cardinal Innitzer's meeting with Hitler is drawn from G. Brook-Shepherd, *Anschluss* (London: Macmillan, 1963), pp 201–2. That meeting is also discussed in Berkley, *Vienna and Its Jews*, p. 323, and in Kershaw, *Hitler*, pp. 81–82.

Carl Zuckmayer's description of Vienna in 1938 is from his autobiography, *Als Wärs ein Stück von Mir* (Frankfurt: Fischer Tochenbuch Verlag, 1966), p. 84; my translation. An English-language version was published as *A Part of Myself: Portrait of an Epoch*, trans. Richard and Clara Winston (New York: Carroll & Graf, 1984).

For Hitler's aspirations and accomplishments as an artist, see P. Schjeldahl, "The Hitler show," *The New Yorker*, April 1, 2002, p. 87.

For the taking of neighbors' property, see T. Walzer and S. Templ, *Unser Wien: "Arisierung" auf Österreichisch* (Berlin: Aufbau-Verlag, 2001), p. 110.

For the role of the Catholic Church in the promulgation of anti-Semitism, see F.

Schweitzer, *Jewish-Christian Encounters over the Centuries: Symbiosis, Prejudice, Holocaust, Dialogue*, ed. M. Perry (New York: P. Lang, 1994), especially pp. 136–37.

Other information for this chapter was drawn from my father's file at the Kultusgemeinde in Vienna and from the following:

Applefeld, A. "Always, darkness visible." *New York Times*, January 27, 2005, p. A25.

Beller, S. *Vienna and the Jews, 1867–1938: A Cultural History*. Cambridge: Cambridge University Press, 1989.

Clare, G. *Last Waltz in Vienna*. New York: Avon, 1983, especially pp. 176–77.

Freud, S. *The Psychopathology of Everyday Life*. Translated by James Strachey. 1901. Reprint, New York: W. W. Norton, 1989.

Gedye, G. E. R. *Betrayal in Central Europe: Austria and Czechoslovakia, The Fallen Bastions*. New York: Harper & Brothers, 1939, especially p. 284.

Kamper, E. "Der schlechte Ort zu Wien: Zur Situation der Wiener Juden um Anschluss zum Novemberprogrom 1938." In *Der Novemberprogrom 1938: Die "Reichkristallnacht" in Wien*. Vienna: Wienkultur, 1988, especially p. 36.

Lee, A. "La ragazza," *The New Yorker*, February 16–23, 2004, pp. 174–87, especially p. 176.

Lesky, E. *The Vienna Medical School of the Nineteenth Century*. Baltimore: Johns Hopkins University Press, 1976.

McCragg, W. O., Jr. *A History of the Hapsburg Jews, 1670–1918*. Bloomington: Indiana University Press, 1992.

Neusner, J. *A Life of Yohanan ben Zaggai: Ca. 1–80 C.E.* 2nd ed. Leiden: Brill, 1970.

Pulzer, P. *The Rise of Political Anti-Semitism in Germany and Austria*. Cambridge, Mass: Harvard University Press, 1988.

Sachar, H. M. *Diaspora: An Inquiry into the Contemporary Jewish World*. New York: Harper & Row, 1985.

Schütz, W. "The medical faculty of the University of Vienna sixty years following Austria's annexation." *Perspectives in Biology and Medicine* 43 (2000):389–96.

Spitzer, L. *Hotel Bolivia*. New York: Hill & Wang, 1998.

Stern, F. *Einstein's German World*. Princeton, N.J.: Princeton University Press, 1999.

Weiss, D. W. *Reluctant Return: A Survivor's Journey to an Austrian Town*. Bloomington: Indiana University Press, 1999.

Zweig, S. *World of Yesterday*. New York: Viking, 1943.

3: An American Education

For a discussion of the academic motivation of Viennese émigrés, see G. Holton and G. Sonnert, "What happened to Austrian refugee children in America?" in *Österreichs Umgang mit dem Nationalsozialismus* (Vienna: Springer Verlag, 2004).

The Yeshivah of Flatbush is now the largest and still one of the best Jewish day schools in the United States. In 1927 the founding parent body asked Dr. Joel Braverman, an exceptional educational leader, to head the school. He recruited an outstanding Hebrew-speaking faculty from what was then Palestine and from Europe and initiated a radical change in Jewish education in the United States. This change had three components. First, rather than conduct the religious studies—fully half the curriculum—in English or Yiddish, the common language among Jewish immigrants of the day, Braverman insisted on carrying out these classes exclusively in Hebrew, a language that was then rarely spoken outside of Palestine. The Yeshivah of Flatbush was the first school in the country practicing the principle of "Hebrew in Hebrew." Second, the secular curriculum received equal emphasis, and was taught in English by an excellent faculty. Finally, the Yeshivah was modern and enrolled an almost equal number of girls and boys. Later, many other day schools followed in the footsteps of the Yeshivah of Flatbush. For a history of this institution, see Jodi Bodner DuBow, ed., *The Yeshivah of Flatbush: The First Seventy-five Years* (Brooklyn: Yeshivah of Flatbush, 2002).

Erasmus Hall High School was founded in 1787. With an initial enrollment of twenty-six boys, it was the first secondary school to be chartered by the Regents of the University of the State of New York. Often called the "mother of high schools," it generated the development of the secondary school system in New York State. The original building, which still stands at the center of the campus, was built in the year of the school's founding with money contributed by John Jay, Aaron Burr, and Alexander Hamilton. For a history of Erasmus, see Rita Rush, ed., *The Chronicles of Erasmus Hall High School* (New York: Board of Education, 1987). The yearbook of my high school class of 1948, *The Arch*, was also an invaluable source for this section.

Harvard College was founded in Cambridge, Massachusetts, in 1636. In the years I was at Harvard, it was led by James Bryant Conant. A first-rate chemist, Conant introduced four major initiatives that further assured Harvard's intellectual preeminence. The first was a system of ad hoc committees made up of independent scholars to evaluate the eligibility for tenure of each academic appointment. This step ensured that tenure was based on scholarly accomplishment rather than social status or other unrelated factors. The second initiative was the National Scholars Program, which guaranteed a full scholarship for two deserving students from each state in the union, thereby ensuring geographic diversity as well as excellence in the student body. Third, Conant established a program of general education that required students to take courses in both the sciences and the humanities, ensuring that they received a liberal arts education. Fourth, he signed an agreement with Radcliffe College that gave its women students free access to classes at Harvard. See H. Hawkins, *Between Harvard and America: The Educational Leadership of Charles W. Eliot* (New York: Oxford University Press, 1972); and R. A. McCaughey, "The transformation of American academic life: Harvard University 1821–1892," *Perspectives in American History* 8 (1974):301–5.

For a discussion of Freud, see P. Gay, *Freud: A Life for Our Time* (New York: W. W. Norton, 1988); and E. Jones, *The Life and Work of Sigmund Freud*, 3 vols. (New York: Basic Books, 1952–1957).

For a discussion of behaviorism, see E. Kandel, *Cellular Basis of Behavior: An Introduction to Behavioral Neurobiology* (San Francisco: Freeman, 1976); J. A. Gray, *Ivan Pavlov* (New York: Penguin Books, 1981); and G. A. Kimble, *Hilgard and Marquis' Conditioning and Learning*, 2nd ed. (New York: Appleton-Century-Crofts, 1961).

Other information for this chapter was drawn from the following:

Freud, S. *Beyond the Pleasure Principle*. Translated by James Strachey. 1922. Reprint, New York: Liveright, 1950; quotation on p. 83.

Kandel, E. "Carl Zuckmayer, Hans Carossa, and Ernst Jünger: A study of their attitude toward National Socialism." Senior thesis, Harvard University, June 1952.

Stern, F. *Dreams and Delusions*. New York: Alfred A. Knopf, 1987.

———. *Einstein's German World*. Princeton, N.J.: Princeton University Press, 1999.

Vietor, K. *Georg Buchner*. Bern: A. Francke AG Verlag, 1949.

———. *Goethe*. Bern: A. Francke AG Verlag, 1949.

———. *Der Junge Goethe*. Bern: A. Francke AG Verlag, 1950.

4: One Cell at a Time

For psychoanalysis and brain function, see L. S. Kubie, "Some implications for psychoanalysis of modern concepts of the organization of the brain," *Psychoanalytic Quarterly* 22 (1953):21–68; M. Ostow, "A psychoanalytic contribution to the study of brain function. I: The frontal lobes," *Psychoanalytic Quarterly* 23 (1954):317–38; and M. Ostow, "A psychoanalytic contribution to the study of brain function II: The temporal lobes," *Psychoanalytic Quarterly* 24 (1955):383–423.

For a history of the cell theory and the neuron doctrine, see E. Mayr, *The Growth of Biological Thought: Diversity, Evolution and Inheritance* (Cambridge, Mass.: Belknap, 1982); P. Mazzarello, *The Hidden Structure: The Scientific Biography of Camillo Golgi* (Oxford: Oxford University Press, 1999); and G. M. Shepherd, *Foundations of the Neuron Doctrine* (New York: Oxford University Press, 1991).

Sherrington wrote about Cajal in an essay called "A memorial on Ramón y Cajal," which appeared originally in D. F. Cannon, ed., *Explorers of the Human Brain: The Life of Santiago Ramón y Cajal* (New York: Henry Schuman, 1949). It is reprinted in J. C. Eccles and W. C. Gibson, *Sherrington: His Life and Thought* (Berlin: Springer Verlag, 1979); "in describing what the microscope showed . . ." is from p. 204; "the intense anthropomorphic descriptions . . ." is from pp. 204–5; and "Is it too much to say of him . . ." is from p. 203.

Cajal's memoir, *Recollections of My Life*, was translated in 1937 by E. H. Craigie and J. Cano, and appeared in *Am Philos. Soc. Mem.* 8; he compares cells to a "full grown forest" on pp. 324–25 and likens himself and Golgi to "Siamese twins" on p. 553. Golgi's Nobel lecture was reprinted in his *Opera Omnia*, ed. L. Sala, E. Veratti, and G. Sala, vol. 4 (Milan: Hoepl, 1929); quotation from p. 1259; it was translated into English as "The neuron theory: Theory and facts," in *Nobel Lectures: Physiology or Medicine, 1901–1921*, ed. Nobel Foundation (Amsterdam: Elsevier, 1967).

Hodgkin wrote about scientific jealousy in his "Autobiographical essay," in *The History of Neuroscience in Autobiography*, ed. L. R. Squire, vol. 1 (Washington, D.C.: Society for Neuroscience, 1996); quotation from p. 254. Darwin's remark on the same theme is from R. K. Merton, "Priorities in scientific discovery: A chapter in the sociology of science," *Am. Soc. Rev.* 22 (1957):635–59.

For more on Sherrington's life and research, see C. Sherrington, *The Integrative Action of the Nervous System* (New Haven: Yale University Press, 1906); and R. Granit, *Charles Scott Sherrington: A Biography of the Neurophysiologist* (Garden City, N.Y.: Doubleday, 1966).

Robert Holt's remarks about Freud are from p. 17 of F. J. Sulloway, *Freud, Biologist of the Mind* (New York: Basic Books, 1979). Freud himself is quoted about this happy period in W. R. Everdell, *The First Moderns* (Chicago: University of Chicago Press, 1997), p. 131.

Other information for this chapter was drawn from the following:

Cajal, S. R. "The Croonian Lecture: La fine structure des centres nerveux." *Proc. R. Soc. London Ser. B* 55 (1894):444–67.

———. *Histologie du Systeme Nerveux de l'Homme et des Vertebres.* 2 vols. Madrid: Consejo Superior de Investigaciones Cientificas, 1909–1911. (English translation, *Histology of the Nervous System.* Translated by N. Swanson and L. W. Swanson. 2 vols. New York: Oxford University Press, 1995.)

———. *Neuron Theory or Reticular Theory: Objective Evidence of the Anatomical Unity of Nerve Cells.* Translated by M. U. Purkiss and C. A. Fox. Madrid: Consejo Superior de Investigaciones Cientificas, 1954.

———. "History of the synapse as a morphological and functional structure." In *Golgi Centennial Symposium: Perspectives in Neurobiology*, edited by M. Santini, 39–50. New York: Raven Press, 1975.

Freud, S. *New Introductory Lectures on Psychoanalysis.* Translated by James Strachey. 1933. Reprint, New York: W.W. Norton, 1965.

Kandel, E. R., J. H. Schwartz, and T. M. Jessell. *Principles of Neural Science.* 4th ed. New York: McGraw-Hill, 2000.

Katz, B. *Electrical Excitation of Nerve.* London: Oxford University Press, 1939.

Reuben, J. P. "Harry Grundfest—January 10, 1904–October 10, 1983." *Biog. Mem. Natl. Acad. Sci.* 66 (1995): 151–66.

5: The Nerve Cell Speaks

Adrian wrote elegantly about impulses in his *The Basis of Sensation: The Action of the Sense Organs* (London: Christopher, 1928). Motor discharges are discussed in E. D. Adrian and D. W. Bronk, "The discharge of impulses in motor nerve fibers. Part I: Impulses in single fibers of the phrenic nerve," *J. Physiol.* 66 (1928):81–101; "the motor fibers . . ." is from p. 98. Adrian's praise of Sherrington appears in J. C. Eccles and W. C. Gibson, *Sherrington: His Life and Thought* (Berlin: Springer Verlag, 1979), p. 84.

For a discussion of Hermann Helmholtz's remarkable set of contributions to conduction of the nervous impulse, to perception, and to unconscious inference, see E. G. Boring, *A History of Experimental Psychology*, 2nd ed. (New York: Appleton-Century-Crofts, 1950).

For a discussion of Julius Bernstein's contribution, see A. L. Hodgkin, *The Conduction of the Nervous Impulse* (Liverpool: Liverpool University Press, 1967); A. Huxley, "Electrical activity in nerve: The background up to 1952," in *The Axon: Structure, Function and Pathophysiology*, ed. S. G. Waxman, J. D. Kocsis, and P. K. Stys,

3–10 (New York: Oxford University Press, 1995); B. Katz, *Nerve, Muscle, Synapse* (New York: McGraw-Hill, 1966); and S. M. Schuetze, "The discovery of the action potential," *Trends in Neuroscience* 6 (1983):164–68.

Other information for this chapter was drawn from the following:

Adrian, E. D. *The Mechanism of Nervous Action: Electrical Studies of the Neuron*. (London: Oxford University Press, 1932).

Bernstein, J. "Investigations on the thermodynamics of bioelectric currents." *Pflügers Arch* 92 (1902):521–62. (English translation in *Cell Membrane Permeability and Transport*, edited by G. R. Kepner, 184–210. Stroudsburg, Pa.: Dowden, Hutchinson & Ross, 1979.)

Doyle, D. A., J. M. Cabral, R. A. Pfuetzner, A. Kuo, J. M. Gulbis, S. L. Cohen, B. T. Chait, and R. MacKinnon. "The structure of the potassium channel: Molecular basis of K^+ conduction and selectivity." *Science* 280 (1998):69–77.

Galvani, L. *Commentary on the Effect of Electricity on Muscular Motion*. Translated by Robert Montraville Green. Cambridge, Mass.: E. Licht, 1953. (A translation of Luigi Galvani's 1933 *De Viribus Electricitatis in Motu Musculari Commentarius*.)

Hodgkin, A. L. *Chance and Design*. Cambridge: Cambridge University Press, 1992.

———. "Autobiographical essay." In *The History of Neuroscience in Autobiography*, edited by L. R. Squire. Vol. 1., 253–92. Washington, D.C.: Society for Neuroscience, 1996.

Hodgkin, A. L., and A. F. Huxley. "Action potentials recorded from inside a nerve fibre." *Nature* 144 (1939):710–11.

Young, J. Z. "The functioning of the giant nerve fibers of the squid." *J. Exp. Biol.* 15 (1938):170–85.

6: Conversation Between Nerve Cells

Grundfest remained a sparker for a very long time, even after Eccles and most other neurophysiologists had become convinced about the chemical nature of synaptic transmission. It was only in September 1954, a year before I arrived in his laboratory, that Grundfest, in an important symposium on nerve impulses, shifted his view. He wrote, "Eccles has recently adopted the position that this [nerve cell to nerve cell] transmission is chemically mediated. Some of us opposed the view. . . . We may have been in error." (D. Nachmansohn and H. H. Merrit, eds., *Nerve Impulses; Transactions* [New York: Josiah Macy Jr. Foundation, 1956], p.184).

For a history of synaptic transmission, see W. M. Cowan and E. R. Kandel, "A brief history of synapses and synaptic transmission," in *Synapses*, ed. W. M. Cowan, T. C. Südhof, and C. F. Stevens (Baltimore: Johns Hopkins University Press, 2000), 1–87.

Bernard Katz recounts his arrival in Britain in "To tell you the truth, sir, we do it because it's amusing!" in *The History of Neuroscience in Autobiography*, ed. L. R. Squire, vol. 1 (Washington, D.C.: Society for Neuroscience, 1996): 348–81; quotation from p. 373.

For Eccles on Popper, see his "Under the spell of the synapse," in *The Neurosciences: Paths of Discovery*, ed. F. G. Worden, J. P. Swazey, and G. Adelman (Cambridge, Mass.: MIT Press, 1976), 159–80; quotations from pp. 162 and 163. For other reminiscences on the history of the synapse and on the soup and spark controversy, see S. R. Cajal, *Recollections of My Life*, translated by E. H. Craigie and J. Cano, *Am. Philos. Soc. Mem.* 8 (1937); H. H. Dale, "The beginnings and the prospects of neurohumoral transmission," *Pharmacol. Rev.* 6 (1954):7–13; O. Loewi, *From the Workshop of Discoveries* (Lawrence: University of Kansas Press, 1953). Paul Fatt reviewed synaptic transmission in "Biophysics of junctional transmission," *Physiol. Rev.* 34 (1954):674–710; quotation from p. 704.

Other information for this chapter was drawn from the following:

Brown, G. L., H. H. Dale, and W. Feldberg. "Reactions of the normal mammalian muscle to acetylcholine and eserine." *J. Physiol.* 87 (1936):394–424.

Eccles, J. C. *Physiology of the Synapses*. Berlin: Springer Verlag, 1964.

Furshpan, E. J., and D. D. Potter. "Transmission at the giant motor synapses of the crayfish." *J. Physiol.* 145 (1959):289–325.

Grundfest, H. "Synaptic and ephaptic transmission." In *Handbook of Physiology.* Section I: *Neurophysiology*, 147–97. Washington, D.C.: American Physiological Society, 1959.

Kandel, E. R., J. H. Schwartz, and T. M. Jessell. *Principles of Neural Science.* 4th ed. New York: McGraw-Hill, 2000.

Katz, B. *Electric Excitation of Nerve.* Oxford: Oxford University Press, 1939.

———. *The Release of Neural Transmitter Substances*. Liverpool: University Press, 1969.

———. "Stephen W. Kuffler." In *Steve: Remembrances of Stephen W. Kuffler*, edited by O. J. McMahan. Sunderland, Mass.: Sinauer Associates, 1990.

Loewi, O., and E. Navratil. "On the humoral propagation of cardiac nerve action. Communication X: The fate of the vagus substance." In *Cellular Neurophysiology: A Source Book*, edited by I. Cooke and M. Lipkin Jr., 478–85. New York: Holt, Rinehart & Winston, 1972. (Original German-language publication 1926.)

Palay, S. L. "Synapses in the central nervous system." *J. Biophys. Biochem. Cytol.* 2 (Suppl.) (1956):193–202.

Popper, K. R., and J. C. Eccles. *The Self and Its Brain*. Berlin: Springer Verlag, 1977.

7: Simple and Complex Neuronal Systems

Visual experiences in response to LSD are described in A. L. Huxley, *The Doors of Perception* (New York: Harper & Brothers, 1954); J. H. Jaffe, "Drugs of addiction and drug abuse," in *The Pharmacological Basis of Therapeutics*, 7th ed., ed. L. S. Goodman and A. Gilman (New York: Macmillan, 1985); and D. W. Woolley and E. N. Shaw, "Evidence for the participation of serotonin in mental processes." *Annals N. Y. Acad. of Sci.* 66 (1957):649–65; discussion, 665–67.

In reconstructing my recollections of Wade Marshall for this chapter, I benefited from discussions with William Landau, Stanley Rappaport, and Tom Marshall, Wade Marshall's son.

Marshall's first seminal papers were: R. W. Gerard, W. H. Marshall, and L. J. Saul, "Cerebral action potentials," *Proc. Soc. Exp. Biol. and Med.* 30 (1933): 1123–25; and R. W. Gerard, W. H. Marshall, and L. J. Saul, "Electrical activity of the cat's brain," *Arch. Neurol. and Psychiat.* 36 (1936): 675–735. His later classic papers include: W. H. Marshall,

C. N. Woolsey, and P. Bard, "Observations on cortical somatic sensory mechanisms of cat and monkey," *J. Neurophysiol.* 4 (1941):1–24; and W. H. Marshall and S. A. Talbot, "Recent evidence for neural mechanisms in vision leading to a general theory of sensory acuity," in *Visual Mechanisms*, ed. H. Kluver, 117–64 (Lancaster, Pa.: Cattell, 1942).

Other information for this chapter was drawn from the following:

Eyzaguirre, C., and S. W. Kuffler. "Processes of excitation in the dendrites and in the soma of single isolated sensory nerve cells of the lobster and crayfish." *J. Gen. Physiol.* 39 (1955):87–119.

———. "Further study of soma, dendrite and axon scitation in single neurons." *J. Gen. Physiol.* 39 (1955):121–53.

Jackson, J. H. *Selected Writings of John Hughlings Jackson*. Edited by J. Taylor. Vol. 1. London: Hodder & Stoughton, 1931.

Katz, B. "Stephen W. Kuffler." In *Steve: Remembrances of Stephen W. Kuffler*. Edited by O. J. McMahan. Sunderland, Mass.: Sinauer Associates, 1990.

Kuffler, S. W., and C. Eyzaguirre. "Synaptic inhibition in an isolated nerve cell." *J. Gen. Physiol.* 39 (1955):155–84.

Penfield, W., and E. Boldrey. "Somatic motor and sensory representation in the cerebral cortex of man as studied by electrical stimulation." *Brain* 60 (1937):389–443.

Penfield, W., and T. Rasmussen. *The Cerebral Cortex of Man: A Clinical Study of Localization of Function*. New York: Macmillan, 1950.

Purpura, D. P., E. R. Kandel, and G. F. Gestrig. "LSD-serotonin interaction on central synaptic activity." Cited in D. P. Purpura. "Experimental analysis of the inhibitory action of lysergic acid diethylamide on cortical dendritic activity in psychopharmacology of psychotomimetic and psychotherapeutic drugs." *Annals N. Y. Acad. of Sci.* 66 (1957):515–36.

Sulloway, F. J. *Freud: Biologist of the Mind*. New York: Basic Books, 1979.

8: Different Memories, Different Brain Regions

For a discussion of Gall, see A. Harrington, *Medicine, Mind, and the Double Brain: A Study in Nineteenth-Century Thought* (Princeton, N.J.: Princeton University Press, 1987); and R. M. Young, *Mind, Brain and Adaptation in the 19th Century* (Oxford: Clarendon Press, 1970).

Broca's 1864 announcement that the left hemisphere governed speech was reprinted in "Sur le siège de la faculté du langue articulé," *Bull. Soc. Antropol.* 6 (1868):337–93; quotation from p. 378. This article has been translated into English by E. A. Berker, A. H. Berker, and A. Smith as "Localization of speech in the third left frontal convolution." *Arch. Neurol.* 43 (1986):1065–72.

Milner talked about H.M. in P. J. Hills, *Memory's Ghost* (New York: Simon & Schuster, 1995), p. 110.

Other information in this chapter was drawn from the following:

For a discussion of Broca and Wernicke, see N. Geschwind, *Selected Papers on Language and the Brain*, Boston Studies in the Philosophy of Science 16 (Norwell, Mass.:

Kluwer, 1974); and T. F. Feinberg and M. J. Farah, *Behavioral Neurology and Neuropsychology* (New York: McGraw Hill, 1997).

Bruner, J. S. "Modalities of memory." In *The Pathology of Memory*, edited by G. A. Talland and N. C. Waugh. New York: Academic Press, 1969.

Flourens, P. *Recherches Expérimentales sur les Propriétes et les Fonctions du Système Nerveux, dans les Animaux Vertébrés*. Paris: Chez Crevot, 1824.

Gall, F .J., and G. Spurzheim. *Anatomie et Physiologie du Système Nerveux en Général, et du Cerveau en Particulier, avec des Observations sur la Possibilité de Reconnaître Plusiers Dispositions Intellectuelles et Morales de l'Homme et des Animaux, par la Configuration de leurs Têtes*. Paris: Schoell, 1810.

James, W. *The Works of William James: The Principles of Psychology*. Edited by F. Burkhardt and F. Bowers. 3 vols. 1890. Reprint, Cambridge, Mass.: Harvard University Press, 1981.

Lashley, K. S. "In search of the engram." *Soc. Exp. Biol.* 4 (1950):454–82.

Milner, B, L. R. Squire, and E. R. Kandel. "Cognitive neuroscience and the study of memory." Review. *Neuron* 20 (1998):445–68.

Ryle, G. *Concept of Mind*. New York: Barnes & Noble, 1949.

Schacter, D. *Searching for Memory: The Brain, the Mind and the Past*. New York: Basic Books, 1996.

Scoville, W. B., and B. Milner. "Loss of recent memory after bilateral hippocampal lesion." *J. Neurol. Neurosurg. Psychiat.* 20 (1957):411–21.

Searle, J. R. *Mind: A Brief Introduction*. London: Oxford University Press, 2004.

Spurzheim, J. G. *A View of the Philosophical Principles of Phrenology*, 3rd ed. London: Knight, 1825.

Squire, L. R. *Memory and Brain*. New York: Oxford University Press, 1987.

Squire, L. R., and E. R. Kandel. *Memory: From Mind to Molecules*. New York: Scientific American, 1999.

Squire, L. R., P. C. Slater, and P. M. Chace. "Retrograde amnesia: Temporal gradient in very long term memory following electroconvulsive therapy." *Science* 187 (1975):77–79.

Warren, R. M. *Helmholtz on Perception: Its Physiology and Development*. New York: John Wiley & Sons, 1968.

Wernicke, C. *Der Aphasische Symptomencomplex*. Breslau: Cohn & Weigert, 1874.

9: Searching for an Ideal System to Study Memory

Alden Spencer and I published several papers together on the hippocampus. See E. R. Kandel, W. A. Spencer, and F. J. Brinley Jr., "Electrophysiology of hippocampal neurons. I: Sequential invasion and synaptic organization," *J. Neurophysiol.* 24 (1961):225–42; E. R. Kandel and W. A. Spencer, "Electrophysiology of hippocampal neurons. II: After-potentials and repetitive firing," *J. Neurophysiol.* 24 (1961):243–59; W. A. Spencer and E. R. Kandel, "Electrophysiology of hippocampal neurons. III: Firing level and time constant," *J. Neurophysiol.* 24 (1961):260–71; and W. A. Spencer and E. R.

Kandel, "Electrophysiology of hippocampal neurons. IV: Fast prepotentials." *J. Neurophysiol.* 24 (1961):272–85; E. R. Kandel and W. A. Spencer, "The pyramidal cell during hippocampal seizure." *Epilepsia* 2 (1961):63–69; and W. A. Spencer and E. R. Kandel, "Hippocampal neuron responses to selective activation of recurrent collaterals of hippocampofugal axons," *Exptl. Neurol.* 4 (1961):149–61.

The experiments on learning memory and the perforant pathway were performed in 2004 and published as: M. F. Nolan, G. Malleret, J. T. Dudman, D. L. Buhl, B. Santoro, E. Gibbs, S. Vronskaya, G. Buzsáki, S. A. Siegelbaum, E. R. Kandel, and A. Morozov, "A behavioral role for dendritic integration: HCN1 channels constrain spatial memory and plasticity at inputs to distal dendrites of CA1 pyramidal neurons." *Cell* 119 (2004):719–32.

The advantages and biology of *Aplysia* are described in E. R. Kandel, *Cellular Basis of Behavior: An Introduction to Behavioral Neurobiology*, (San Francisco: Freeman, 1976); and in *The Behavioral Biology of Aplysia: A Contribution to the Comparative Study of Opisthobranch Molluscs* (San Francisco: Freeman, 1979).

Other information for this chapter was drawn from the following:

Brenner, S. *My Life in Science*. London: Biomed Central, 2002. "What you need . . ." is from pp. 56–60.

———. "Nature's gift to science." In *Les Prix Nobel/The Nobel Prizes*, edited by Nobel Foundation, 268–83. Stockholm: Almquist & Wiksell International, 2002.

Hilgard, E. *Theories of Learning*. New York: Appleton-Century-Crofts, 1956.

10: Neural Analogs of Learning

For an earlier discussion of the Massachusetts Mental Health Center, see E. R. Kandel, "A new intellectual framework for psychiatry," *Am. J. Psych.* 155 (1998):457–69. The study I carried out as a resident is E. R. Kandel, "Electrical properties of hypothalamic neuroendocrine cells." *J. Gen. Physiol.* 47 (1964):691–717.

For a discussion of behaviorism, see I. P. Pavlov, *Conditioned Reflexes: An Investigation of the Physiological Activity of the Cerebral Cortex*, trans. G. V. Anrep (London: Oxford University Press, 1927); B. F. Skinner, *The Behavior of Organisms* (New York: Appleton-Century-Crofts, 1938); E. G. Boring, *A History of Experimental Psychology*, 2nd ed. (New York: Appleton-Century Crofts, 1950); G. A. Kimble, *Hilgard and Marquis' Conditioning and Learning*, 2nd ed. (New York: Appleton-Century-Crofts, 1961); and J. Kornorski, *Conditioned Reflexes and Neuron Organization* (Cambridge: Cambridge University Press, 1948; quotation from pp. 79–80).

The quote from Max Perutz about Jim Watson is from H. F. Judson, *The Eighth Day of Creation* (New York: Simon & Schuster, 1979), p. 21.

The quote from Eccles is in J. C. Eccles, "Conscious experience and memory," in *Brain and Conscious Experience*, ed. J. C. Eccles (New York: Springer, 1966):314–44; quotation from p. 330.

Other information for this chapter was drawn from the following:

Cajal, S. R. "The Croonian Lecture. La fine structure des centres nerveux." *Proc. R. Soc. London Ser. B* 55 (1894):444–67. "Mental exercise facilitates . . ." is from p. 466.

Doty, R. W., and C. Guirgea. "Conditioned reflexes established by coupling electrical excitation to two cortical areas." In *Brain Mechanisms and Learning*, edited by A. Fessard, R. W. Gerard, and J. Kornoski, 133–51. Oxford: Blackwell, 1961.

Kimble, G. A. *Foundations of Conditioning and Learning.* New York: Appleton-Century-Crofts, 1967.

11: Strengthening Synaptic Connections

The studies of the analogs of habituation and sensitization were carried out in cell R2, earlier called the giant cell of *Aplysia*. This was published as E. R. Kandel and L. Tauc, "Mechanism of heterosynaptic facilitation in the giant cell of the abdominal ganglion of *Aplysia depilans*," *J. Physiol.* (London) 181 (1965):28–47. The studies on classical conditioning were carried out on nearby cells that were smaller; see E. R. Kandel and L. Tauc, "Heterosynaptic facilitation in neurons of the abdominal ganglion of *Aplysia depilans*," *J. Physiol.* (London) 181 (1965):1–27; quotation ("The fact that the connections . . .") from p. 24.

Konrad Lorenz is quoted on earthworms in Y. Dudai, *Memory from A to Z* (Oxford: Oxford University Press, 2002), p. 225.

Katz's comment on Hill is also described in his "To tell the you truth, sir, we do it because it's amusing!" in *The History of Neuroscience in Autobiography* , ed. L. R. Squire, vol. 1, 348–81 (Washington, D.C.: Society for Neuroscience, 1996).

For the excellent discussion of learning paradigms that influenced me, see E. Hilgard, *Theories of Learning* (New York: Appleton-Century-Crofts, 1956); and G. A. Kimble, *Foundations of Conditioning and Learning* (New York: Appleton-Century-Crofts, 1967).

On historical anti-Semitism in France, see I. Y. Zingular and S. W. Bloom, eds. *Inclusion and Exclusion: Perspectives on Jews from the Enlightenment to the Dreyfus Affair* (Leiden and Boston: Brill, 2003).

Other information for this chapter was drawn from the following:

Kandel, E. R. *Cellular Basis of Behavior: An Introduction to Behavioral Neurobiology*. San Francisco: Freeman, 1976.

Kandel, E. R., and L. Tauc. "Mechanism of prolonged heterosynaptic facilitation." *Nature* 202 (1964):145–47.

———. "Heterosynaptic facilitation in neurons of the abdominal ganglion of *Aplysia depilans*." *J. Physiol.* (London) 181 (1965):1–27.

———. "Mechanism of heterosynaptic facilitation in the giant cell of the abdominal ganglion of *Aplysia depilans*." *J. Physiol.* (London) 181 (1965):28–47.

12: A Center for Neurobiology and Behavior

The environment at Harvard during the Kuffler period is well described in O. J. McMahan, ed., *Steve: Remembrances of Stephen W. Kuffler* (Sunderland, Mass.: Sinauer

Associates, 1990); and in D. H. Hubel and T. N. Wiesel, *Brain and Visual Reception* (Oxford: Oxford University Press, 2005).

The quote from Per Andersen is from P. Andersen, "A prelude to long-term potentiation," in *LTP: Long-Term Potentiation*, edited by T. Bliss, G. Collingridge, and R. Morris (Oxford: Oxford University Press, 2004). The review that Alden Spencer and I wrote is in E. R. Kandel and W. A. Spencer, "Cellular neurophysiological approaches in the study of learning," *Physiol. Rev.* 48 (1968):65–134.

13: Even a Simple Behavior Can Be Modified by Learning

Mapping of connections between identified cells is based on W. T. Frazier, E. R. Kandel, I. Kupfermann, R. Waziri, and R. E. Coggeshall, "Morphological and functional properties of identified neurons in the abdominal ganglion of *Aplysia californica*." *J. Neurophysiol.* 30 (1967):1288–1351; E. R. Kandel, W. T. Frazier, R. Waziri, and R. E. Coggeshall, "Direct and common connections among identified neurons in *Aplysia*," *J. Neurophysiol.* 30 (1967):1352–76; I. Kupfermann and E. R. Kandel, "Neuronal controls of a behavioral response mediated by the abdominal ganglion of *Aplysia*," *Science* 164 (1969):847–50. In the initial experiments we often used shock to the head instead of tail for a strong unconditioned stimulus in sensitization experiments.

Other information for this chapter was drawn from the following:

Arvanitaki, A., and N. Chalazonitis. "Configurations modales de l'activité, propres à différents neurons d'un même centre," *J. Physiol.* (Paris) 50 (1958):122–25.

Byrne, J., V. Castellucci, and E. R. Kandel. "Receptive fields and response properties of mechanoreceptor neurons innervating siphon skin and mantle shelf of *Aplysia*." *J. Neurophysiol.* 37 (1974):1041–64.

———. "Contribution of individual mechanoreceptor sensory neurons to defensive gill-withdrawal reflex in *Aplysia*." *J. Neurophysiol.* 41 (1978):418–31.

Cajal, S. R. "The Croonian Lecture: La fine structure des centres nerveux." *Proc. R. Soc. London Ser. B* 55 (1894):444–67.

Carew, T. J., R. D. Hawkins, and E. R. Kandel. "Differential classical conditioning of a defensive withdrawal reflex in *Aplysia californica*." *Science* 219 (1983):397–400.

Goldschmidt, R. "Das nervensystem von Ascaris lumbricoides und megalocephala: Ein versuch in den aufhaus eines einfaches nervensystem enzudringen. Erster Teil. Z. Wiss." *Zool.* 90 (1908):73–126.

Hawkins, R. D., V. F. Castellucci, and E. R. Kandel. "Interneurons involved in mediation and modulation of the gill-withdrawal reflex in *Aplysia*. II: Identified neurons produce heterosynaptic facilitation contributing to behavioral sensitization." *J. Neurophysiol.* 45 (1981):315–26.

Kandel, E. R. *Cellular Basis of Behavior: An Introduction to Behavioral Neurobiology*. San Francisco: Freeman, 1976.

———. *The Behavioral Biology of Aplysia: A Contribution to the Comparative Study of Opisthobranch Molluscs*. San Francisco: Freeman, 1979.

Köhler, W. *Gestalt Psychology. An Introduction to New Concepts of Modern Psychology.* Denver: Mentor Books/New American Library, 1947.

Pinsker, H., I. Kupfermann, V. Castellucci, and E. R. Kandel. "Habituation and dishabituation of the gill-withdrawal reflex in *Aplysia.*" *Science* 167 (1970):1740–42.

Thorpe, W. H. *Learning and Instinct in Animals.* Rev. ed. Cambridge, Mass.: Harvard University Press, 1963.

14: Synapses Change with Experience

For a discussion of Freud's theories of synaptic plasticity and memory, see S. Freud, "Project for a scientific psychology," in *Standard Edition,* trans. and ed. James Strachey et al., vol. 1, 281–397 (New York: W. W. Norton, 1976); K. H. Pribram and M. M. Gill, *Freud's "Project" Re-assessed: Preface to Contemporary Cognitive Theory and Neuropsychology* (New York: Basic Books, 1976); and F. J. Sulloway, *Freud: Biologist of the Mind* (New York: Basic Books, 1979).

My colleagues and I also analyzed the mechanisms of classical conditioning. In 1983, Hawkins, Carew, and I delineated a presynaptic component, an enhancement of the mechanisms that contribute to sensitization. In 1992 Nicholas Dale and I found that the sensory neuron uses glutamate as its transmitter. In 1994, my former student David Glanzman and subsequently Robert Hawkins and I made the important observation that there is also an important postsynaptic component. See X. Y. Lin and D. L. Glanzman, "Long-term potentiation of *Aplysia* sensorimotor synapses in cell culture regulation by postsynaptic voltage," *Biol. Sci.* 255 (1994):113–18; and I. Antonov, I. Antonova, E. R. Kandel, and R. D. Hawkins, "Activity-dependent presynaptic facilitation and Hebbian LTP are both required and interact during classical conditioning in *Aplysia,*" *Neuron* 37 (2003):135–47.

For alternative views of the mechanisms of learning, see R. Adey, "Electrophysiological patterns and electrical impedance characteristics in orienting and discriminative behavior," *Proc. Int. Physiol. Soc.* (Tokyo) 23 (1965):324–29; quotation from p. 235; B. D. Burns, *The Mammalian Cerebral Cortex* (London: Arnold, 1958); quotation from p. 96; S. R. Cajal, "The Croonian Lecture. La Fine structure des centers nerveux," *Proc. R. Soc. London Ser. B* 55 (1894):444–67; and D. O. Hebb, *The Organization of Behavior: A Neuropsychological Theory* (New York: John Wiley, 1949).

Other information for this chapter was drawn from the following:

Castellucci, V., H. Pinsker, I. Kupfermann, and E. R. Kandel. "Neuronal mechanisms of habituation and dishabituation of the gill-withdrawal reflex in *Aplysia.*" *Science* 167 (1970):1745–48. "The data indicate . . ." is from p. 1748.

Hawkins, R. D., T. W. Abrams, T. J. Carew, and E. R. Kandel. "A cellular mechanism of classical conditioning in *Aplysia*: Activity-dependent amplification of presynaptic facilitation." *Science* 219 (1983):400–405.

Kandel, E. R. *A Cell-Biological Approach to Learning.* Grass Lecture Monograph I. Bethesda, Md.: Society for Neuroscience, 1978.

Kupfermann, I., V. Castellucci, H. Pinsker, and E. R. Kandel. "Neuronal correlates of habituation and dishabituation of the gill-withdrawal reflex in *Aplysia*." *Science* 167 (1970):1743–45.

Pinsker, H., I. Kupfermann, V. Castellucci, and E. R. Kandel. "Habituation and dishabituation of the gill-withdrawal reflex in *Aplysia*." *Science* 167 (1970):1740–43. "The analysis of the neural mechanisms . . ." is from p. 1740.

15: The Biological Basis of Individuality

The discussion of Helmholtz's work on unconscious inference is based on C. Frith, "Disorders of cognition and existence of unconscious mental processes: An introduction," in E. Kandel et al., *Principles of Neural Science*, 5th ed. (New York: McGraw-Hill, forthcoming); R. M. Warren and R. P. Warren, *Helmholtz on Perception: Its Physiology and Development* (New York: John Wiley & Sons, 1968); R. J. Herrnstein and E. Boring, eds., *A Source Book in the History of Psychology* (Cambridge, Mass.: Harvard University Press, 1965), especially pp. 189–93; and R. L. Gregory, ed., *The Oxford Companion to the Mind* (Oxford: Oxford University Press, 1987), pp. 308–9.

For a discussion of Ebbinghaus, see H. Ebbinghaus, *Memory: A Contribution to Experimental Psychology*, trans. H. A. Ruger and C. E. Bussenius (New York: Teacher's College / Columbia University, 1913); original German-language publication 1885.

For structural changes in *Aplysia*, see C. H. Bailey and M. Chen, "Long-term memory in *Aplysia* modulates the total number of varicosities of single identified sensory neurons," *Proc. Natl. Acad. Sci. USA* 85 (1988):2373–77 and C. H. Bailey and M. Chen, "Time course of structural changes at identified sensory neuron synapses during long-term sensitization in *Aplysia*," *J. Neurosci.* 9 (1989):1774–80; C. H. Bailey and E. R. Kandel, "Structural changes accompanying memory storage," *Annu. Rev. Physiol.* 55 (1993):397–426.

Other information for this chapter was drawn from the following:

Cajal, S. R. "The Croonian Lecture: La fine structure des centres nerveux." *Proc. R. Soc. London Ser. B* 55 (1894):444–67.

Dudai, Y. *Memory from A to Z*. Oxford: Oxford University Press, 2002.

Duncan, C. P. "The retroactive effect of electroshock on learning." *J. Comp. Physiol. Psychol.* 42 (1949):32–44.

Ebert, T., C. Pantev, C. Wienbruch, B. Rockstroh, and E. Taub. "Increased cortical representation of the fingers of the left hand in string players." *Science* 270 (1995):305–7.

Flexner, J. B., L. B. Flexner, and E. Stellar. "Memory in mice as affected by intracerebral puromycin." *Science* 141 (1963):57–59.

Jenkins, W. M., M. M. Merzenich, M. T. Ochs, T. Allard, and E. Guic-Robles. "Functional reorganization of primary somatosensory cortex in adult owl monkeys after behaviorally controlled tactile stimulation." *J. Neurophysiol.* 63 (1990):83–104.

16: Molecules and Short-Term Memory

For background of cyclic AMP, see R. J. DeLange, R. G. Kemp, W. D. Riley, R. A. Cooper, and E. G. Krebs. "Activation of skeletal muscle phosphorylase kinase by adenosine triphosphate and adenosine 3',5'-monophosphate." *J. Biol. Chem.* 243, no. 9 (1968):2200–2208; E. G. Krebs, "Protein phosphorylation and cellular regulation, I" in *Les Prix Nobel (The Nobel Prizes)*, ed. Nobel Foundation (Stockholm: Almquist & Wiksell International, 1992); T. W. Rall and E. W. Sutherland, "The regulatory role of adenosine 3',5'-phosphate. Cold Spring Harbor Symp.," *Quant. Biol.* 26 (1961):347–54; A. E. Gilman, "Nobel lecture. G Proteins and regulation of adenylyl cyclase," *Biosci. Reports* 15 (1995):65–97; P. Greengard, "The neurobiology of dopamine signaling," in *Les Prix Nobel (The Nobel Prizes)*, ed. Nobel Foundation, 262–81 (Stockholm: Almquist & Wiksell International, 2000).

For cAMP in *Aplysia*, see J. H. Schwartz, V. F. Castellucci, and E. R. Kandel, "Functioning of identified neurons and synapses in abdominal ganglion of *Aplysia* in absence of protein synthesis," *J. Neurophysiol.* 34 (1971):939–53; H. Cedar, E. R. Kandel, and J. H. Schwartz, "Cyclic adenosine monophosphate in the nervous system of *Aplysia californica:* Increased synthesis in response to synaptic stimulation." *J. Gen. Physiol.* 60 (1972):558–69; M. Brunelli, V. Castellucci, and E. R. Kandel, "Synaptic facilitation and behavioral sensitization in *Aplysia:* Possible role of serotonin and cyclic AMP." *Science* 194 (1976):1178–81; also, V. F. Castellucci, E. R. Kandel, J. H. Schwartz, F. D. Wilson, A. C. Nairn, and P. Greengard. "Intracellular injection of the catalytic subunit of cyclic AMP-dependent protein kinase simulates facilitation of transmitter release underlying behavioral sensitization in *Aplysia.*" *Proc. Natl. Acad. Sci. USA* 77 (1980):7492–96.

For cAMP in *Drosophila,* see S. Benzer, "Behavioral mutants of *Drosophila* isolated by counter current distribution," *Proc. Natl. Acad. Sci.* 58 (1967): 1112–19; D. Byers, R. L. Davis, and J. R. Kiger, Jr., "Defect in cyclic AMP phosphodiesterase due to the dunce mutation of learning in *Drosophila melanogaster,*" *Nature* 289 (1981):79–81; Y. Dudai, Y. N. Jan, D. Byers, W. G. Quinn, and S. Benzer. "Dunce, a mutant of *Drosophila* deficient in learning." *Proc. Natl. Acad. Sci. USA* 73, no. 5 (1976): 1684–88.

Other information in this chapter was drawn from the following:

Castellucci, V., and E. R. Kandel. "Presynaptic facilitation as a mechanism for behavioral sensitization in *Aplysia.*" *Science* 194 (1976):1176–78.

Dale, N., and E. R. Kandel. "L-glutamate may be the fast excitatory transmitter of *Aplysia* sensory neurons." *Proc. Nat. Acad. Sci. USA* 90 (1993):7163–67.

Jacob, F. *The Possible and the Actual.* New York: Pantheon, 1982; quotation from pp. 33–35.

———. *The Statue Within.* Translated by F. Philip. New York: Basic Books, 1988.

Kandel, E. R. *Cellular Basis of Behavior: An Introduction to Behavioral Neurobiology.* San Francisco: Freeman, 1976.

Kandel, E. R., M. Klein, B. Hochner, M. Shuster, S. Siegelbaum, R. Hawkins, D. Glanzman, V. F. Castellucci, and T. Abrams. "Synaptic modulation and learning: New insights into synaptic transmission from the study of behavior." In *Synaptic*

Function, edited by G. M. Edelman, W. E. Gall, and W. M. Cowan, 471–518. New York: John Wiley & Sons, 1987.

Kistler, H. B., Jr., R. D. Hawkins, J. Koester, H. W. M. Steinbusch, E. R. Kandel, and J. H. Schwartz. "Distribution of serotonin-immunoreactive cell bodies and processes in the abdominal ganglion of mature *Aplysia*." *J. Neurosci.* 5 (1985):72–80.

Kriegstein, A., V. F. Castellucci, and E. R. Kandel. "Metamorphosis of *Aplysia californica* in laboratory culture." *Proc. Nat. Acad. Sci. USA* 71 (1974):3654–58.

Kuffler, S., and J. Nicholls. *From Neuron to Brain: A Cellular Approach to the Function of the Nervous System*. Sunderland, Mass.: Sinauer Associates, 1976.

Siegelbaum, S., J. S. Camardo, and E. R. Kandel. "Serotonin and cAMP close single K^+ channels in *Aplysia* sensory neurons." *Nature* 299 (1982):413–17.

17: Long-Term Memory

François Jacob writes about day versus night science in *The Statue Within*, trans. F. Philip (New York: Basic Books, 1988), pp. 296–97.

For a discussion of Thomas Hunt Morgan, see two biographies: G. E. Allen, *Thomas Hunt Morgan: The Man and His Science* (Princeton, N.J.: Princeton University Press, 1978); and A. H. Sturtevant, *Thomas Hunt Morgan* (New York: National Academy of Sciences, 1959). See also E. R. Kandel, "Thomas Hunt Morgan at Columbia: Genes, chromosomes, and the origins of modern biology," pp. 29–35, and E. R. Kandel, "An American century of biology," pp. 36–39, both in *Living Legacies: Great Moments in the Life of Columbia for the 250th Anniversary*, fall 1999 issue of *Columbia: The Magazine* of *Columbia University*.

Watson and Crick first announced their findings in "Molecular structure of nucleic acids: A structure of deoxyribose nucleic acid," *Nature* 171 (1953):737–38; quotation from p. 738. See also J. D. Watson and F. H. C. Crick, "Genetical implications of the structure of deoxyribonucleic acid," *Nature* 171 (1953):964–67; J. D. Watson, *The Double Helix* (1968; reprint, New York: Touchstone / Simon & Schuster, 2001); and J. D. Watson and A. Berry, *DNA: The Secret of Life* (New York: Alfred A. Knopf, 2003). The latter book is the source for Watson's reflections (p. 88). Schrödinger's essay appeared in E. Schrödinger, *What Is Life? The Physical Aspect of the Living Cell*. 1944. (Reprint, Cambridge: Cambridge University Press, 1947).

Other information for this chapter was drawn from the following:

Avery, O. T., C. M. MacLeod, and M. McCarty. "Studies on the chemical nature of the substance inducing transformation of pneumococcal types: Induction of transformation by a desoxyribonucleic acid fraction isolated from Pneumococcus Type III." *J. Exp. Med.* 79 (1944):137–58.

Chimpanzee Genome. Special issue on chimpanzees. *Nature* 437, September 1, 2005.

Cohen, S. N., A. C. Chang, H. W. Boyer, and R. B. Helling. "Construction of biologically functional bacterial plasmids *in vitro*." *Proc. Natl. Acad. Sci. USA* 70, no. 11 (1973):3240–44.

Crick, F. H., L. Barnett, S. Brenner, and R. J. Watts-Tobin. "General nature of the genetic code for proteins." *Nature* 192 (1961):1227–32.

Gilbert, W. "DNA sequencing and gene structure." *Science* 214 (1981):1305–12.

Jackson, D. A., R. H. Symons, and P. Berg. "Biochemical method for inserting new genetic information into DNA Simian Virus 40: circular SV40 DNA molecules containing lambda phage genes and the galactose operon of *Escherichia coli*." *Proc. Nat. Acad. Sci. USA* 69 (1972):2904–09.

Jessell, T. M., and E. R. Kandel. "Synaptic transmission: A bidirectional and a self-modifiable form of cell-cell communication." *Cell 72/Neuron 10* (Suppl.) (1993):1–30.

Matthaei, H., and M. W. Nirenberg. "The dependence of cell-free protein synthesis in *E. coli* upon RNA prepared from ribosomes." *Biochem. Biophys. Res. Commun.* 4 (1961):404–8.

Sanger, F. "Determination of nucleotide sequences in DNA." *Science* 214 (1981):1205–10.

18: Memory Genes

Jacob and Monod's classic paper is F. Jacob and J. Monod, "Genetic regulatory mechanisms in the synthesis of proteins," *J. Molec. Biol.* 3 (1961):318–56.

Other information in this chapter was drawn from the following:

Buck, L., and R. Axel. "Novel multigene family may encode odorant receptors: A molecular basis for odor recognition." *Cell* 65, no. 1 (1991):175–87.

Jacob, F. *The Statue Within*. Translated by F. Philip. New York: Basic Books, 1988.

Kandel, E. R., A. Kriegstein, and S. Schacher. "Development of the central nervous system of *Aplysia* in the terms of the differentiation of its specific identifiable cells." *Neurosci.* 5 (1980):2033–63.

Scheller, R. H., J. F. Jackson, L. B. McAllister, J. H. Schwartz, E. R. Kandel, and R. Axel. "A family of genes that codes for ELH, a neuropeptide eliciting a stereotyped pattern of behavior in *Aplysia*." *Cell* 28 (1982):707–19; quotation from p. 707.

Weinberg, R. A. *Racing to the Beginning of the Road: The Search for the Origin of Cancer*. San Francisco: Freeman, 1998; quotation from pp. 162–63.

19: A Dialogue Between Genes and Synapses

The two reviews by Phillip Goelet are P. Goelet, V. F. Castellucci, S. Schacher, and E. R. Kandel, "The long and short of long-term memory—a molecular framework," *Nature* 322 (1986):419–22; and P. Goelet and E. R. Kandel, "Tracking the flow of learned information from membrane receptors to genome," *Trends Neurosci.* 9 (1986):472–99.

In the experiments on the translocation of the cAMP-dependent protein kinase, we collaborated with Roger Tsien, a Howard Hughes investigator at the University of California, San Diego, who developed the method we used to visualize the movement of the cAMP-dependent protein kinase to the nucleus. This work is described in B. J. Bacskai, B. Hochner, M. Mahaut-Smith, S. R. Adams, B.-K. Kaang, E. R. Kandel, and R. Y. Tsien, "Spatially resolved dynamics of cAMP and protein kinase A subunits in *Aplysia* sensory neurons," *Science* 260 (1993):222–26.

The development of tissue culture methods for the *Aplysia* neuron was initiated by Sam Schacher in collaboration with my students Stephen Rayport, Pier Giorgio Montarolo, and Eric Proshansky.

The initial evidence for CREB in learning-related plasticity is in P. K. Dash, B. Hochner, and E. R. Kandel, "Injection of cAMP-responsive element into the nucleus of *Aplysia* sensory neurons blocks long-term facilitation," *Nature* 345 (1990):718–21.

The finding of a repressor in *Aplysia* is described in D. Bartsch, M. Ghirardi, P. A. Skehel, K. A. Karl, S. P. Herder, M. Chen, C. H. Bailey, and E. R. Kandel, "*Aplysia* CREB-2 represses long-term facilitation: Relief of repression converts transient facilitation into long-term functional and structural change," *Cell* 83 (1995):979–92.

For the new protocol to study memory in *Drosophila,* see T. Tully, T. Preat, S. C. Boynton, and M. Del Vecchio, "Genetic dissection of consolidated memory in *Drosophila melanogaster,*" *Cell* 79 (1994): 35–47.

The studies in *Drosophila* that pointed to the role of CREB repressor in blocking long-term memory and activator over-expressed in enhancing memory storage for learned fear is in J. C. P. Yin, J. S. Wallach, M. Del Vecchio, E. L. Wilder, H. Zhuo, W. G. Quinn, and T. Tully, "Induction of a dominant negative CREB transgene specifically blocks long-term memory in *Drosophila,*" *Cell* 79 (1994):49–58; J. C. P. Yin, M. Del Vecchio, H. Zhou, and T. Tully, "CREB as a memory modulator: Induced expression of a dCREB2 activator isoform enhances long-term memory in Drosophila." *Cell* 81 (1995): 107–15.

For evidence of CREB in the honeybee, see D. Eisenhardt, A. Friedrich, N. Stollhoff, U. Müller, H. Kress, and R. Menzel, "The *AmCREB* gene is an ortholog of the mammalian CREB/CREM family of transcription factors and encodes several splice variants in the honeybee brain," *Insect Molecular Biol.* 12 (2003):373–82.

The evidence for CREB in learned fear in the mouse is in P. W. Frankland, S. A. Josselyn, S. G. Anagnostaras et al., "Consolidation of CS and US representations in associative fear conditioning," *Hippocampus* 14 (2004):557–69; and S. Kida, S. A. Josselyn, S. P. de Ortiz et al., "CREB required for the stability of new and reactivated fear memories," *Nature Neurosci.* 5 (2002):348–55.

For the evidence for CREB in human learning, see J. M. Alarcon, G. Malleret, K. Touzani, S. Vronskaya, S. Ishii, E. R. Kandel, and A. Barco, "Chromatin acetylation, memory, and LTP are impaired in CBP$^{+/-}$ mice: A model for the cognitive deficit in Rubinstein-Taybi Syndrome and its amelioration," *Neuron* 42 (2004):947–59.

Other information for this chapter was drawn from the following:

Bailey, C. H., P. Montarolo, M. Chen, E. R. Kandel, and S. Schacher. "Inhibitors of protein and RNA synthesis block structural changes that accompany long-term heterosynaptic plasticity in *Aplysia*." *Neuron* 9 (1992):749–58.

Bartsch, D., A. Casadio, K. A. Karl, P. Serodio, and E. R. Kandel. "CREB-1 encodes a nuclear activator, a repressor, and a cytoplasmic modulator that form a regulatory unit critical for long-term facilitation." *Cell* 95 (1998):211–23.

Bartsch, D., M. Ghirardi, A. Casadio, M. Giustetto, K. A. Karl, H. Zhu, and E. R. Kandel. "Enhancement of memory-related long-term facilitation by ApAF, a novel transcription factor that acts downstream from both CREB-1 and CREB-2." *Cell* 103 (2000):595–608.

Casadio, A., K. C. Martin, M. Giustetto, H. Zhu, M. Chen, D. Bartsch, C. H. Bailey, and E. R. Kandel. "A transient neuron-wide form of CREB-mediated long-term facilitation can be stabilized at specific synapses by local protein synthesis." *Cell* 99 (1999):221–37.

Chain, D. G., A. Casadio, S. Schacher, A. N. Hegde, M. Valbrun, N. Yamamoto, A. L. Goldberg, D. Bartsch, E. R. Kandel, and J. H. Schwartz. "Mechanisms for generating the autonomous cAMP-dependent protein kinase required for long-term facilitation in *Aplysia*." *Neuron* 22 (1999):147–56.

Dale, N., and E. R. Kandel. "L-glutamate may be the fast excitatory transmitter of *Aplysia* sensory neurons." *Proc. Natl. Acad. Sci. USA* 90 (1993):7163–67.

Glanzman, D. L., E. R. Kandel, and S. Schacher. "Target-dependent structural changes accompanying long-term synaptic facilitation in *Aplysia* neurons." *Science* 249 (1990):799–802.

Kaang, B.-K., E. R. Kandel, and S. G. N. Grant. "Activation of cAMP-responsive genes by stimuli that produce long-term facilitation in *Aplysia* sensory neurons." *Neuron* 10 (1993):427–35.

Lorenz, K. Z. *The Foundations of Ethology*. New York: Springer Verlag, 1981.

Martin, K. C., D. Michael, J. C. Rose, M. Barad, A. Casadio, H. Zhu, and E. R. Kandel. "MAP kinase translocates into the nucleus of the presynaptic cell and is required for long-term facilitation in *Aplysia*." *Neuron* 18 (1997):899–912.

Martin, K. C., A. Casadio, H. Zhu, E. Yaping, J. Rose, C. H. Bailey, M. Chen, and E. R. Kandel. "Synapse-specific transcription-dependent long-term facilitation of the sensory to motor neuron connection in *Aplysia*: A function for local protein synthesis in memory storage." *Cell* 91 (1997):927–38.

Mayford, M., A. Barzilai, F. Keller, S. Schacher, and E. R. Kandel. "Modulation of an NCAM-related adhesion molecule with long-term synaptic plasticity in *Aplysia*." *Science* 256 (1992):638–44.

Montarolo, P. G., P. Goelet, V. F. Castellucci, J. Morgan, E. R. Kandel, and S. Schacher. "A critical period for macromolecular synthesis in long-term heterosynaptic facilitation in *Aplysia*." *Science* 234 (1986):1249–54.

Montminy, M. R., K. A. Sevarino, J. A. Wagner, G. Mandel, and R. H. Goodman. "Identification of a cyclic-AMP-responsive element within the rat somatostatin gene." *Proc. Natl. Acad. Sci. USA* 83, no. 18 (1986):6682–86.

Prusiner, S. B. "Prions." *Les Prix Nobel/The Nobel Prizes*, edited by Nobel Foundation. Stockholm: Almquist & Wiksell International, 1997.

Rayport, S. G., and S. Schacher. "Synaptic plasticity *in vitro*: Cell culture of identified *Aplysia* neurons mediating short-term habituation and sensitization." *J. Neurosci.* 6 (1986):759–63.

Schacher, S., V. F. Castellucci, and E. R. Kandel. "cAMP evokes long-term facilitation in *Aplysia* sensory neurons that requires new protein synthesis." *Science* 240 (1988):1667–69.

Si, K., M. Giustetto, A. Etkin, R. Hsu, A. M. Janisiewicz, M. C. Miniaci, J.-H. Kim, H.

Zhu, and E. R. Kandel. "A neuronal isoform of CPEB regulates local protein synthesis and stabilizes synapse-specific long-term facilitation in *Aplysia*." *Cell* 115 (2003):893–904.

Si, K., S. Lindquist, and E. R. Kandel. "A neuronal isoform of the *Aplysia* CPEB has prion-like properties." *Cell* 115 (2003):879–91.

Steward, O., and E. M. Schuman. "Protein synthesis at synaptic sites on dendrites." *Annu. Rev. Neurosci.* 24 (2001):299–325.

20: A Return to Complex Memory

Virginia Woolf wrote about memories of her mother in "Sketches of the Past," which was reprinted in J. Schulkind, ed., *Moments of Being* (New York: Harcourt Brace, 1985), p. 98; and is cited in S. Nalbation, *Memory in Literature: Rousseau to Neuroscience* (New York: Palgrave Macmillan, 2003).

Christof Koch quotes Tennessee Williams's *The Milk Train Doesn't Stop Here Anymore* on p. 187 of *The Quest for Consciousness: A Neurobiological Approach* (Englewood, Col.: Roberts, 2004).

The first description of place cells is in J. O'Keefe and J. Dostrovsky. "The hippocampus as a spatial map. Preliminary evidence from unit activity in the freely-moving rat." *Brain Res.* 34, no. 1 (1971):171–75.

For an excellent review of long-term potentiation, see T. Bliss, G. Collingridge, and R. Morris, eds., *LTP: Long-Term Potentiation* (Oxford: Oxford University Press, 2003). Among the many illuminating articles in this volume are P. Andersen, "A prelude to long-term potentiation"; R. Malinow, "AMPA receptor trafficking and long-term potentiation"; R. G. M. Morris, "Long-term potentiation and memory"; and R. A. Nicoll, "Expression mechanisms underlying long-term potentiation: a postsynaptic view."

Other information for this chapter was drawn from the following:

Baudry, M., R. Siman, E. K. Smith, and G. Lynch. "Regulation by calcium ions of glutamate receptor binding in hippocampal slices." *Euro. J. Pharmacol.* 90, no. 2–3 (1983):161–68.

Bliss, T. V., and T. Lømo. "Long-lasting potentiation of synaptic transmission in the dentate gyrus of the anesthethized rabbit following stimulation of the perforant path." *J. Physiol.* 232 (1973):331–56.

Collingridge, G. L., S. J. Kehl, and H. McLennan. "Excitatory amino acids in synaptic transmission in the Schaffer collateral-commissural pathway of the rat hippocampus." *J. Physiol.* (London) 334 (1983):33–46.

Curtis, D. R., J. W. Phillis, and J. C. Watkins. "The chemical excitation of spinal neurons by certain acidic amino acids." *J. Physiol.* 150 (1960):656–82.

Eccles, J. C. *The Physiology of Synapses*. Berlin: Springer Verlag, 1964.

Hebb, D. O. *The Organization of Behavior: A Neuropsychological Theory.* New York: Wiley, 1949; quotation from p. 62.

Nowak, L., P. Bregestovski, P. Ascher, A. Herbet, and A. Prochiantz. "Magnesium gates glutamate-activated channels in mouse central neurons." *Nature* 307 (1984):462–65.

O'Dell, T. J., S. G. N. Grant, K. Karl, P. M. Soriano, and E. R. Kandel. "Pharmacological and genetic approaches to the analysis of tyrosine kinase function in long-term potentiation." Cold Spring Harbor Symp. *Quant. Biol.* 57 (1992):517–26.

Roberts, P. J., and J. C. Watkins. "Structural requirements for inhibition for L-glutamate uptake by glia and nerve endings." *Brain Res.* 85, no. 1 (1975):120–25.

Schacter, D. L. *Searching for Memory: The Brain, the Mind and the Past.* New York: Basic Books, 1996.

Spencer, W. A. and E. R. Kandel. "Electrophysiology of hippocampal neurons. IV: Fast prepotentials." *J Neurophysiol.* 24 (1961): 272–85.

Westbrook, G. L., and M. L. Mayer. "Glutamate currents in mammalian spinal neurons resolution of a paradox." *Brain Res.* 301, no. 2 (1984):375–79.

21: Synapses Also Hold Our Fondest Memories

Methods for developing genetically modified mice are described in R. L. Brinster and R. D. Palmiter. "Induction of foreign genes in animals." *Trends Biochem. Sci.* 7 (1982):438–40; and M. R. Capecchi, "High-efficiency transformation by direct microinjection of DNA into cultured mammalian cells." *Cell* 22, no. 2 (1980):479–88.

The first reports of the effects of gene knockout in LTP and spatial memory are in S. G. N. Grant, T. J. O'Dell, K. A. Karl, P. L. Stein, P. Soriano, and E. R. Kandel, "Impaired long-term potentiation, spatial learning, and hippocampal development in fyn mutant mice." *Science* 258 (1992):1903–10; A. J. Silva, R. Paylor, J. M. Wehner, and S. Tonegawa, "Impaired spatial learning in alpha-calcium-calmodulin kinase II mutant mice." *Science* 257 (1992):206–11.

The collaborative experiments with Steven Siegelbaum, also referred to in chapter 9, were carried out by Matt Nolan and Josh Dudman. The experiments are described in: M. F. Nolan, G. Malleret, J. T. Dudman, D. Buhl, B. Santoro, E. Gibbs, S. Vronskaya, G. Buzsáki, S. A. Siegelbaum, E. R. Kandel, and A. Morozov, "A behavioral role for dendritic integration: HCN1 channels constrain spatial memory and plasticity at inputs to distal dendrites of CA1 pyramidal neurons." *Cell* 119 (2004):719–32.

Other information for this chapter was drawn from the following:

Mayford, M., T. Abel, and E. R. Kandel. "Transgenic approaches to cognition." *Curr. Opin. Neurobiol.* 5 (1995):141–48.

Mayford, M., M. E. Bach, Y.-Y. Huang, L. Wang, R. D. Hawkins, and E. R. Kandel. "Control of memory formation through regulated expression of a CaMLIIα transgene." *Science* 274 (1996):1678–83.

Mayford, M., D. Baranes, K. Podyspanina, and E. R. Kandel. "The 3'-untranslated region of CaMLIIα is a cis-acting signal for the localization and translation of mRNA in dendrites." *Proc. Natl. Acad. Sci. USA* 93 (1996):13250–55.

Silva, A. J., C. F. Stevens, S. Tonegawa, and Y. Wang. "Deficient hippocampal long-term potentiation in alpha-calcium-calmodulin kinase-II mutant mice." *Science* 257 (1992):201–6.

Tsien, J. Z., D. F. Chen, D. Gerber, C. Tom, E. H. Mercer, D. J. Anderson, M. Mayford, E. R. Kandel, and S. Tonegawa. "Subregion and cell-type restricted gene knockout in mouse brain." *Cell* 87 (1996):1317–26.

Tsien, J. Z., P. T. Huerta, and S. Tonegawa. "The essential role of hippocampal CA1 NMDA receptor-dependent synaptic plasticity in spatial memory." *Cell* 87 (1996):1327-38.

22: The Brain's Picture of the External World

For a neurologist's perspective on cognition, see S. Freud, *The Interpretation of Dreams,* 1900 (reprint, London: Hogarth, 1953); and O. Sacks, *The Man Who Mistook His Wife for a Hat* (New York: Alfred A. Knopf, 1985).

For a perspective on cognitive psychology, see G. A. Miller, *Psychology: The Science of Mental Life* (New York: Harper & Row, 1962); and U. Neisser, *Cognitive Psychology* (New York: Appleton-Century-Crofts, 1967), quotation from p. 3.

For reviews of the work of Mountcastle, Hubel, and Wiesel, see D. H. Hubel and T. N. Wiesel, *Brain and Visual Perception* (New York: Oxford University Press, 2005); V. B. Mountcastle, "Central nervous mechanisms in mechanoreceptive sensibility," in *Handbook of Physiology.* Section 1, *The Nervous System.* Vol. 3, *Sensory Processes,* Part 2, 789–878, ed. I. Darian Smith (Bethesda, Md.: American Physiological Society, 1984); and V. B. Mountcastle, "The view from within: Pathways to the study of perception," *Johns Hopkins Med J.* 136, no. 3 (1975): 109–31, quotation from p. 109 (original italics).

Other information for this chapter was drawn from the following:

Evarts, E. V. "Pyramidal tract activity associated with a conditioned hand movement in the monkey." *J. Neurophysiol.* 29 (1966):1011–27.

Gregory, R. L., ed. *The Oxford Companion to the Mind.* Oxford: Oxford University Press, 1987.

Marshall, W. H., C. N. Woolsey, and P. Bard. "Observations on cortical somatic sensory mechanisms of cat and monkey." *J. Neurophysiol.* 4 (1941):1–24.

Marshall, W. H., and S. A. Talbot. "Recent evidence for neural mechanisms in vision leading to a general theory of sensory acuity." In *Visual Mechanisms*, edited by H. Kluver, 117–64. Lancaster, Pa.: Cattell, 1942.

Movshon, J. A. "Visual processing of moving images." In *Images and Understanding: Thoughts About Images; Ideas About Understanding,* edited by H. Barlow, C. Blakemore, and M. Weston-Smith, 122-37. New York: Cambridge University Press, 1990.

Tolman, E. C. *Purposive Behavior in Animals and Men.* New York: Century, 1932

Wurtz, R. H., M. E. Goldberg, and D. L. Robinson. "Brain mechanisms of visual attention." *Sci. Am.* 246, no. 6 (1982):124.

Zeki, S. M. *A Vision of the Brain.* Oxford: Oxford University Press, 1993; quotation from pp. 295–96 (original italics).

23: Attention Must Be Paid!

For a detailed discussion of the hippocampus and space, see J. O'Keefe and L. Nadel, *The Hippocampus as a Cognitive Map* (Oxford: Clarendon Press, 1978), quotation from p. 5.

For discussion of attention, see W. James, *The Works of William James. The Principles of Psychology*, ed. F. Burkhardt and F. Bowers, 3 vols. (1890) (reprint, Cambridge, Mass.: Harvard University Press, 1981), quotation from I: pp. 380–81, italics in original.

For attention, space, and memory, see F. A. Yates, *The Art of Memory* (Chicago: University of Chicago Press; London: Routledge & Kegan Paul, 1966).

Gender differences are discussed in E. A Maguire, N. Burgess, and J. O'Keefe, "Human spatial navigation: Cognitive maps, sexual dimorphism and neural substrates," *Current Opin Neurobiol.* 9, no.2 (1999):171–77.

Other information in this chapter was drawn from the following:

Agnihotri, N. T., R. D. Hawkins, E. R. Kandel, and C. G. Kentros. "The long-term stability of new hippocampal place fields requires new protein synthesis." *Proc. Natl. Acad. Sci. USA* 101 (2004):3656–61.

Bushnell, M. C., M. E. Goldberg, and D. L. Robinson."Behavioral enhancement of visual responses in monkey cerebral cortex. 1: Modulation in posterior parietal cortex related to selective visual attention." *J. Neurophysiol.* 46, no. 4 (1981):755–72.

Kentros, C. G., N. T. Agnihotri, S. Streater, R. D. Hawkins, and E. R. Kandel. "Increased attention to spatial context increases both place field stability and spatial memory." *Neuron* 42 (2004):283–95.

McHugh, T. J., K. I. Blum, J. Z. Tsien, S. Tonegawa, and M. A. Wilson. "Impaired hippocampal representation of space in CA1-specific NMDAR1 knockout mice." *Cell* 87 (1996):1339–49.

O'Keefe, J., and J. Dostrovsky. "The hippocampus as a spatial map: Preliminary evidence from unit activity in the freely-moving rat." *Brain Res.* 34, no. 1 (1971):171–75.

Rotenberg, A., M. Mayford, R. D. Hawkins, E. R. Kandel, and R. U. Muller. "Mice expressing activated CaMKII lack low frequency LTP and do not form stable place cells in the CA1 region of the hippocampus." *Cell* 87 (1996):1351–61.

Theis, M., K. Si, and E. R. Kandel. "Two previously undescribed members of the mouse CPEB family of genes and their inducible expression in the principal cell layers of the hippocampus." *Proc. Natl. Acad. Sci. USA* 100 (2003):9602-7.

Zeki, S. M. *A Vision of the Brain.* Oxford: Oxford University Press, 1993.

24: A Little Red Pill

For a discussion of Pasteur's contribution to science and industry, see R. J. Dubos, *Louis Pasteur* (Boston: Little, Brown, 1950); and M. Perutz, "Deconstructing Pasteur," in *I Wish I'd Made You Angry Earlier: Essays on Science, Scientists and Humanity* (Plainview, N.Y.: Cold Spring Harbor Laboratory Press, 1998), pp. 119–30.

For a discussion of Dale's interaction with academic and industrial life, see H. H. Dale, *Adventures in Physiology* (London: Pergamon, 1953).

For a discussion of the early history of biotechnology, see S. Hall, *Invisible Frontiers: The Race to Synthesize a Human Gene* (New York: Atlantic Monthly Press, 1987); and J. D. Watson and A. Berry, *DNA: The Secret of Life* (New York: Alfred A. Knopf., 1987). Hall (p. 94) is the source for "sin" and "purist heaven."

Kenney, M. *Biotechnology. The University-Industrial Complex.* New Haven: Yale University Press, 1986.

For a discussion of neuroethics, see M. J. Farah, J. Illes, R. Cook-Deegan, H. Gardner, E. R. Kandel, P. King, E. Parens, B. Sahakian, and P. R. Wolpe. "Science and society: Neurocognitive enhancement: What can we do and what should we do?" *Nat. Rev. Neurosci.* 5 (2004):421–25; S. Hyman, "Introduction: The brain's special status," *Cerebrum* 6, no. 4 (2004):9–12, quotation from p. 9. S. J. Marcus, ed. *Neuroethics: Mapping the Field* (New York: Dana Press, 2004).

Other information in this chapter was drawn from the following:

Bach, M. E., M. Barad, H. Son, M. Zhuo, Y.-F. Lu, R. Shih, I. Mansuy, R. D. Hawkins, and E. R. Kandel. "Age-related defects in spatial memory are correlated with defects in the late phase of hippocampal long-term potentiation *in vitro* and are attenuated by drugs that enhance the cAMP signaling pathway." *Proc. Natl. Acad. Sci. USA* 96 (1999):5280–85.

Barad, M., R. Bourtchouladze, D. Winder, H. Golan, and E. R. Kandel. "Rolipram, a type IV-specific phosphodiesterase inhibitor, facilitates the establishment of long-lasting long-term potentiation and improves memory." *Proc. Natl. Acad. Sci. USA* 95 (1998):15020–25.

25: Mice, Men, and Mental Illness

Another major impetus behind the rise of molecular neurology was the early emergence of patient advocacy groups. Groups of patients and their families and friends have formed around particular diseases since at least the 1930s, when the Infantile Paralysis Foundation initiated the March of Dimes, spurred by President Franklin D. Roosevelt, who had contracted poliomyelitis in 1921. The foundation supported both the basic and the clinical research that led to the development of the polio vaccines, which ultimately eradicated the disease. This was a remarkable process and it was based on the ability of the foundation to raise substantial sums of money and to select scientific advisors who supported imaginative, rigorous research.

In the 1960s a similar approach was taken toward genetic diseases of the nervous system. As the historian Alice Wexler, herself a member of a patient advocate group, has written: "The decade of the 1960s, with its blossoming of social activism, also helped foster a political atmosphere favorable to mobilizing families directly affected by the illness. Civil rights activism, the feminist health movement, and the patient rights movement of the sixties and seventies all created an environment that encouraged families [of patients with genetic diseases of the nervous system] . . . to act on their behalf" (A. Wexler, *Mapping Fate: A Memoir of Family, Risk, and Genetic Research* [New York: Times Books/Random House, 1995], p. xv).

In 1967 songwriter and poet Woody Guthrie died of Huntington's disease. This terrible illness mobilized his former wife, the dancer Marjorie Guthrie, to organize families of people with the disease into a group called the Committee to Combat Huntington's, which later evolved into the Huntington's Disease Society of America. This advocacy group lobbied Congress to speed up the development of effective therapies and to obtain greater support for efforts to alleviate the consequences of the disease by educating family members and training health professionals.

The same year that Woody Guthrie died, Leonore Wexler was diagnosed with Huntington's disease, as her two siblings had been before her. Leonore's husband, Milton Wexler, a gifted and farsighted psychoanalyst with a successful practice in Los Angeles, realized that having one parent with Huntington's disease gave his daughters, Alice, the historian, and Nancy, a psychologist who later became a friend and colleague of mine at Columbia, a 50-50 chance of inheriting the disease. Worried about his daughters and pained by his ex-wife's illness, Wexler formed the Hereditary Disease Foundation. This foundation had a different orientation from Guthrie's and produced a paradigm shift not only in patient advocacy, but also in how to carry out effective research on a genetic disorder.

Wexler decided not to focus on treatment of the disease because too little was known about it to make such an effort productive. Rather, he turned to basic science and raised money for research targeted at finding and characterizing the mutant gene that causes the disease. Wexler did not simply provide scientists with resources, however. He set up and led working groups of the best scientists to debate alternative strategies and to delineate the ones most likely to succeed. He then recruited and supported scientists with those strategic skills, and he met with them frequently to review progress and plan the next steps.

This strategy, initiated by Milton and carried forward for the next thirty years by Nancy, proved amazingly successful. People with Huntington's disease were identified, their families' medical histories established, and tissue banks organized. The scientific community was kept informed of these efforts, so each step the foundation took—from locating the gene (by Nancy Wexler and Jim Gusella) to cloning it and to developing animal models of the disease—was cause for general celebration for the whole scientific community.

This is described in A. Wexler, *Mapping Fate: A Memoir of Family, Risk, and Genetic Research* (New York: Times Books/Random house, 1995).

Indeed, the success of the Hereditary Disease Foundation did not go unnoticed by the relatives of the mentally ill. A number of patient-interest groups related to mental illness have now been formed, of which the most influential has been the National Association for Research in Schizophrenia and Depression (NARSAD). Founded in 1986 by Connie and Steve Lieber and Herbert Pardes, former director of the National Institute of Mental Health, NARSAD provided major direction and support for research in mental illness. Now several other foundations based on patient interest groups also are having an important impact in mental health research including the National Alliance for Mental Illness, the Fragile-X Foundation, and Cure Autism Now.

For a general review of the biology of emotional states, see C. Darwin, *The Expression of Emotion in Man and Animals* (New York: Appleton, 1873); W. B. Cannon, "The James-Lange theory of emotions: A critical examination and an alternative theory," *Am. J. Psychol.* 39 (1927):106–24; W. B. Cannon, *The Wisdom of the Body* (New York: W. W. Norton, 1932); A. R. Damasio, *The Feeling of What Happens: Body and Emotion in the Making of Consciousness* (New York: Harcourt Brace, 1999); M. Davis, "The role of the amygdala in fear and anxiety," *Annu. Rev. Neurosci.* 15 (1992):353–75; J. E. LeDoux, *The Emotional Brain* (New York: Simon & Schuster, 1996); J. Panskseep, *Affective Neuroscience: The Foundations of Human and Animal Emotions* (New York: Oxford University Press, 1998); W James, "What is an emotion?" *Mind* 9 no. 34 (1884):188–205; and C. G. Lange, *Om Sindsbe Vaegelser et Psycho* (Copenhagen: Kromar, 1885). James republished Lange's theory in his *Principles of Psychology*, now available in a definitive three-volume set, *The Works of William James*, ed. F. Burkhardt and F. Bowers (1890; reprint, Cambridge, Mass.: Harvard University Press, 1981).

Other information for this chapter was drawn from the following:

Cowan, W. M., and E. R. Kandel. "Prospects for neurology and psychiatry," *JAMA* 285 (2001):594–600. Huang, Y.-Y., K. C. Martin, and E. R. Kandel. "Both protein kinase A and mitogen-activated protein kinase are required in the amygdala for the macromolecular synthesis-dependent late phase of long-term potentiation." *J. Neurosci.* 20 (2000):6317–25.

Kandel, E. R. "Disorders of mood: Depression, mania and anxiety disorders," in *Principles of Neural Science,* 4th ed., E. R. Kandel, J. H. Schwartz, and T. M. Jessell, eds. New York: McGraw Hill, 2000, pp. 1209–26.

Rogan, M. T., M. G. Weisskopf, Y.-Y. Huang, E. R. Kandel, and J. E. LeDoux. "Long-term potentiation in the amygdala: Implications for memory." Chapter 2 in *Neuronal Mechanisms of Memory Formation: Concepts of Long-Term Potentiation and Beyond,* edited by C. Hölscher, 58–76. Cambridge: Cambridge University Press, 2001.

Rogan, M. T., K. S. Leon, D. L. Perez, and E. R. Kandel. "Distinct neural signatures for safety and danger in the amygdala and striatum of the mouse." *Neuron* 46 (2005):309–20.

Shumyatsky, G. P., E. Tsvetkov, G. Malleret, S. Vronskaya, M. Hatton, L. Hampton, J. F. Battey, C. Dulac, E. R. Kandel, and V. Y. Bolshakov. "Identification of a signaling network in lateral nucleus of amygdala important for inhibiting memory specifically related to learned fear." *Cell* 111 (2002):905–18.

Snyder, S. H. *Drugs and the Brain.* New York: Scientific American Books, 1986.

Tsvetkov, E., W. A. Carlezon, Jr., F. M. Benes, E. R. Kandel, and V. Y. Bolshakov. "Fear conditioning occludes LTP-induced presynaptic enhancement of synaptic transmission in the cortical pathway to the lateral amygdala." *Neuron* 34 (2002):289–300.

26: A New Way to Treat Mental Illness

Information for this chapter was drawn from the following:

Abi-Dargham, A., D. R. Hwang, Y. Huang, Y. Zea-Ponce, D. Martinez, I. Lombardo, A. Broft, T. Hashimoto, M. Slifstein, O. Mawlawi, R. VanHeertum, and M. Laruelle. "Quantitative analysis of striatal and extrastriatal D_2 receptors in humans with [^{18}F]fallypride: Validation and reproducibility." In preparation.

Ansorge, M. S., M. Zhou, A. Lira, R. Hen, and J. A. Gingrich. "Early-life blockade of the 5-HT transporter alters emotional behavior in adult mice." *Science* 306 (2004):879–81.

Baddeley, A. D. *Working Memory*. Oxford: Clarendon Press, 1986.

Carlsson, M. L., A. Carlsson, and M. Nilsson. "Schizophrenia: From dopamine to glutamate and back." *Curr. Med. Chem.* 11, no. 3 (2004):267–77.

Fuster, J. M. "The prefrontal cortex—an update: Time is of the essence." *Neuron* 30, no. 2 (2001):319–33.

Goldman-Rakic, P. "The 'psychic' neuron of the cerebral cortex." *Ann. N. Y. Acad. Sci.* 868 (1999):13–26.

Huang, Y.-Y., E. Simpson, C. Kellendonk, and E. R. Kandel. "Genetic evidence for the bi-directional modulation of synaptic plasticity in the prefrontal cortex by D1 receptors." *Proc. Natl. Acad. Sci. USA* 101 (2004):3236–41.

Jacobsen, C. F. *Studies of Cerebral Function in Primates*. Baltimore: Johns Hopkins University Press, 1936.

Kandel, E. R. "Disorders of thought: Schizophrenia." In *Principles of Neural Science.* 3rd ed. Edited by E. R. Kandel, J. H. Schwartz, and T. M. Jessell, 853–68. New York: Elsevier, 1991.

Lawford, B. R., R. M. Young, E. P. Noble, B. Kann, L. Arnold, J. Rowell, and T. L. Ritchie. "D2 dopamine receptor gene polymorphism: Paroxetine and social functioning in posttraumatic stress disorder." *Euro. Neuropsychopharm.* 13, no. 5 (2003):313–20.

Santarelli, L., M. Saxe, C. Gross, A. Surget, F. Battaglia, S. Dulawa, N. Weisstaub, J. Lee, R. Duman, O. Arancio, C. Belzung, and R. Hen. "Requirement of hippocampal neurogenesis for the behavioral effects of antidepressants." *Science* 301 (2003):805–9.

Seeman, P., T. Lee, M. Chau-Wong, and K. Wong. "Antipsychotic drug doses and neuroleptic/dopamine receptors." *Nature* 261 (1976):717–19.

Snyder, S. H. *Drugs and the Brain*. New York: Scientific American Books, 1986.

Schwartz, J. M., P. W. Stoessel, L. R. Baxter, K. M. Martin, and M. E. Phelps. "Systematic changes in cerebral glucose metabolic rate after successful behavior modification treatment of obsessive-compulsive disorders." *Arch Gen Psychiatry* 53 (1996):109–13.

27: Biology and the Renaissance of Psychoanalytic Thought

For an introduction to psychoanalysis, see C. Brenner, *An Elementary Textbook of Psychoanalysis*, rev. ed. (New York: International University Press, 1973).

For an introduction to Aaron Beck's work, see J. S. Beck, *Cognitive Therapy: Basics and Beyond* (New York: Guilford, 1995).

For a constructive critique of empirically supported psychotherapies, see D. Westen, C. M. Novotny, and H. Thompson Brenner, "The empirical status of empirically supported psychotherapies: Assumptions, findings, and reporting in controlled clinical trials," *Psychol. Bull.* 130 (2004):631–63.

Other information for this chapter was drawn from the following:

Etkin, A., K. C. Klemenhagen, J. T. Dudman, M. T. Rogan, R. Hen, E. R. Kandel, and J. Hirsch. "Individual differences in trait anxiety predict the response of the basolateral amygdala to unconsciously processed fearful faces." *Neuron* 44 (2004):1043–55.

Etkin, A., C. Pittenger, H. J. Polan, and E. R. Kandel. "Towards a neurobiology of psychotherapy: Basic science and clinical applications." *J. Neuropsychiatry Clin. Neurosci.* 17 (2005):145–58.

Jamison, K. R. *An Unquiet Mind*. New York: Alfred A. Knopf, 1995; quotation from pp. 88–89.

Kandel, E. R. "A new intellectual framework for psychiatry." *Am. J. Psych.* 155, no. 4 (1998):457–69.

———. "Biology and the future of psychoanalysis: A new intellectual framework for psychiatry revisited." *Am. J. Psych.* 156, no. 4 (1999):505–24 (see in particular the references cited in this paper).

———. *Psychiatry, Psychoanalysis and the New Biology of Mind*. Arlington, Va.: APA Publishing, 2005.

28: Consciousness

For a discussion of mind-brain dualism, see P. S. Churchland, *Brain Wise Studies in Neurophilosophy* (Cambridge, Mass.: MIT Press, 2002); A. R. Damasio, *Descartes: Error, Emotion, Reason and the Human Brain* (New York: Putman, 1994); R. Descartes, *The Philosophical Writings of Descartes*, trans. E. S. Haldane and G. R. T. Ross, vol. 1 (New York: Cambridge University Press, 1972); J. C. Eccles, *Evolution of the Brain: Creation of the Self* (London/New York: Routledge, 1989); and M. S. Gazzaniga and M. S. Steven, "Free will in the twenty-first century: A discussion of neuroscience and the law," in *Neuroscience and the Law*, ed. B. Garland (New York: Dana Press, 2004), p. 57, citing V. Ramachandran.

For a discussion of unconscious processes in perception, see C. Frith, "Disorders of cognition and existence of unconscious mental processes: An introduction," in E. R. Kandel et al., *Principles of Neural Science*, 5th ed. (New York: McGraw-Hill, forthcoming).

For a discussion of free will, see ibid.; S. Blackmore, *Consciousness: An Introduction* (Oxford/New York: Oxford University Press, 2004); L. Deecke, B. Grozinger, and H. H. Kornhuber, "Voluntary finger movement in man: Cerebral potential and theory," *Biol.*

Cyber. 23 (1976):99–119; B. Libet, "Autobiography," in *History of Neuroscience in Autobiography*, ed. L. R. Squire, vol. 1, 414–53 (Washington, D.C.: Society for Neuroscience, 1996); B. Libet, C. A. Gleason, E. W. Wright, and D. K. Pearl, "Time of conscious intention to act in relation to onset of cerebral activity (readiness-potential): The unconscious initiation of a freely voluntary act," *Brain* 106 (1983):623–42; and M. Wegner, *The Illusion of Conscious Will* (Cambridge, Mass.: MIT Press, 2002).

The Academy in Athens, which Plato founded, still exists today. I was inducted as a foreign member in 2005!

Other information in this chapter was drawn from the following:

Bloom, P. "Dissecting the right brain." Book review of *The Ethical Brain*, by M. Gazzaniga. *Nature* 436 (2005):178–79; quotation from p. 178.

Crick, F. C., and C. Koch. "What is the function of the claustrum?" *Philos. Trans. R. Soc. Lond. B Biol. Sci.*, June 30, 2005:1271–79.

Durnwald, M. "The psychology of facial expression." *Discover* 26 (2005):16–18.

Edelman, G. *Wider than the Sky: The Phenomenal Gift of Consciousness*. New Haven: Yale University Press, 2004.

Etkin, A., K. C. Klemenhagen, J. T. Dudman, M. T. Rogan, R. Hen, E. R. Kandel, and J. Hirsch. "Individual differences in trait anxiety predict the response of the basolateral amygdala to unconsciously processed fearful faces." *Neuron* 44 (2004):1043–55.

Kandel, E. R. "From nerve cells to cognition: The internal cellular representation required for perception and action." In *Principles of Neural Science*, 4th ed., edited by E. R. Kandel, J. H. Schwartz, and T. M. Jessell. New York: McGraw-Hill, 2000, pp. 381–403.

Koch, C. *The Quest for Consciousness: A Neurobiological Approach*. Denver, Col.: Roberts, 2004.

Lumer, E. D., K. J. Friston, and G. Rees. "Neural correlates of perceptual rivalry in the human brain." *Science* 280 (1998):1930–34.

Miller, K. "Francis Crick, 1916–2004." *Discover* 26 (2005):62.

Nagel, T. "What is the mind-brain problem?" In *Experimental and Theoretical Studies of Consciousness*, 1–13. CIBA Foundation Symposium Series 174. New York: John Wiley & Sons, 1993.

Polonsky, A., R. Blake, J. Braun, and D. J. Heeger. "Neuronal activity in human primary visual cortex correlates with perception during binocular rivalry." *Nature Neuroscience* 3 (2000):1153–59.

Ramachandran, V. "The astonishing Francis Crick." *Perception* 33 (2004):1151–54; quotation from p. 1154.

Searle, J. R. *Mind: A Brief Introduction*. Oxford: Oxford University Press, 2004.

———. "Consciousness: What we still don't know." Review of *The Quest for Consciousness*, by Christof Koch. *New York Review of Books* 52 (2005):36–39.

Stevens, C. F. "Crick and the claustrum." *Nature* 435 (2005):1040–41.

Watson, J. D. *The Double Helix*. 1968. Reprint, New York: Touchstone, 2001); quotation from p. 115.

Zimmer, C. *Soul Made Flesh: The Discovery of the Brain and How It Changed the World.* New York: Free Press, 2004.

29: Rediscovering Vienna via Stockholm

There are several good biographies of Alfred Nobel. For example, see T. Frängsmyr's short portrait, *Alfred Nobel*, trans. J. Black (Stockholm: Swedish Institute, 1996); and the book by Ragnar Sohlman, Nobel's executor, *The Legacy of Alfred Nobel: The Story Behind the Nobel Prize*, trans. E. Schubert (London: Bodley Head, 1983).

For a discussion of the Nobel Prize including a brief history of Nobel and his will, see B. Feldman, *The Nobel Prize* (New York: Arcade, 2000); and I. Hargittai, *Nobel Prizes, Science, and Scientists* (Oxford: Oxford University Press, 2002).

A scholarly discussion of American laureates from a sociological perspective is in H. Zuckerman, *Scientific Elite: Nobel Laureates in the United States* (New York: Free Press, 1977).

The fate of the Jewish academic physicians is discussed in a special issue (February 27, 1998) of the *Wiener Klinische Wucheschrift*—Vienna's most significant medical journal—entitled *On the Sixtieth Anniversary of the Dismissal of the Jewish Faculty Members from the Vienna Medical School*. This issue also has a discussion of Eduard Pernkopf by Peter Malina, pp. 193–201. See also G. Weissman's essay "Springtime for Pernkopf," *Hospital Practice* 30 (1985):142-68.

George Berkley's *Vienna and Its Jews: The Tragedy of Success, 1880s–1980s* (Cambridge, Mass.: Abt Books, 1988) was an invaluable source for this chapter. The figures on the role of Austrians in the Holocaust come from p. 318; the Hans Tietze quotation comes from p. 41.

The volume that emerged from the symposium of the summer of 2003 is F. Stadtler, E. R. Kandel, W. Kohn, F. Stern, and A. Zeilinger, eds. *Osterreichs Umgang mit dem National Socialism Springer Wien* (Vienna: Springer Verlag, 2004).

Other information in this chapter was drawn from the following:

Bettauer, H. *The City Without Jews: A Novel of Our Time.* Translated by S. N. Brainin. New York: Bloch, 1926; quotation from p. 130.

Sachar, H. M. *Diaspora: An Inquiry into the Contemporary Jewish World.* New York: Harper & Row, 1985.

Wistrich, R. *The Jews of Vienna in the Age of Franz Joseph.* Oxford: Oxford Univeristy Press, 1989; quotation from p. viii.

Young, J. E. *The Texture of Memory: Holocaust Memorials and Meaning.* New Haven: Yale University Press, 1993.

30: Learning from Memory: Prospects

For a discussion of Leonardo da Vinci's training in Andrea del Verrochio's studio, see E. T. DeWald, *History of Italian Painting, 1200–1600* (New York: Holt Rinehart & Winston, 1961), especially pp. 356–57.

Other information for this chapter was drawn from the following:

De Bono, M., and C. I. Bargmann. "Natural variation in a neuropeptide Y receptor homolog modifies social behavior and food responses in *C. elegans*." *Cell* 94 (1998):679–89.

Demir, E., and B. J. Dickson. "Fruitless splicing specifies male courtship behavior in *Drosophila*." *Cell* 121 (2005):785–94.

Insel, T. R., and L. J. Young. "The neurobiology of attachment." *Nat. Rev. Neurosci.* 2 (2001):129–36.

Kandel, E. R. *Psychiatry, Psychoanalysis and the New Biology of Mind*. Arlington, Va.: APA Publishing, 2005.

Rizzolatti, G., L. Fadiga, V. Gallese, and L. Fogassi. "Premotor cortex and the recognition of motor actions." *Cogn. Brain Res.* 3 (1996):131–41.

Stockinger, P., D. Kvitsiani, S. Rotkopf, L. Tirian, and B. J. Dickson. "Neural circuitry that governs *Drosophila* male courtship behavior." *Cell* 121 (2005):795–807.

ACKNOWLEDGMENTS

During the course of my career I have had the privilege of working with and learning from many gifted collaborators, fellows, and students, and I have tried throughout this book to acknowledge their contributions. Beyond individual collaborators my science has benefited enormously from the interactive environment created by the Center for Neurobiology and Behavior at the College of Physicians and Surgeons of Columbia University. It would be hard to find a more ideal environment in which to mature as a scientist. Specifically, I have benefited greatly from my long-standing friendships with Richard Axel, Craig Bailey, Jane Dodd, Robert Hawkins, Michael Goldberg, Samuel Schacher, John Koester, Thomas Jessell, James H. Schwartz, Steven Siegelbaum, and Gerald Fischbach, the current dean of the College of Physicians and Surgeons. I am further grateful to John Koester for his excellent leadership of the Center for Neurobiology and Behavior.

My research has been generously supported by the Howard Hughes Medical Institute and by the NIH. I am particularly indebted to the leadership of the Howard Hughes Medical Institute: Donald Fredrickson, George Cahill, Purnell Chopin, Max Cowan, Donald Harter, and, more recently, Tom Cech and Gerry Rubin. Their farsighted vision has encouraged Hughes investigators to adopt a long-term perspective on research and to tackle challenging problems. Research on learning and memory certainly meets both of these criteria!

I am grateful to the Sloan Foundation for a grant that helped me start on this book and to my agents, John Brockman and Katinka Matson, who aided me in framing the proposal for this book and in guiding it through the editorial process.

Many people have read part or all of the several earlier versions of this book. Professor Edward Timms, a historian of modern Austria at the University of Sussex in England, and Dieter Kuhl, a student of Viennese culture, kindly read and commented on chapters 2 and 24. David Olds, an academic psychoanalyst and colleague at Columbia, commented on chapters 3, 22, and 27. Several of my scientific colleagues read one or more versions of the whole text. I am particularly grateful to Tom Jessell, Jimmy Schwartz, Tom Carew, Jack Byrne, Yadin Dudai, Tamas Bartfei, Roger Nicoll, Sten Grillner, David Olds, Rod MacKinnon, Michael Bennett, Dominick Purpura, Dusan Bartsch, Robert Wurtz, Tony Movshon, Chris Miller, Anna Kris Wolfe, Marianne Goldberger, Christof Koch, and Bertil Hille for their thoughtful comments. I have also benefited from insightful readings of an early draft by several nonscientists, Connie Casey, Amy Bednick, June Bingham Birge, Natalie Lehman Haupt, Robert Kornfeld, and Sarah Mack, who pointed out difficulties posed by certain technical discussions.

Jane Nevins, the editor-in-chief of the DANA Foundation, and Sibyl Golden read later versions of the manuscript and helped me make some of the more technical portions more understandable for the general reader. Howard Beckman, my longtime friend who has edited several versions of *Principles of Neural Science*, generously read and commented on the text, and the superb science writer Geoffrey Montgomery worked with me on several chapters to help make them come more alive. Most of all, I am enormously indebted to my excellent editor Blair Burns Potter, who read almost all versions of the text and the figures, and in each case improved their overall clarity and coherence. Before I began this book, I had heard about Blair's talents but had met her only briefly. Through our extensive e-mail correspondence, I have come to value her as a wonderful friend.

I am fortunate to have had help on the art program from Maya Pines, a longtime friend and a science editor at the Howard Hughes Medical Institute, and from Sarah Mack, my colleague at Columbia and the art director of *Principles of Neural Science*. I am grateful to Sarah and to Charles Lam, who also executed the art program, animating what were originally rather vague ideas. In addition, I want to thank my associates at Columbia—Aviva Olsavsky for help with the glossary and text, Shoshana Vasheetz for word processing assistance, Seta Izmirly, Millie Pellan, Arielle Rodman, Brian Skorney, and Heidi Smith for proofreading the galleys, and especially Maria Palileo for her assiduous organization of the numerous versions of the manuscript.

My book editor at Norton, Angela von der Lippe, helped me to rethink and reorganize portions of the book and thereby improve it in numerous ways. I am also indebted to Angela's colleagues at Norton, in particular Vanessa Levine-Smith, Winfrida Mbewe, and Trent Duffy, my copyeditor. All of them caringly helped the book achieve its present form. All deserve my deepest gratitude.

INDEX

Page numbers in *italics* refer to illustrations.

Aß peptide, 331
acetylcholine, 91, 93, 94, 96, 97, 98, 100–101
action potentials, 60, *83*, 74–89, 90, 91, 92, 98, *102*, 143, 155, 194, 201, 204
  as all-or-none, 77, *78*, 79, 80, 109
  amplitude of, 82, 85, 86, 87–88, 92–93
  connection specificity and, 79
  dendrites and, 106–7, 140, 293
  and differentiating sensory information, 78–79
  frequency of, 78, 80
  generation of, 76, 85–88, 158
  in ionic hypothesis, 83–89
  in membrane hypothesis, 80–83, 85, 87–88
  of motor neurons, 140
  of place cells, 282
  of plant cells, 165
  propagation of, 75–76, 77, 80, 84, 86, 88, *99*, 100
  in pyramidal cells, 139, 140, 309
  recordings of, *76*, 77, *78*, 80, 93, 107, 108–9, 138, 139–40
  refractory period and, 158
  sensory duration and, 77–78
  sensory intensity and, 77, 78, 79, 80
  in sensory neurons, 77–80
  shape of, *76*, 77, *78*, 86, *87*, 232
  speed of, 75–76, 77, 80
  synaptic potentials vs., 92–93
adenine, 243, 244
adenylyl cyclase, 227
Adey, Ross, 199–200
Adler, Fred, 324
adrenaline, 91
Adrian, Edgar Douglas, *76*, 77–80, *78*, 83, 93, 94, 108
  Nobel Prize of, 79–80
age-related memory loss (benign senescent forgetfulness), xiii, 11, 266, 327–31
  drug treatment of, 326, 330, 331, 332
  in explicit vs. implicit memory, 328–29
  hippocampus and, 328–30
  in mice, 326, 328–30, *329*, 335
  onset of, 327, 329
agnosias, 303
alcoholism, 99, 336
  amentias and, 154
Alden Spencer Lectureship and Award, 220
alleles, 289
allocentric spatial coordinates, 308
Alzheimer's disease, xiii, 11, 289, 323–24, 331–32, 336
  ß-amyloid plaques in, 327, 331
  early-onset, 337
  genetic component of, 337
  hippocampus and, 331
  progression of, 327
  Rolipram treatment of, 330, 331
ambient (basal) attention, 312
amino acids, 99, 244, 245, 250, 272, 320

AMPA receptor, 283, 284, 292
amygdala, 44, 45, *130*, 132, 186, 343–48, *345*, *347*, 356, 383
basolateral nucleus of, 387, 388
central nucleus of, 387
damage to, 342
inhibition of, 348, 350
lateral nucleus of, 344, 345, 348, 349, 350
long-term potentiation in, 345–46, 348, 349
amyotrophic lateral sclerosis (ALS), 67, 218–20, 289, 325, 337
Andersen, Per, 186, 282, 287
antidepressant drugs, 325, 359–61 369
antipsychotic drugs, 355–58
anti-Semitism, 5, 26
cultural vs. racial, 30–31
in France, 178–79
at University of Vienna, 18, 30
in Vienna, 15–18, 20, *21*, 27–31, 152, 412–14; *see also* Kristallnacht
anxiety, 11, 338–51, 355, 388
anticipatory, 343
background, 386–87, 388
instinctive, 339–40
learned, 339, 340
pathological, 339–40
see also fear
anxiety disorders, 336, 337, 338–39, 342–43, 350–51, 352, 365, 368, 388, 421
cognitive behavioral therapy for, 369
incidence of, 340
aphasias, 120, 121–24, *121*
Aplisa, The (Minouche), *206*, 207
Aplysia, 145–49, 159, 160–62, 165–72, *166*, 173, 181, 182, 183, 187–97, 198–207, 221–32, 234, 287, 293, 294, 418
abdominal ganglion of, 165, 166, 188, *190*, 192, 193, *193*, 194, *195*, 225
anatomical changes in, 212–16, *214*
bag cells of, 250
behavioral repertoire of, 187–88, *190*, 248, 250, 251
biological principles derived from, 204–5
brain size of, 146–47, *147*
cell R2 of, 165, 166, 169, 171–72, *193*
classical conditioning of, 192, 201, 202, 204, 205, 343, 346
connection specificity in, 193–96, *196*, 200
egg-laying behavior of, 250, 251
experimental strategy for, 252–53
ganglia of, 147, *147*, 161, 188
genes of, 245
genome of, 419
gill-withdrawal reflex of, 188–92, *191*, 194–97, *195*, *196*, 200–203, 204, 205, 215–16, 254, 273–74, 308, 310, 354
habituation of, 189, 191–92, *191*, 200, 201, 202, 203, 204, 205, 212, 213–14, *214*, 215, 222
inking of, 146, 188, *190*, 204, 216
interneurons of, 196, 214, 222–23, *229*, 230, 254
large cells of, 171–72, 184
life cycle of, 253, *254*
locomotor activity of, 188, *189*
long-term memory of, 191–92, 212–15, 259
mantle cavity of, 188–89, *191*
mantle shelf of, 189
memory retrieval in, 215–16
memory storage in, 204–5
Minouche's poem about, *206*, 207
motor neurons of, 194–95, *195*, 196, *196*, 200–201, 204, 205, 213, 214, 253–56
neural circuits of, 166, 169, 187, 192–96, 200, 203, 262–63, 281
neuronal uniqueness of, 192–93
neuron size in, 147
with Nobel Prize medal, 399–400, *400*
presynaptic terminals of, 213–15
reflex responses of, 147
reports on, 196–97, 200, 203, 251
research strategy for, 200
seaweed needed by, 253, *254*
sensitization of, 189–91, *191*, 200, 201, 202, 203, 204, 205, 212, 213–14, *214*, 215, 224, 230, 259, 314, 346
sensory neurons of, 195–96, *196*, 200–201, 204, 205, 213–14, 222–23, 253–56, 259
sexual behavior of, 188, *190*
short-term memory of, 189, 192, 204, 213–14, 292
siphon of, 189, *191*, 201
suitable characteristics of, 146–48
synaptic potentials in, 148, 200–201, 222–23, 230, 231
tissue cultures of, 253–56, *255*, 259, 262–63
Aplysia californica, 146, *146*
Aplysia depilans, 146
a priori knowledge, 202, 203, 296, 309
Arancio, Ottavio, 331
Aristotle, 40, 41, 285
Arvanitaki-Chalazonitis, Angelique, 145, 146–47, 148, 193
Aryanization of property, 28
Ascaris, 145, 193
attention, 7, 44–45, 59, 128, 295, 298, 305, 306, 307, 309, 311–15, 376–77, 385, 423
basal (ambient), 312
dopamine and, 312–13, 314, 315, *315*, 424
involuntary, 313–14
mechanism of, 312–15
salience in, 313–14, *315*
selective, 311, 312, 335, 382, 424
spatial memory and, 311, 312, 328
voluntary, 313–14, *315*
auditory cortex, 123, 142, 344
Auerwald, Wilhelm, 236–37
Aunt Minna and Uncle Srul, 24, 24–25, 32

Austria, 12–32, 102, 118, 405–15, 417, 428
Allied occupation forces in, 406, 408
concentration camps in, 408
Dollfuss's assassination in, 25
inconstant political allegiance in, 14–15, 27
military defeats of, 19, 22
Nazi annexation of, 138–18, 27, 92, 96
Nazi collaboration denied by, 405–6, 408–9, 410–11
Nazi party of, 20, *21*, 25, 26–27, 30, 92, 405, 407–8, *407*
Nazi Reich functionaries from, 405
racial anti-Semitism in, 31
"rehabilitated" citizens of, 408
reparations unpaid by, 406, 410–11
war criminals law of, 406, 408
see also Vienna
Austrian Physiological Society, 236, 408
"Austria's Response to National Socialism: Implications for Scientific and Humanistic Scholarship" symposium, 405, 409–10, 411, 412, 414
Austro-Hungarian Empire, 5, 19
autonomic nervous system, 45, 90–92, 93, 94, 136, 167, 387
emotions and, 340–42
functions of, 90–91
Avery, Oswald, 242
Axel, Richard, 151–52, 247–51, *248*, 291, 332, 398, 427, 428
as biotechnology consultant, 323–25
biotechnology patent of, 323
co-transfection developed by, 249
Nobel Prize of, *248*, 251, 404–5
personality of, 248–49
sense of smell studied by, *248*, 251, 405
serotonin receptor cloned by, 325
"Axel syndrome," 248–49
axons, 44, *45*, 63–64, *63*, 67, 74–84
crayfish's large, 108, 145
electrical signaling of, *see* action potentials
as processes, 62, 63
signaling function of, 65
squid's giant, *83*, 84–85, 86, 88, 107, 108, 422
surface membrane of, 80–83
terminals of, 64, 65, 88, 91, 100–101, 226, 259; *see also* presynaptic axon terminals
Aymard, Alfred and Louise, 174–75

Babylonian Talmud, 266
bacteria, 246, 256, 257, 322
E. coli, 225, 234, 258–59, 320
in genetic engineering, 320
Baddeley, Alan, 354
Bailey, Craig, 213–14, 259
ß-amyloid plaques, 327, 331
Barad, Mark, 330
Bard, Philip, 110–11, 114, 299
Bargmann, Cori, 425
Barnes Maze, 328–29, *329*
Bartsch, Dusan, 263, 264, 418
basal (ambient) attention, 312
basal ganglia, 45
obsessive-compulsive disorder and, 371
Baxter, Lewis R., Jr., 371
Beck, Aaron, 367–69, 370
Beckmann, Max, 428
Bedik, Amy, 398
behaviorist psychology, 7, 9, 41–42, 132, 135, 159, 188, 198, 208, 295, 296, 298, 305, 422
Benzer, Seymour, 232–34, 286, 287, 419, 422
Berg, Paul, 246
Berkley, George, 17, 20, 27
Berlin, 12, 28
Berman, Ronald, 36, *36*, 37
Bernstein, Julius, 80–83, *82*, *83*, 85, 86, 87–88
Beth Israel Hospital, Boston, 181
Bettauer, Hugo, 412
Beyond the Pleasure Principle (Freud), 46, 51
Bibring, Grete, 181
binding problem, 303–4, 308, 382
binocular rivalry, 384–85
biochemical signaling pathways, 225–34
bioethics, 333–34, 390
Biogen, 321, 324
Biotechnology General, 323–24, 327
biotechnology industry, 246, 319–34
academic biologists in, 319, 321–33, 324
cognitive enhancement as ethical issue in, 333–34
genetic engineering and, 319–21
hormone synthesis in, 320–21, 323
molecular biologists in, 323–33
monoclonal antibodies and, 323
pharmaceutical industry and, 319, 321, 322, 323, 324
university patents in, 323
bipolar disorder, 372
Birnstiel, Max, 408
Bliss, Tim, 282–83, 284, 286, 309, 345
Bolshakov, Vadim, 346
Boston Veterans Administration Hospital, 152–53
botulism toxin, 67
bovine spongiform encephaly (mad cow disease), 273
Bowlby, John, 374
Boyer, Herbert, 246, 319–21
brain, 6, 10, 44–46, *45*, 53–59, 75, 100, 101–2, 144, 237, 250–51, 295–306
clay models of, 44, 58
computational power of, 268, 297–98
dualist view of, 117–18
Freud's structural theory and, 43, 45, 53–55, *54*, 116, 376
gender differences in, 316
individuality of, 218

brain (*continued*)
mind as one with, xii, 8–9, 117–18, 120
nerve net view of, 66
number of cells in, 146–47
pain receptors lacked by, 125
proteins in, 236
psychoanalysis and, 45–47
psychoanalytic view of, 202
regions of, 79, 116–34; *see also specific regions*
brain damage, 119, 120, 217–33, 337
to amygdala, 342
aphasias and, 120, 121–24, *121*
of concussions, 211
to hippocampus, 127–33, 282, 291, 328, 342
to prefrontal cortex, 354, 355, 385
to visual cortex, 303
brain imaging, 7–8, 113, 305, 306, 316, 423
depression and, 371
of emotionally-charged visual perception, 385–89
psychoanalysis and, 424–25
psychotherapy and, 370–71, 372, 425
schizophrenia and, 354
brain slices, 283, 292
brain stem, 44–45, 66, 90, 113, 300
Brenner, Sydney, 144, 244, 261, 317, 419
brief dynamic therapy, 370
Brinley, Jack, 136–37, 140
Broca, Pierre-Paul, 120–24, *121*, 131, 305
Broca's area, *121*, 122, 123, *123*, 126
Brunelli, Marcello, 230
Bruner, Jerome, 130
Buck, Linda, *248*, 251, 404–5
Burns, B. Delisle, 199
Byers, Duncan, 234
Byrne, Jack, 195–96, 400
Bystryn, Denise, *see* Kandel, Denise Bystryn
Bystryn, Iser, 48–49, 174–75
Bystryn, Jean-Claude, 174–75, 238, *399*
Bystryn, Sara, 48–49, 174–75, 177, 182–83, 238–39, *239*, 428

Caenorhabditis elegans, 144, 245, 419
social behavior of, 425
Cajal, Santiago Ramón y, *60*, 61–70, *63*, 72, 79, 107, 158, 194, 225, 301
anthropomorphic imagery of, 61
background of, 61
Golgi's disagreement with, 67–68
learning theory of, 158, 159–60, 198, 199, 204
newborn animal cells studied by, 62
Nobel Prize of, 67, 68
Sherrington's memorial to, 69
silver staining method of, 62, 67
see also neuron doctrine
calcium ions, 100, 101, *102*, 232, 284, 292
Campagna, John, 36, 37, 115
Canetti, Elias, 12
Carew, Tom, 191–92, 201, 212–15
Carl XVI Gustaf, King of Sweden, 400, 401, 402, 404
Carlsson, Arvid, 356–57
Nobel Prize of, 229, 230, 356, 393–405
Carossa, Hans, 37–38
Castellucci, Vincent, 195–96, 213
caudate nucleus, 337, 371
cell biology, 58, 88, 108, 140, 149, 184–85, 204, 218, 236, 419
see also molecular biology
cell division, 242
cells, 7, 225, 257
biochemical signaling pathways in, 225
components of, 58–59
effector, 66
nucleus of, 58, 241, 257, 258, 262, 263, 264, 272
specialized functions of, 59
surface membrane of, 58, 226
cell theory, 58, 63
Cellular Basis of Behavior (Kandel), 236
"Cellular Neurophysiological Approaches in the Study of Learning" (Spencer and Kandel), 185–86
central nervous system, 92, 136
structure of, 44–45, *45*, *46*
cerebellum, 44, *130*, 132
cerebral cortex, 45, 103–7, *104*, 116–34, 138, 192, 301, 315, 342, 387
damage to, 120
divisions of, 109
electrical stimulation of, 122, 125–27, *125*, 160–61
Gall's view of, 117–20, *118*
lobes of, 109, *111*
psychedelic drugs and, 103–5; *see also* LSD
size of, 109–10
spreading cortical depression of, 114, 136–37
voluntary attention initiated in, 313, 314, *315*
see also specific cortical regions
cerebral hemispheres, 45, 120–21, *121*, 342
in split-brain patients, 379
channelopathies, 89
chemical signaling, 60, 90–102, 205
see also neurotransmitters
Chen, Mary, 213–14
child development, 365
attachment system in, 374
habituation and, 168
maternal deprivation in, 373–74
psychoanalytic studies of, 43, 365, 372–74
in schizophrenia, 357–58
chloride ions, 80, 81
chlorpromazine, 355–56
chromosomes, 58, 245, 257, 320
of *Drosophila*, 232
paired, 241, 242

cingulate cortex, 342, 345
City Without Jews, The: A Novel About the Day After Tomorrow (Bettauer), 412
classical conditioning, 40, 41, 160–61, *167,* 186, 279, 422
of *Aplysia,* 192, 201, 202, 204, 205, 343, 346
aversive, 170, 171, 232–34, 343–45, *344, 345,* 346
involuntary attention in, 313
in neural analogs of learning, 160, 161, 166, 170, 171
sensory stimuli in, 160–61, 166, 170
claustrum, 383–84, 385
cognitive behavioral therapy, 369–70, 371
cognitive enhancers, xiii, 333–34
cognitive maps, 298, 310
see also sensory map of body; spatial maps
cognitive neuroscience, 7–8, 298–99, 304, 382, 390, 423
cognitive psychology, 7, 9, 130, 135, 159, 198, 208, 307, 301–2, 305, 368
assumptions of, 296–98
development of, 296
psychoanalysis and, 43, 296, 298, 368
cognitive styles, 368, 369
Cohen, David, 186
Cohen, Gerson D., 260
Cohen, Stanley, 246
Cold Spring Harbor Laboratory, 266, 291
Collingridge, Graham, 283
Columbia University, xiv, 47, 48, 49, 50, 53, *56,* 57, 103, *138,* 208, 231, 261, 272, 290, 325, 331, 364, 398–99, 416
biotechnology patent of, 323
Center for Neurobiology and Behavior at, 220, 252
College of Physicians and Surgeons of, 57, 247–51, *248,* 399
Howard Hughes Medical Institute at, 251–52, 326, 332, 336
press conference at, 394, 395
computers, 297–98
connection specificity, 64, 65–66, 79, 193–96, *196,* 200, 225
conscious memory, *see* explicit memory
consciousness, 43, 53, 291, 295, 374, 376–90, 418, 423
binding problem of, 382
binocular rivalry and, 384–85
as biological process, 8–9
evolution of, 8
free will and, 7, 11, 375, 389–90
Freud's view of, 53–55, *54,* 133
historical views of, 377–78
philosophers' view of, 377, 378–82, 389–90
psychoanalysis and, 385, 388, 389, 424–25
reductionism and, 381
of split-brain patients, 379
subjectivity of, 379–82
unconscious awareness and, 340–42, 344, 376, 384, 385–90, 424–25
unity of, 303–4, 311, 313, 375, 379, 380, 382–85
working definition of, 376–77
see also attention
co-transfection, 249
Cowan, Max, 336
CPEB (cytoplasmic polyadenylation element-binding protein), 272–75, *274,* 314, 315
Crain, Stanley, 106, 108–9, 115, 139
crayfish, 100, 108–9, 136, 139, 186
dendrites of, *95,* 107
Freud's study of, 107
large axon of, 108, 145
CREB (cyclic AMP response element-binding protein), 263–66, 268, 270, 422
activator vs. repressor, 264–66, 275–76
in hippocampus, 293, 294, 346
Creutzfeldt-Jacob disease, 273, 320
Crick, Francis, xi, 242–44
career of, 377, 383–84
claustrum as focus of, 383–84
consciousness and, 377, 378, 380, 382, 383–84, 385, 388, 424
cultural anti-Semitism, 30–31
Curtis, David, 213
cyclic AMP, 226–35, *229, 233,* 238, 240–41, 245, 284, 422
Alzheimer's disease and, 331
D1 dopamine receptor and, 358
genes and, 234, 263
in hippocampus, 293, 330, 331, 346
in long-term memory genetics, 258–59, 263, 264, 267
Rolipram treatment and, 330, 331
cyclic AMP response element, 263
see also CREB
cytoplasm, 58, 62, 80–81, 258
cytosine, 243, 244

Dahl, Hartvig, 367
Dale, Henry, 91–92, 93, 94, 98
in biotechnology industry, 322
Damasio, Antonio, 341
DANA Foundation, 334
Darwin, Charles, 6–7, 8, 40, 68, 339, 384, 386
Dash, Pramod, 263
Davanloo, Habib, 370
Davis, Michael, 343
defenses, psychological, 40, *54,* 55
defensive responses, 160, 167–68, 170
to fear, *339,* 340–41, 343, *344,* 346, *350,* 387
Delay, Jean, 355
Demir, Ebru, 425
dendrites, 63–65, *63,* 66, 67, 69, *101,* 145
action potentials and, 106–7, 140, 293

dendrites (*continued*)
of crayfish, *95*, 107
electrical properties of, 106–7
intracellular recordings of, 107
postsynaptic site on, 65
signaling function of, 65
synaptic potentials of, 106
Deniker, Pierre, 355
Dennett, Daniel, 378–79
depression, 11, 153, 242, 336, 349, 358–61
antidepressant drug treatment of, 325, 359–61, 369
brain imaging of, 371
clinical features of, 358–59
cognitive behavioral therapy for, 369
as emotional disorder, 337–38
genetic components of, 338
hypothalamic neuroendocrine cells and, 155
incidence of, 359
memory impairment in, 331, 335, 361
norepinephrine in, 359–60
onset of, 359
prefrontal cortex and, 371
as self-directed anger, 368
serotonin in, 359–60
systematic negative bias of, 368–69
Descartes, René, 117, 135, 378
developmental processes, 202–3, 223, 272
see also child development
Dickson, Barry, 425
DNA (deoxyribonucleic acid), xi, 6, 58, 242–46, 324, 377, 419
double helix of, 242–43
nucleotide bases of, 8, 243–45, 246
promoter sites of, 257–58, 263
sequencing of, 245, 320, 324
transcription of, 258–59, 267, 275
see also recombinant DNA
Doctrine of Deicide, 31
Dodd, Jane, 394, 398
Dollfuss, Engelbert, 25, 27
domains, 245
D1 dopamine receptor, 358
Doors of Perception, The (Huxley), 103
dopamine, 230
age-related memory loss and, 328
attention and, 312–13, 314, 315, *315*, 424
in hippocampus, 203, 313, 314, 315, 328, 330
receptors for, 325, 330, 356, 357–58, 361
schizophrenia and, 356–58, 361
Doty, Robert, 160
Double Helix, The (Watson), 377
Down's syndrome, 11
Drosophila, 245, 252, 292, 294, 332, 419, 422
genetics of, 232–34, 241, 286, 287
learned fear in, 266, 308
sexual behavior of, 425
drug abuse, 48
drug addiction, 349, 351
drug treatments, 154, 249, 331–32, 334, 348, 350–51, 388, 401
of age-related memory loss, 326, 330, 331, 332
of Alzheimer's disease, 331
antidepressant, 325, 359–61, 369
antipsychotic, 355–58
psychotherapy combined with, 372
tranquilizing drugs in, 99, 355–56
D2 dopamine receptor, 357–58, 361
dualism, 117–18, 135, 378
Dudai, Yadin, 232–34
Duman, Ronald, 360–61
Duncan, C. P., 211–12
dynamic polarization, 64, 66, 68

Ebbinghaus, Hermann, 208–11, 214
forgetting curve plotted by, 210
nonsense words of, 209–10, 211
Ebert, Thomas, 217–18
Eccles, John, *95*, 98–99, 107, 136, 140, 141, 144, 149, 159, 252, 294
in Australia, 94–96
dualism of, 378
falsification of theory by, 96–97, 99
in New Zealand, 96
Nobel Prize of, 179
in "soup versus spark" controversy, 92, 93, 94–97
Edelman, Gerald, 383, 385
Edelson, Marshall, 420–21
effector cells, 66
effector genes, 257–58, 259, 262, 270, 293
effector proteins, 257
ego, 43, 45, 53–55, *54*, 105, 116, 248
definition of, 54–55
egocentric spatial coordinates, 308
Ehrenzeichen für Wissenschaft und Kunst, 413
Eichmann, Adolf, 405
Eisenberg, Paul Chaim, 413
Ekman, Peter, 386, *386*
electrical signaling, 56, 74–89, 90–98, 100, 205, 227
in "soup versus spark" controversy, 91–98
see also action potentials; synaptic potentials
electricity, passive propagation of, 75–76
electrodes, 105, 107, 108, 136, 139, 147, 172, 194, 389
electron microscope, 69, 101
electrophysical recording systems, 108–9
"Electrophysiological Patterns of Serotonin and LSD Interaction on Afferent Cortical Pathways" (Kandel), 224
emotional disorders, 337–38
see also specific disorders
emotions, 45, 132, 141, 264–65, 280, 383
conscious and unconscious, 340–42, 385–89

in long-term memory, 342–43
neural circuits of, 338, 342
positive, *347,* 349–51
visual perception and, 385–89
see also fear
empiricism, 202–3, 285
endocrine system, 45
entorrhinal cortex, 140, 293
environmental factors, 338, 358
environmental stimuli, 79, 170, 257, 258–59
genes and, 262–64, 276
enzymes, 225, 242, 245, 257, 258, 293, 330
adenylyl cyclase, 227
see also protein kinase A
epilepsy, 125–34, 142, 289
familial idiopathic, 89
of H.M., 127–34
Jacksonian sensory march in, 113–14
Penfield's neurosurgical work on, 125–27, *125*
retrograde amnesia in, 211–12
Erasmus Hall High School, 33, 36–37
escape responses, 160, 168, 169
Escherichia coli, 225, 234, 258–59
in genetic engineering, 320
ethology, 144, 186, 422
Etkin, Amit, 385–89
Evarts, Ed, 304–5
evolution, xii–xiii, 144, 162, 374, 417, 422
cultural, 10
Darwinian, 6–7, 8, 40
fear conserved in, 339, *339*
implicit memory conserved in, 234, 294
learning as conserved in, 144, 162, 186, 192, 266, 267, 276, 422
mutations in, 234–36
natural selection in, 8, 235–36, 287
Ewalt, Jack, 152–53
experimental animals, 40–41, 119, 131, 319, 338–39
cats, 70, 105, 139, 143, 147, 183, 299
frogs, 75, 91, 101
goldfish, 155
invertebrate, *see* invertebrate experimental animals
newborn, 62
pigeons, 186
rats, 109, 124–25, 129, 168, 186, 225, 282, 283, 308–9
selection of, 107–9, 144–45
see also mice; monkeys; rabbits
explicit memory, *128,* 129–33, *130,* 134, 213, 279–85, 286, 289, 292–94, 295, 310, 311, 353
age-related memory loss and, 328–29
emotionally charged, 342
salience in, 313–14, *315*
storage of, *130,* 132, 280; *see also* hippocampus
voluntary attention in, 313–14
see also spatial maps; spatial memory
Expression of the Emotions in Man and Animals, The (Darwin), 339

face recognition, 128, 298, 303, 381
facial expressions, 386–89, *386*
Fatt, Paul, 99–100
fear, 6, 338–51
amygdala and, 342, 343–48, *345, 347,* 349, 350, 387
conscious and unconscious components of, 340–42, 344, 385–89, *386*
as conserved in evolution, 339, *339*
defensive responses to, *339,* 340–41, 343, 344, 346, *350,* 387
genetic predisposition to, 287, 339
inhibition of, 348, 350
instinctive, 339, 343, 348
learned, *see* learned fear
pathological, 339–40
survival value of, 339
see also anxiety
Fechner, Gustav, 208
Feldberg, William, 93, 102
fight-or-flight response, 345, 387
first-messenger signaling, 227, *228,* 231
Fischbach, Ruth and Gerry, 398–99, *399*
Fischer, Heinz, 414
flash-bulb memories, 265, 266
Fleming, Jonathan, 326
Flexner, Louis, 212, 225
Flourens, Pierre, 119–20, 124, 133
Forbes, Alexander, 199
forgetting, 275
phases of, 210
forward genetics, 287–88
fragile X syndrome, 67, 337
France, 26, 148, 150, 151 165–79, 183, 231, 355, 428
anti-Semitism in, 178–79
Bystryn family in, 48–49, 174–75
Institut Pasteur of, 256, 322
Nobel's fortune claimed by, 397
science in, 178, 417
Frank, Karl, 136, 140
Franklin, Rosalind, 242
Frederick II "the Great," King of Prussia, 412
Frederickson, Donald, 251
free association, 74, 364
free will, 7, 11, 375, 389–90
Freud, Anna, 42, 373
Freud, Sigmund, 12, 13, 42, 51, 53–56, 61, 74, 135, 155, 305, 340, 371, 375, 421, 427
on agnosias, 303
on anxiety, 339
crayfish neurons studied by, 107
depression as viewed by, 368
empirical psychology as aim of, 366

Freud, Sigmund (*continued*)
English residence of, 102
learning mechanism proposed by, 198
neuroanatomical research of, 45–46, 55–56, 72–73, 107
prose style of, 39
psychic determinism discovered by, 39–40, 367, 389
psychotherapeutic method of, 364–65
structural theory of, 43, 45, 53–55, *54*, 56, 116, 376
unconscious mind as viewed by, 39–40, 53–54, *54*, 55, 56, 133, 388, 424–25
as Viennese emigrant, 42, 102, 409
Fritsch, Gustav Theodor, 122
frogs, 75, 91, 101, 422
development of, 272
From Neuron to Brain (Kuffler and Nicholls), 181, 236
frontal lobe, 105, 109, *111*, 355
Broca's area of, *121*, 122, 123, *123*, 126
fruit flies, 7, 143, 145, 148
see also Drosophila
fruitless protein, 425
functional magnetic resonance imaging (fMRI), 8, 305, 306, 385
Furshpan, Edwin, 100, 108
Fuster, Joaquin, 354

GABA (gamma-aminobutyric acid), 99, 348
Gage, Phineas, 354
Gall, Franz Joseph, 117–20, *118*
localization of function theory of, 118–20, *118*
phrenology developed by, *118*, 119
Galvani, Luigi, 75, 83
Gasser, Herbert, 56–57, 84
gastrin-releasing peptide, 348
gateway hypothesis, 48
Gaucher's disease, 67
Gazzaniga, Michael, 390
Gehrer, Elisabeth, 405
gender differences, xiii–xiv, 315–16, 352, 359
gene cloning, 245–46, 248, 321, 325, 333
Genentech, 319–21, 322, 332
genes, 58, 238–49, 241–46, 247–59, 306, 320, 418
activator, 258
alleles, 289
for *Aplysia* egg-laying, 250
cyclic AMP and, 234, 263
domains of, 245
effector, 257–58, 259, 262, 270, 293
environmental stimuli and, 262–64, 276
knockout, 288, 289, 291
learning and, 202–3, 287
mutant, 232–34, 266, 337, 338
particular proteins encoded by, 243–45, 246, 257–58, 261–62, 291
of prions, 273
regulatory, 257–58, 259
regulatory mechanisms of, 256–59, 261–62, 263–66, 268, 275–76, 293–94, 310, 314, 324
replication of, 243, 244
repressor, 258
of spinal neurons, 252
in tissue culture, 323
transcription of, 258–59
see also chromosomes; DNA
genetic code, 243–45, 246, 377, 418
genetic disease components, 320–21, 361, 337
alleles, 289
of depression, 338
ion channels in, 89
of schizophrenia, 153, 155, 242, 338, 355, 357–58
genetic engineering, 250, 319–21, 425
bacteria in, 320
disease-causing alleles and, 289
in hormone synthesis, 320–21, 323
of mice, 186, 281, 286–94, 295, 306, 319, 348, 357, 358, 374
transgenesis in, 288, 289
see also Drosophila; recombinant DNA
genetic predispositions, 30, 338
to fear, 287, 339
"Genetic Regulatory Mechanisms in the Synthesis of Protein" (Jacob and Monod), 256
genetics, xi, 6, 7, 148, 153, 186, 223, 252, 418, 419
of *Drosophila*, 232–34, 241, 286, 287
forward, 287–88
in ion channel defects, 89
molecular, 241–46, 286–94, 336, 348, 427; *see also* long-term memory, genetic studies of
racial anti-Semitism and, 31
reverse, 288
genomes, 245, 246, 257, 288, 289, 320, 418, 419
Germany, 5, 6, 14, 17, 26, 30, 37–38, 102, 217–18, 417, 428
post-World War II soul-searching in, 405, 406, 409, 412
see also Nazis, Nazism
Gestalt psychology, 199, 296, 301–2
Gilbert, Walter, 244–45, 321, 322, 324–25, 326
glucose, 258–59
glutamate, 99, 222, 224, 226, 227, *229*, 230, 231, 323, 240–41, 255, 348
in hippocampus, 283, 284, 292
Goelet, Philip, 261–62, 267
Goldberg, Michael, 304–5, 312
Goldberger, Robert, 44, 47–48
Goldman-Rakic, Patricia, 353, 354, 358
Goldschmidt, Richard, 192–93

Golgi, Camillo, 62, 79, 100
neuron doctrine rejected by, 67–68
Nobel Prize of, 67, 68
Gombrich, Ernst, 42
Goodman, Corey, 332
Grant, Seth, 289, 290
Greengard, Paul, 229–30, 231
Nobel Prize of, 229, 356, 393–405
on Synaptic Pharmaceuticals advisory board, 325
Gregory, Richard, 390
Grundfest, Harry, 47, 53–59, 74–75, 84, 108, 109, 115, 126, 140, 149, 221, 248, 364, 376, 416, 417
career of, 56–58
chemical transmission doubted by, 91, 92
at McCarthy hearings, 57
Purpura and, 103–7, *104*, 427
retirement of, 247
Grynszpan, Herschel, 28
guanine, 243, 244

habituation, 41, 160, *167*, 279, 422
of *Aplysia*, 189, 191–92, *191*, 200, 201, 202, 203, 204, 205, 212, 213–14, *214*, 215, 222
defensive responses and, 167–68
in infant development research, 168
in neural analogs of learning, 160, 161, 166–69
sensory stimuli in, 160, 166–67, 168–69
sexual behavior and, 168
hallucinations, 106, 154, 357
auditory, 353
of epileptic seizures, 126
visual, 103, 104, 105
Hapsburg dynasty, 22, 150, 406, 409, 414
Harlow, Harry, 373–74
Harlow, John, 354
Hartmann, Ernst, 153, 154
Hartmann, Heinz, 43, 55, 368, 373
Harvard University, 33, 36–39, 40, 43, 116, 152, 180, 183, 184–85, 193, 244, 321, 324, 334, 363
see also Massachusetts Mental Health Center
Häupel, Michael, 411
Hausman, Louis, 44, 47
Hawkins, Robert, 201–2
head trauma, 154
retrograde amnesia caused by, 211
Hebb, D. O., 199, 284
Helmholtz, Hermann von, 75–76, 77, 80, 83
perception studied by, 133, 208–9
unconscious inference concept of, 209, 305, 424
Hen, Rene, 360
Henneman, Elwood, 153, 156
Hermissenda, 186
heterosynaptic facilitation, 169–70, 205, 283
Hill, A. V., 84, 94, 172
Hilten, Kathy, 398
hippocampus, 45, *63*, *128*, *130*, 136–43, 149, 152, 156, 172, 213, 280, 281–85, 286–94, 308, 316, 356, 422
age-related memory loss and, 328–30
Alzheimer's disease and, 331
AMPA receptor in, 283, 284, 292
antidepressant drugs and, 360–61
of birds, 306
coincidence detector in, 284
CREB in, 293, 294, 346
cyclic AMP in, 293, 330, 331, 346
damage to, 127–33, 282, 291, 328, 342
dentate gyrus of, 360–61
dopamine in, 293, 313, 314, 315, 328, 330
glutamate in, 283, 284, 292
ionotropic receptors in, 283–85
of London taxi drivers, 306
long-term potentiation in, 281, 283–85, 286–87, 288–94, 309, 330, 331, 346
multisensory representation in, 308
neural circuitry of, 139, 142–43
NMDA receptor in, 283–85, 286, 290, 292
perforant pathway and, 293
place cells of, 282, 286, 308–9, 310, 424
pyramidal cells of, 139–40, 142, 147, 172, 281, 282, 308–9, 330
recorded action potentials of, 139–40
Schaffer collateral pathway of, 291, 330
second-messenger signaling in, 284
sensory information and, 142, 143, 282, 308
slices of, 283, 292
synapses of, 142
see also spatial map; spatial memory
Hippocrates, 358, 377–78
Hirsch, Joy, 385
Hirsh, David, 395
histamine, 355
Historia Naturalis (Pliny the Elder), 145
Hitler, Adolf, 18, 25, 28, 29, 38, 48, 92, 93, 96, 102, 398, 405, 406, 407, 410
Schuschnigg's meeting with, 26–27
in Vienna, 13–17, *16*
Hitzig, Edward, 122
H.M. (epileptic patient), 127–34, *128*, *131*, 291, 328
Hobson, Alan, 153
Hochner, Benjamin, 263
Hodgkin, Alan, 68, 75, 83–89, *83*, 90, 94, 98, 108, 143, 422
dissertation of, 84
Nobel Prize of, 88, 179
squid's giant axon studied by, *83*, 84–85, 86, 88, 107
Holt, Robert, 72
Holton, Gerald, 33
homosynaptic depression, 168–69, 205, 283

honeybees, 7, 145, 186
hormones, 141, 226–27, 229, 355
 of *Aplysia*, 250, 251
 epinephrine, 226
 growth, 320
 insulin, 246, 320, 321, 322
 from neuroendocrine cells, 155–56
 peptide, 250, 251
 somatostatin, 320–21
 synthesis of, 320–21, 323
Howard Hughes Medical Institute, 251–52, 326, 332, 336
Hubel, David, 180, 237, 299, 301, 308
Hughes, Howard, 251–52
humors, theory of, 358
Huntington's disease, 242, 289, 336, 337, 338
Huxley, Aldous, 103
Huxley, Andrew, 75, 83–89, *83*, 90, 94, 98, 107, 108, 422
 Nobel Prize of, 88, 179
Hyman, Steven, 334
hypothalamus, 44, 45, 141
 in learned fear, 342, 345
 neuroendocrine cells of, 155–56

id, 45, 53–55, *54*, 105, 116, 248
 definition of, 55
implicit memory, *128*, 129–33, *130*, 134, *167*, 213, 279, 281, 292, 293, 295, 311
 as conserved in evolution, 234, 294
 CREB in, 266
 emotionally charged, 342
 involuntary attention in, 313–14
 memory disorders and, 328–29
 salience in, 313–14, *315*
 storage of, *130*, 132
 see also classical conditioning; habituation; sensitization
In Bluebeard's Castle (Steiner), 1
Innitzer, Theodor Cardinal, 15
Institut Pasteur, 256, 322
instrumental conditioning, 41
insulin, 246, 320, 321, 322
Integrative Action of the Nervous System, The (Sherrington), 69
interneurons, 66–67, *67*, 71, 348
 of *Aplysia*, 196, 214, 222–23, *229*, 230, 254
 inhibitory, 348
 modulatory, 223–24, *224*, 230, 254–55, *255*, 264
interpersonal psychotherapy, 369–70
Interpretation of Dreams, The (Freud), 133
invertebrate experimental animals, 132, 143–48, 185–86, 192, 283
 cockroaches, 186
 crayfish, *95*, 100, 107, 108–9, 136, 139, 145, 186
 fruit flies, 7, 143, 145, 148; *see also Drosophila*
 honeybees, 7, 145, 186
 leeches, 186
 lobsters, 145, 186
 pill bugs, 188
 selection of, 107–8, 143, 144–45
 snails, 7, 143, 145, 185, 186; *see also Aplysia*
 squids, *83*, 84–85, 86, 88, 101, 107
 see also bacteria; worms
investigator bias, 366
involuntary attention, 313–14
ion channels, 81, *82*, 85, 86–89, *87*, *102*, 225, 227, 231, 245, 252, 257, 283, 293
 of acetylcholine receptors, 97–98
 genetic defects of, 89
 non-gated potassium, 87–88, 89
 S, 231
 transmitter-gated, 97–98, 100
 voltage-gated, 86–88, 89, 97–98, 100, 140
ionic hypothesis, 60, 68, 83–89
ionotropic receptors, 227, *228*
 in hippocampus, 283–85
ion pore, 88
ions, 80–83, *83*
 calcium, 100, 101, *102*, 232, 284, 292
 chloride, 80, 81
 potassium, 80, 81, *82*, *83*, 85, 86–88, *87*, 89, 97–98, 136–37, 231, 232
 sodium, 80, 81, *83*, 97–98, 140

Jackson, John Hughlings, 113, 305, 378
Jacksonian sensory march, 113–14
Jacob, François, 163, 235–36, 324, 417
 on day vs. night science, 240
 gene regulation studied by, 256–59, 261, 264
Jacobson, Carlyle, 354
James, William, 133, 210, 311, 313–14, 340–41, 344, 427
Jamison, Kay, 372
Jessell, Tom, 252, 325, 394, 398
Jewish Theological Seminary, 260
Jews, 4–5, 12, 14, 15–18, *17*, 20, 30–31, 34, 42, 49, 92, 93–94, 174, 177, 405
 Austrian reparations unpaid to, 406, 410–11
 as conspiring to control the world, 178–79
 cultural importance of, 406–7, 409–10, 412
 Holocaust of, 5, 10, 48, 405, 410, 415; *see also* Nazis, Nazism
 present-day Viennese community of, 405, 409, 410–13, 414
 values of, 22
 Yom Kippur holiday of, 393, 394, 412–13
 see also anti-Semitism
Jörnvall, Hans, 393–94, 398
Journal of General Physiology, 107
Journal of Neurophysiology, 226
Journal of Physiology, 171
Journal of the American Medical Association, 336
Jünger, Ernst, 37–38

Kafka, Franz, 39
Kallman, Franz, 155
Kaltenbrunner, Ernst, 405
Kandel, Allison, 157, 398, *399, 402*
Kandel, Billy, 176–77
Kandel, Charlotte Zimels, 3–5, 13, *13,* 14, 18–22, 24–25, 31, 34–36, 177, 182–83, 260
background of, 18–19
death of, 36
forced emigration of, 28–29, 34
Kandel, Denise Bystryn, 47–50, *50,* 138, 148, 149, 150–52, 155–58, 162, 181, 182–83, 209, 238, 324–25, 326, 364, 420
art collected by, 428
career of, 48, 49, 247, 395–96, 425
in France, 150, 165, 173–79, 183
Nobel Prize award and, 393, 394, 395, 396, 398–99, *399,* 402, 404
World War II experiences of, 48–49, 174, 400
Kandel, Elise Wilker, 176–79
Kandel, Emily, 157, 398, *399,* 402
Kandel, Eric R.:
art collected by, 173–74, 183, 428
biotechnology companies of, 323–33
birth of, 13, 19
childhood in Vienna, 3–5, 23–25
college education of, 5–6, 33, 36–39, 40, 43, 363
elementary education of, 13, 18, 33–34, 36
forced emigration of, 29, 31–32, 33, 408, 413, 414
high school education of, 33, 36–37
honors awarded to, 259–60, 395
internship of, 108
in medical school, 6, 43–47, *56,* 183, 416–17, 418
Nobel Prize of, xiv, 229, 251, 356, 393–405, 428
opera enjoyed by, 183, 249
as parent, 157–58, 220
personal psychoanalysis of, 158, 182, 420
psychiatric residency of, 148, 152–56, 336, 364, 366, 367, 420, 426
Kandel, Hermann, 3–5, *13,* 22, 24–25, 31, 34–36, 37, 151, 182–83, 408
background of, 18–19
Brooklyn clothing store of, 35
death of, 35–36, 220
forced emigration of, 28–29, 34
Kultusgemeinde file of, 411
Nazi arrests of, 4–5, 17, 28, 34
toothbrush factory job of, 34–35
toy store of, 3, 13, *14,* 28
upstanding character of, 29, 35
Kandel, Lewis (Ludwig), 4, 5, 13–14, *14, 15,* 25, 33, 37, 173–79, 183
academic achievements of, 13
career of, 175–78
death of, 179, 220
forced emigration of, 29, 31–32, 33
in World War II, 176
Kandel, Libby, 157, 398, *399, 402*
Kandel, Minouche, *see* Sheinfeld, Minouche Kandel
Kandel, Paul, 156–58, 162, 165, 175, 177, 181, 220
Nobel Prize and, 394, 398, *399,* 402
Kant, Immanuel, 202, 203, 223, 296, 302, 307, 309
Karolinska Institute, 393–94, 399–400, 401
Katz, Bernard, 86, 93–96, *95,* 97–98, 99, 100–101, 102, *102,* 143, 172
ionotropic receptors found by, 226–27
squid's synapse studied by, 101, 107, 232, 422
Katzman, Irene, 398
Kellendonk, Christoph, 353, 357
Kentros, Cliff, 312
Kernberg, Otto, 370
Kety, Seymour, 155
Khorana, Har Gohind, 244
kinases, 228, 245, 293
calcium, calmodulin-dependent protein, 284
MAP, 263, 264–65
see also protein kinase A
Klerman, Gerald, 369–70
Klestil, Thomas, 405, 409, 414
Klimt, Gustav, 19, 150, 428
knockout genes, 288, 289, 291
Koch, Christof, 377, 380, 382, 383, 385, 388, 424
Koester, John, 398
Kohn, Walter, 409, 410
Kokoschka, Oskar, 19, 150, 428
Kornhuber, Hans, 389
Kornorski, Jerzy, 158
Kraus, Karl, 23, 363–64
Krebs, Ed, 227–28
Krenek, Ernst, 20, *21*
Kriegstein, Arnold, 253
Kris, Anna, 38–39, 42–43, 47, 363
Kris, Ernst, 39, 40, 42–43, 46–47, 55, 126, 181, 363, 368, 373
Kris, Marianne Rie, 39, 42–43, 46–47, 181, 363
Kris, Tony, 153, 154–55
Kristallnacht, 4–5, 27–31, 280, 414
motivations for, 29–31
pretext for, 28
Kubie, Lawrence, 47, 55, 126
Kuffler, Stephen, *95,* 115, 143, 145, 149, 153, 180–81, 236–38, 417
in Australia, 94–96, *95*
in Austrian Physiological Society, 236, 408
crayfish dendrites studied by, *95,* 107
death of, 237
forced emigration of, 94, 102, 237
memorial of, 237
neurobiology department established by, 184–85

Kuffler, Stephen (*continued*)
retinal cells studied by, 300–301
Kupfermann, Irving, 188–92, 194–95, 196–97, 250, 273, 320

Laborit, Henri, *355*
Lange, Carl, 340–41, 344
language, 45, *111*, 120–24, *121*, *123*, 126
Larousse Encyclopedia of Animal Life, The, 207
Lashley, Karl, 116, 124–25, 135, 192, 199, 200, 203
mass action theory of, 124, 129, 133
Lasker Award, 259–60
Last Days of Mankind, The (Kraus), 363–64
Launsky-Tieffenthal, Peter, 410–11
Laurencia pacifica, 253, *254*
L-DOPA, 401
learned fear, 186, 339, 342–48
in anxiety disorders, 342–43
in aversive classical conditioning, 170, 171, 232–34, 343–45, *344*, *345*, 346
behavioral test for, 346
in *Drosophila,* 232–34, 266, 308
molecular genetics of, 348
neural circuitry of, 343–35, *344*, *345*, *347*
phobias and, 342–43
sensitization as, 169
learned safety, *347*, 349–50, *350*
learning, 6, 10, 59, 116, 132, 135, 136, *138*, 141, 142, 143, *147*, 148, 165, 185, 208–18, 238, 257, 276, 279, 421
anatomical changes produced by, 212–18, *214*, *217*
behaviorist, 208
Cajal's theory of, 158, 159–60, 198, 199, 204
chemical vs. electrical transmission in, 205
as conserved in evolution, 144, 162, 186, 192, 266, 267, 276, 422
critical period in, 262
developmental processes in, 202–3
in early years of life, 218
genes and, 202–3, 287
genetic studies of, 232–34
in human individuality, 218
instrumental conditioning in, 41
maze, 124–25, 290–91, 328–29, *329*
of nonsense words, 209–10, 211
one-trial, 265–66
reductionist approach to, 144–49, *147*, 161, 186, 201, 203–4, 236
reflexive, 132
relearning of, 210, 214
repetition in, 132, 209–10, 215, 263, 264–65, 309
spatial maps and, 309
strength of synapses in, *see* synaptic strength
survival value of, 186
synaptic plasticity in, 158–62
see also classical conditioning; habituation; long-term memory; neural analogs of learning; sensitization; synaptic strength
LeDoux, Joseph, 186, 343–44, 346
Lepus marinus, 145
Lettvin, Jerry, 152–53
Libet, Benjamin, 389–90
Lichtenberger, Elisabeth, 413
Lilian, Princess, 401
Limax, 186
local protein synthesis, 270, *271*, 272–75, *274*
Locke, John, 40, 41, 202, 203, 223
Loewi, Otto, 91–92, 94, 102
Lømo, Terje, 282–83, 284, 286, 309, 345
London taxi drivers, 306
"Long and Short of Long-Term Memory, The" (Kandel and Goelet), 262
long-term facilitation, 238, 263, 264–66, *265*, 267, 268, 270, 283, 287, 293, 328, 346
long-term memory, 129, 136, 142–43, 208–18, 238, 328, 401, 424
anatomical changes required for, 212–18, *214*, *217*, 221, 225, 255–56, *256*, 261, 294
of *Aplysia,* 191–92, 212–15, 259
consolidation of, 210–12, 213, 218, 261–62, 266
duration of, 206
emotional, 342–43
of H.M., 127–28, 130–31
protein synthesis required by, 212, 213, 218, 225, 241, 255–56, *256*, 259, 262, 267, 292, 310
retrograde amnesia vs., 211–12
as secondary memory, 210
short-term memory converting to, 128, 129, *130*, 192, 206, 208, 209–10, 213–15, 241, 259, 261–62, 266, 275–76, 314, 328, 422
storage of, 129, *130*, 142–43, 212–13, 215
vulnerability to disruption of, 211–12
see also hippocampus; learning
long-term memory, genetic studies of, 238, 241, 247–48, 251, 252–56, 261–76
CPEB in, 272–75, *274*
CREB in, 263–66, 268, 270, 275–76
cyclic AMP in, 258–59, 263, 264, 267
gene regulatory proteins in, 257–58, 261–62, 263–66, 268, 275–76
local protein synthesis in, 270, *271*, 272–75, *274*
maintenance mechanism of, 270–75, 276
MAP kinase in, 263, 264–65
modulatory interneurons in, 254–55, *255*, 264
mRNA in, 268–70, 272–75, *274*
prions in, 272–75, *274*
protein kinase A in, 263, 264–65, 268, 270, 310
serotonin in, 254–56, *255*, 262–63, 264, 267, 268, *269*, *271*, 272, 274, *274*, 275
synaptic marking in, 267–70
long-term potentiation, 281, 283–85, 286–87, 288–94, 306

age-related memory loss and, 330
Alzheimer's disease and, 331
in amygdala, *345–46, 348, 349*
spatial maps and, 291, 309–10
Lorenz, Konrad, 144, 169
Lowenstein, Rudolph, 43, 55, 368
LSD (lysergic acid diethylamide), 103–5, 106, 107
effects of, 103–4
serotonin and, 104–5, 106
visual hallucinations generated by, 103, 104, 105
Lueger, Karl, 20, 26, 31
Lumer, Eric, 385
Lynch, Gary, 284

McCarthy, Joseph, 57
McCarty, Maclyn, 242
McGinn, Colin, 378
MacKinnon, Roderick, 89
MacLeod, Colin, 242
mad cow disease (bovine spongiform encephaly), 273
Mahler, Gustav, 19
Mann, Thomas, 39
"Man Who Mistook His Wife for a Hat, The" (Sacks), 303
MAP kinase, 263, 264–65
Margolin, Sidney, 47
Marks, Paul, 325
Marshall, Wade, 109–15, 116, 134, 136, 141, 148, 155, 364
mental illness of, 114–15
sensory map of body discovered by, 109–14, *110, 112,* 124, 216, 299, 308
Martin, Kelsey, 267–70, 418
Massachusetts Mental Health Center, 148, 152–56, 180–82, 336, 364, 366, 420
mass action, theory of, 124, 129, 133
maternal deprivation studies, 373–74
Mayford, Mark, 289
Mayr, Ernst, 155
maze learning, 124–25, 290–91, 328–29, *329*
medial temporal lobe, 127–33, *128,* 280
mediating circuits, 223, *224*
Medical Research Council Laboratory, 323
medical sociology, 49
membrane hypothesis, 80–83, 85, 87–88
memories, 238–39, 280–81
core, 281
emotional, 342
flash-bulb, 265, 266
persistent painful, 10, 11, 342–43
traumatic, 10, 342, 343
of Woolf, 280
memorists, 266
memory, 3–6, 9–11, 43, *111,* 116–34, 135, 165, 185, 203, 238, 372, 421, 422, 424
associations in, 209, 215, 285
as distinct mental function, 129
exceptionally good, 265–66
forms of, 198
grading of, 209–10
maze learning and, 124–25
in transmission of culture, 10
see also explicit memory; implicit memory; long-term memory; short-term memory
memory disorders, xii, 10–11, 124, 323–24, 327–33
in depression, 331, 335, 361
drug treatments for, 319, 326, 330, 331–32
of H.M., 127–33, 291
mass action theory of, 124
retrograde amnesia, 211–12
in schizophrenia, 331–32, 335, 353–55, 357, 358
see also age-related memory loss; Alzheimer's disease
Memory Pharmaceuticals, 326–33
memory retrieval, 215–16, 281
memory storage, xiv, 9, 45, 59, *125, 128,* 133–34, 135–36, 142–43, 158, 199, 224–26, 264–66, 327, 423
in *Aplysia,* 204–5
chemical vs. electrical transmission in, 205
of emotional memories, 342
of long-term memory, 129, *130,* 142–43, 212–13, 215
one-stage model of, 212–13, 215
reductionist approach to, 144–49
as sensitive to disruption, 211
of short-term memory, 129, *130,* 212–13, 215, 231, *233*
in specific brain regions, 129, 135, 142, 280; *see also* hippocampus
synaptic plasticity in, 159–60
theories of, 199–200
two-stage model of, 211–12
Mendel, Gregor, 241
"mental facilities," 118–19
mental illness, xiii, 11, 105–6, 141, 153–54, 323, 331–32, 335–51, 352–62, 418, 421
animal models of, 338–39, 358
bioethics and, 334
drug treatments for, 154, 334, 348, 350–51
environmental factors in, 338, 358
functional vs. organic, 336
genetic components of, 242, 338, 361
genetic predisposition in, 338
incidence of, 372
investigating causes of, 337–51
of Marshall, 114–15
neurological disorders vs., 336–37
postmodern examinations and, 336, 337
psychic determinism in, 39–40

mental illness (*continued*)
social stigma of, 336
translational scientific research on, 361–62
see also specific mental disorders
Merzenich, Michael, 216–17
messenger RNA (ribonucleic acid), 244, 258, 268–70, 383–84
active vs. dormant, 272–75, *274*
metabotropic receptors, 227, *228*, 229, 230, 325
mice, 7, 225, 361
age-related memory loss in, 326, 328–30, *329*, 335
anxiety in, 338–39
classical conditioning of, 343
genetically modified, 186, 281, 286–94, 295, 306, 319, 348, 357, 358, 374
instinctive fear response of, 343, *344*, 346, *350*
learned fear in, 343–50, *344, 345, 347, 350*
safety training of, *347,* 349, *350*
schizophrenia modeled in, 353, 357, 358
selective breeding of, 286, 287
sense of smell in, 251
spatial memory in, 290–94, 295, 302, 306, 308, 310, 314
spatial tasks learned by, 290–91, 312, 328–29, *329*
Milk Train Doesn't Stop Here Anymore, The (Williams), 281
Milner, Brenda, 116–17, 124–25, 127–34, 135, 136, 159, 198, 212–13
H.M. as patient of, 127–33, *128*, 291
star-drawing experiment of, 131, *131,* 133
Milstein, Cesare, 323
mind, 7
brain as one with, xii, 8–9, 117–18, 120
empiricist vs. rationalist view of, 202–3
Kantian view of, 202, 203, 223, 296, 302, 307, 309
soul vs., 117, 118, 120, 378
mind, science of, xii–xiii, 6–11, 59, 241, 251, 304, 306, 332, 365, 372, 376, 418
all mental processes as biological in, 336, 421
experimental methods of studying, 40–42
Freud's structural theory of, 53–55
future direction of, 423–26
mind-altering drugs, 103–5
see also LSD
mind-body problem, 420–21
mirror neurons, 425
Mitzi (housekeeper), 23–24, 280
modulatory circuits, 205, 223–24, *224*
modulatory interneurons, 223–24, *224,* 230, 254–55, *255,* 264
modulatory neurotransmitters, 223–24, 252, 283, 313, 314
molecular biology, xii, 6–7, 8, 88, 241–46, *248,* 250–52, 224–26, 237–38, 306, 309, 320, 335, 336, 350, 361–62, 422, 425
biotechnology industry and, 323–33
see also short-term memory, molecular biology of
molecular cognition, 251–52, 295–306
molecular genetics, 241–46, 286–94, 336, 348, 427
see also long-term memory, genetic studies of
molecular sociobiology, 425
monamine oxidase inhibitors (MAOIs), 359
monkeys, 7, 111, 304–5, 353, 354
egocentric coordinates of, 308
maternal deprivation studies of, 373–74
mirror neurons in, 425
sensory map of body in, 216–17, *217*
monoclonal antibodies, 323
Monod, Jacques, 256–59, 261, 264, 324
Montefiore Hospital, 50, 108
Montreal Neurological Institute, 116, 124
Morgan, Thomas Hunt, 232–33, 241
Morris, Richard, 290
motor cortex, 160–61
motor neurons, 66–67, *67,* 71–72, 79, 136, 140, 142, 337
acetylcholine released by, 93, 97
action potentials of, 140
of *Aplysia,* 194–95, *195,* 196, *196,* 200–201, 204, 205, 213, 214, 253–56
in coordinated reflex responses, 71–72
excitatory and inhibitory signals received by, 98–99
synaptic potentials and, 92–93, 97–98
Mountcastle, Vernon, 259, 299, 300, 301, 302, 308
Movshon, Anthony, 305
Mozart, Wolfgang Amadeus, 218, 402–3
Muller, Robert, 310
Mullinex, Kathleen, 325
multiple sclerosis, 67, 325
Murex, 146
muscle cells, 66, 75, 79, 92–93, 96, 97–98, 226, 229
musicians, 217–18
Musil, Robert, 12
Muzicant, Ariel, 410, 411

Nachmansohn, David, 57
Nagel, Thomas, 379–82
National Academy of Sciences, 252
National Institute of Mental Health (NIMH), 109–15, *110,* 116, 135–49, *138,* 155, 156, 364, 426
National Institutes of Health (NIH), 57, 109, *110,* 127, 135, 140, 152, 287, 304, 312
fellowship grant of, 161–62, 203–4
scientific culture of, 141, 145, 148
National Socialism, *see* Nazis, Nazism
natural selection, 235–26, 287
Nature, 242, 262, 267
"Nature of Conduction in Nerve" (Hodgkin), 84

Nature Reviews Neuroscience, 334
Nazis, Nazism, 4–5, 3–18, 26–32, 48, 174
agenda of, 26
Austria annexed by, 13–18, 27, 92, 96
Austrian, 20, *21,* 25, 26–27, 30, 92, 405, 407–8, *407*
Austria's denial of collaboration with, 405–6, 408–9, 410–11
concentration camps of, 48, 405, 408, 414
Hermann Kandel arrested by, 4–5, 28, 34
Pan-German nationalism of, 25, 26, 31
scientists forced to emigrate by, 92, 93–94, 96, 102, 237, 405, 409, 414
Sturm Abteilung of, 34
negative feedback circuits, 348
Neisser, Ulric, 427
nerve cells, *see* neurons
nerves, *45,* 90, 91
neural analogs of learning, 160–62, 165–72, 198–99, 204, 282–83
classical conditioning in, 160, 161, 166, 170, 171
experimental design of, 161
habituation in, 160, 161, 166–69, 171
heterosynaptic facilitation in, 169–70
homosynaptic depression in, 168–69
NIH fellowship grant for, 161–62
recovery from homosynaptic depression in, 169
sensitization in, 160, 161, 166, 169–70, 171
sensory stimulation in, 166, 168–70
synaptic potential in, 168–69
neural circuits (pathways), xii, 68, 69, 93, 113, 123, 138, 158, 166, 185, 198, 279, 422, 423
of *Aplysia,* 166, 169, 187, 192–96, 200, 203, 262–63, 281
connection specificity of, *64,* 65–66, 79, 193–94, 225
of emotions, 338, 342
fornix, 139
of hippocampus, 139, 142–43
of learned fear, 343–45, *344, 345, 347*
mediating, 223, *224*
modulatory, 205, 223–24, *224*
negative feedback, 348
perforant pathway, 140, 293
redundancy of, 124
Schaffer collateral pathway, 291, 330
of simple neuronal systems, 143
and theories of memory storage, 199–200
see also synapses
"Neurocognitive Enhancement: What Can We Do and What Should We Do?" (Kandel et al.), 334
neuroendocrine cells, 155–56
neuroethics, 333–34
Neuroethics: Mapping the Field, 334
neurological disorders, 323–24, 336–37, 401
amyotrophic lateral sclerosis (ALS), 67, 218–20, 289, 325, 337
anatomical location of, 337
channelopathies, 89
Creutzfeldt-Jacob disease, 273, 320
fragile X syndrome, 67, 337
Gaucher's disease, 67
genetic components of, 89, 289, 337, 338
Huntington's disease, 242, 289, 336, 337, 338
multiple sclerosis, 67, 325
myasthenia gravis, 98
Parkinson's disease, 67, 230, 242, 289, 337, 356
poliomyelitis, 67
prion, 272–73, 275, 320
"Neuronal Controls of a Behavioral Response Mediated by the Abdominal Ganglion of *Aplysia*" (Kandel and Kupfermann), 196–97
neuronal systems, simple, 107–9, 143, 144–45, 185–86
neuron doctrine, 56, 59, *60,* 61–68, 72, 73, 88
classes of neurons in, 66–67, *67*
principles of, *64,* 65–66, 79, 107
see also Cajal, Santiago Ramón y
neurons, xii, 9, 55–56, 59–69, *63, 64,* 110, 267, 422
cell body of, 63–65, *63,* 67, 107
classes of, 66–67, *67*
components of, 63–65, *63*
conduction speed along, 209
disease states and, 67
dynamic polarization of, *64,* 66, 68
electrical signaling in, 56
excitatory action of, *70,* 71, 93, 214, 252, 264, 348
inhibitory action of, *70,* 71, 93, 214, 264, 348
integrative functions of, *70,* 71, 93, 264
number of, 59
reaction time of, 209
receptive field of, 300
self-exciting loops of, 199, 200
shapes of, 61–63, *63*
size of, 147
surface membrane of, 62, 63, 80–89, *82, 89,* 91, 101
uniqueness of, 192–93
see also axons; dendrites
neuropsychology, 121
neurotic disorders, 39–40, 365
neurotransmitters, 60, 69, 90–102, *101, 102,* 214, 222–23, 236, 262, 401
acetylcholine, 91, 93, 94, 96, 97, 98, 100–101
excitatory, *95,* 98–99, 105, 222, 283, 348
GABA, 99, 348
inhibitors, *95,* 98–99, 105, 348
ion channels gated by, 97–98
ionotropic receptors gated by, 226–27, *228*
modulatory, 223–24, 252, 283, 313, 314

neurotransmitters (*continued*)
  norepinephrine, 359–60
  peptide, 348
  release of, 100–102, *101, 102*
  in "soup versus spark" controversy, 91–98
  see also dopamine; glutamate; serotonin
New Introductory Lectures on Psychoanalysis (Freud), *54*
Newsome, William, 305
New York University Medical School, 44–47, 59, 92, *138*, 152, 181
  ERK's Student Day talk at, 224
  Neurobiology and Behavior Division of, 181–86, 188, 191, 218–20, 310
Nicholls, John, 181, 236
Nicoll, Roger, 284, 287, 292
Nirenberg, Marshall, 244
NMDA receptor, 283–85, 286, 290, 292
Nó, Rafael Lorente de, 199
Nobel, Alfred, 396–97, 400, 401, 402
Nobel Foundation, 393–94, 396–98
  awards committees of, 397, 398, 401
  endowment of, 397
  establishment of, 396–97
  first awards of, 397–98
Nobel Medal, 399–400, *400*, 402, *402*
Nobel Peace Prize, 397
Nobel Prize, xiv, 34, 57, 252
  of Adrian and Sherrington, 79–80
  of Axel and Buck, *248*, 251, 404–5
  of Cajal and Golgi, 67, 68
  of Carlsson, Greengard and Kandel, 229, 356, 393–405; *see also* Kandel, Eric R., Nobel Prize of
  of Dale and Loewi, 92
  of Eccles, 179
  of Gilbert, 324
  of Hill, 172
  of Hodgkin and Huxley, 88, 179
  of MacKinnon, 89
  of Prusiner, 273
  of Tonegawa, 291
non-gated potassium channels, 87–88, 89
nonsense words, 209–10, 211
norepinephrine, 359–60
nucleotide bases, 89, 243–45, 246
  triplets of, 244
nucleus, cell, 58, 241, 257, 258, 262, 263, 264, 272
Nunberg, Henry, 44, 420
Nunberg, Herman, 44, 420, 421

obsessive-compulsive disorders, 365, 368, 421
  basal ganglia and, 371
  cognitive behavioral therapy for, 369, 371
occipital lobes, 109, 111, *111*
Of Flies, Mice and Men (Jacob), 163
O'Keefe, John, 281–82, 307, 308–9, 310, 315–16
"On a Scientific Psychology" (Freud), 371
"On Narcissism" (Freud), 45–46
operant conditioning, 422
Orden pour le Mérite, 412–14
Organization of Behavior, The: A Neuropsychological Theory (Hebb), 199
Ostow, Mortimer, 47, 55, 260
Oxford Partners, 326

Palade, George, 69, 101
Palay, Sanford, 69, 101
Pan-German nationalism, 25, 26, 31
panic attacks, 369
parent-child relationship, 373–74
parietal lobes, 109, 110–11, *111, 112*, 315, 385
Parkinson's disease, 67, 230, 242, 289
  anatomical location of, 337
  dopamine treatment of, 356, 401
Pasteur, Louis, 322
Pasteur Institute, 256, 322
pathways, neural, *see* neural circuits
Pavlov, Ivan, 40–41, 132, 158, 159, 160, 166, 170, 187, 198, 295, 427
Pelinka, Anton, 408
Penfield, Wilder, 111, *112*, 124, 299
  electrical stimulation of cortex by, 125–27, *125*
perception, 45, 59, 106, *111*, 132, 208–9, 295–306, 311
  see also sensory information; sensory representation; *specific senses*
perforant pathway, 140, 293
peripheral autonomic ganglia, 90
peripheral nervous system, *45*, 90
Pernkopf, Eduard, 407–8, *407*
  Atlas of Anatomy, 408
  concentration camp victims and, 408
Perutz, Max, 162
pharmaceutical industry, 249, 319, 321, 322, 323, 324, 355
philosophy, xii, 9, 40, 202, 333
  consciousness as viewed by, 377, 378–82, 389–90
  empiricism vs. rationalism in, 202–3
phobias, 342–43, 365, 369
  drug treatment of, 348
phosphorylation, 228, 231, 263, 401
phrenology, *118*, 119
Picasso, Pablo, 173–74
Pilzecker, Alfons, 210–11
place cells, 282, 286, 309, 310, 424
Plato, xiii, 377, 378
Pliny the Elder, 145
Polan, Jonathan, 353, 357
Poland, 28, 29, 48
  Nazi death camps in, 405
poliomyelitis, 67
Popper, Karl, 12, 96–97, 99, 378, 409, 414

positive reinforcement, 349
positron-emission tomography (PET), 8, 305
postsynaptic cell, 92, 93, 98, *101, 102*
postsynaptic neurons, 90, 94, 205
post-traumatic stress disorder, 10, 335, 342–43
  cognitive behavioral therapy for, 369
  drug treatment of, 348
  learned fear in, 342, 343
  manifestation of, 343
potassium ions, 80, 81, *82, 83,* 85, 86–88, *87,* 89, 97–98, 136–37, 231, 232
  radioactive, 137
Potter, David, 100, 180
preconscious mind, 54, *54,* 126, 365, 374
prefrontal cortex, *130,* 313, 315, 350, 385
  damage to, 354, 355, 385
  depression and, 371
  functions of, 354
  scientific investigation of, 354–55
  working memory in, 127, 353–55, 357, 359
premotor cortex, 425
presynaptic axon terminals, 65–66, 69, 71, 92, 94, 100, 101, *101,* 102, 223, 224, 230, 231, 259, 267, 348
  change in numbers of, 213–15, *214*
presynaptic neurons, 90, 92, 93, 94, 98, 205
primary memory, 210
Principles of Neural Science (Kandel and Schwartz), xiv
Principles of Psychology, The (James), 133, 311, 313
prions, 272–75, *274,* 314, 320, 424
"Project for a Scientific Psychology" (Freud), 72, 198
promoter sites, 257–58, 263
protein kinase A, 228–30, *229,* 231, *233,* 240–41, 314, 330, 346, 401
  Alzheimer's disease and, 331
  in long-term memory genetics, 263, 264–65, 268, 270, 310
proteins, 7, 88, 240, 256–58, 320–21, 377
  of acetylcholine receptors, 97
  amino acids of, 99, 244, 245, 250, 272, 320
  in brain, 236
  cell manufacture of, 58
  effector, 257
  fruitless, 425
  gene regulatory, 257–58, 261–62, 263–66, 268, 275–76, 324
  genes' encoding of, 243–45, 246, 257–58, 261–62, 291
  kinases and, 228
  local synthesis of, 270, *271,* 272–75, *274*
  long-term memory's required synthesis of, 212, 213, 218, 225, 241, 255–56, *256,* 259, 262, 267, 292, 310
  membrane, 227
  negative charge of, 81
  phosphorylation of, 228, 231, 263, 401
  prions, 272–75, *274,* 314, 320, 424
Prusiner, Stanley, 272–73
psychedelic drugs, 103–5
  see also LSD
psychiatry, 335–37, 362, 372, 421
  academic, 180–81, 364
  antidepressants and, 359
  antipsychotic drugs and, 355–56
  ERK's residency in, 148, 152–56, 336, 364, 366, 367, 420, 426
  medical illnesses treated by, 366
  psychoanalytic, 264–67, 372–75
  in World War II, 366
psychic determinism, 39–40, 367, 389
psychoanalysis, xi, xv, 6, 7, 12, 39–40, 41–43, 44, 53–55, 59, 72–73, 109, 116, 126, 133, 143, 145, 153, 181, 260, 295, 363–75, 416
  behaviorism vs., 41
  brain as viewed by, 202
  brain biology as relevant to, 45–47
  brain imaging and, 424–25
  child development studies in, 43, 365, 372–74
  cognitive psychology and, 43, 296, 298, 368
  consciousness and, 385, 388, 389, 424–25
  depression as viewed by, 368
  of ERK, 158, 182, 420
  expanded therapeutic scope of, 365–66
  interpretation in, 364, 365
  introspection in, 41–42
  mind-body problem in, 420–21
  as nonempirical specialty, 364, 365, 366–67
  objectivity of, 42
  original contributions of, 365
  reductionist approach by, 372–75
  resistance in, 365
  for schizophrenia, 155, 365
  therapeutic method of, 74, 364–65, 366–67; *see also* psychotherapy
  transference in, 153, 365, 370, 420
  see also Freud, Sigmund
Psychoanalytic Study of the Child, The, 373
psychology, xii, 250–51
  see also behaviorist psychology; cognitive psychology
psychopathology, 367, 388
Psychopathology of Everyday Life (Freud), 25, 39–40
psychopharmacology, *see* drug treatments
psychosis, 104–5, 154, 421
  see also schizophrenia
psychosomatic diseases, 366
psychotherapy, 116, 153, 154, 180, 364, 365–72, 388
  brain imaging and, 370–71, 372, 425
  brief dynamic, 370
  cognitive behavioral, 369–70, 371
  controlled clinical trials of, 369

psychotherapy (*continued*)
drug treatment combined with, 372
evidence-based, 367–70
interpersonal, 369–70
optimal duration of, 370
outcome studies of, 366, 370–71
privacy of, 366–67
short-term, 369–71, 372
for situational crises, 369
structural changes produced by, 367, 370–71
transference-focused, 370
Purpura, Dominick, 130–7, *104*, 115, 140, 224, 417, 427
pyramidal cells, 139–40, 142, 147, 172, 281, 282, 308–9, 330
of amygdala, 344, 348

Quanta, 101
Quinn, Chip, 148, 232–34

rabbits, 230, 282
sensory map of body in, 299
racial anti-Semitism, 30–31
Rall, Willifred, 140
Ramachandran, Vilayanur, 384, 390, 425
Rappaport, Judy Livant, 153
rationalism, 202–3
readiness potential, 389
receptors, *45*, 60, 79, 91, *101*, *102*, 222, 225, 236, 425
acetylcholine, 97, 98
AMPA, 283, 284, 292
dopamine, 325, 330, 356, 357–58, 361
GABA, 99
for gastrin-releasing peptide, 348
ionotropic, 227, *228*, 283–85
metabotropic, 227, *228*, 229, 230, 325
NMDA, 283–85, 286, 290, 292
pain, 125
for sense of smell, 251
serotonin, 104, 105, 325
touch, 109, 113
reciprocal control, 71
recombinant DNA, 245, 246, 251, 259, 286–88, 419
co-transfection method for, 249
in synthesizing proteins, 320–21
see also genetic engineering
recordings, intracellular, *76*, 77, *78*, 80, 93, 107, 108–9, 138, 139–40
recovery from homosynaptic depression, 169
refractory period, 158
Reiser, Morton, 420
relearning, 210, 214
repression, *54*, 55
res cogitans vs. *res externa*, 117
resistance, 365
resting membrane potential, 80–83, *82*, *83*, 85, 86, 87–88, 98, 109, 165, 232
retina, 111, 296, 300–301
retrograde amnesia, 211–12
reverberatory circuits, 199
reverse genetics, 288
ribosomes, 258
Rizzolatti, Giacomo, 425
RNA (ribonucleic acid), 377
messenger, *see* messenger RNA
Rockefeller University, 56–57, 69, 84, 89, 104, 225, 242, 325, 425
Rogan, Michael, 346, 349
Roland, Lewis, 325
Rolipram, 330, 331
Roman Catholicism, 15, 49
Doctrine of Deicide of, 31
dualism and, 117–18
Ryle, Gilbert, 133, 378

Sacks, Oliver, 303, 305
safety, learned, 349–50, *350*
Safire, William, 333–34
salience, 223–24, 313–14, *315*
Samuelson, Bengt, 401
Sanger, Frederick, 244–45
Schacher, Samuel, 254
Schacter, Daniel, 131–32
Schaffer collatoral pathway, 291, 330
S channel, 231
Scheller, Richard, 250, 332
Schiele, Egon, 19, 150, 428
Schildkraut, Joseph, 153, 154–55
schizophrenia, 11, 104, 106, 153, 154, 336, 352–58
antipsychotic drug treatment of, 355–58
brain imaging and, 354
cognitive symptoms of, 352, 353–54, 356–58
delusions of, 352–53, 357
derailment in, 353
developmental causes of, 357–58
environmental triggers in, 338, 358
excessive dopamine transmission in, 356–58, 361
flattening of affect in, 353
genetic components of, 153, 155, 242, 338, 355, 357–58
genetic predisposition to, 357
hallucinations of, 353, 357
incidence of, 352
memory impairment in, 331–32, 335, 353–55, 357, 358
negative symptoms of, 352, 353, 356–57
positive symptoms of, 352–53, 356–57
psychotic episodes in, 352, 353, 356
as thought disorder, 337–38, 357
schizotypal personality disorder, 352
Schleiden, Mattias Jakob, 58
Schnitzler, Arthur, 23, 39

Schönberg, Arnold, 19
Schönerer Georg von, 31
Schorske, Carl, 23
Schrödinger, Erwin, 241–42, 243
Schuschnigg, Kurt von, 25, 26–27
Schüssel, Wolfgang, 411
Schwann, Theodor, 58
Schwartz, James H., xiv, 44, 183–84, *184*, 185, 191, 224–30, 241, 247, 251–52, 310, 398, 427
Schweitzer, Frederick, 31
Science, 196–97, 200, 203
science, scientists, 96–97, 105, 172–73, 252, 416–29
  acceptable research questions in, 144–45
  basic vs. clinical, 361–62
  biased vs. unbiased, 288
  commercialization of, *see* biotechnology industry
  day vs. night, 240, 241, 246, 247, 261
  disputes of, 67–69, 419–23
  ethics of, 333–34
  forced to emigrate by Nazis, 42, 92, 93–94, 96, 102, 237, 405, 409, 414
  in France, 178, 417
  freedom of research in, 322, 333
  intellectual context of, 309–10
  investigator bias in, 366
  social context of, 106, 417–18
  taking risks in, 419–20, 427
  training for, 426–28
  see also mind, science of
Scoville, William, 116–17, 127
Searle, John, 379–81, 382
seaweed, 253, *254*
secondary memory, 210
second-messenger signaling, 226–27, *228*, 230, 231, 234, 259, 284, 292, 325, 401
security, sense of, 349–51
Seeman, Philip, 357
selective attention, 311, 312, 335, 383, 424
selective serotonin reuptake inhibitors, 360, 371
self, sense of, 11
Semrad, Elvin, 153–54
sensitization, 41, 160, *167*, 186, 261–63, 279, 422
  of *Aplysia*, 189–91, *191*, 200, 201, 202, 203, 204, 205, 212, 213–14, *214*, 215, 224, 230, 259, 314, 346
  as learned fear, 169
  in neural analogs of learning, 160, 161, 166, 169–70, 171
  sensory stimuli in, 160, 166, 169–70
  survival value of, 169
  see also synaptic strength
sensory homunculus, *112*
sensory information, 44, *45*, 70, 75, 78–79, 120–21, 129, *130*, 209, 311, 424
  hippocampus and, 142, 143, 282, 308
  see also specific senses
sensory map of body, 109–14, *110*, *112*, 124, 216–18, 276, 299, 308
  of monkeys, 216–17, *217*, 299
  of musicians, 217–18
sensory neurons, 66–67, *67*, 77–80, 113, 343–44
  of *Aplysia*, 195–96, *196*, 200–201, 204, 205, 213–14, 222–23, 253–56, 259
  of rabbits, 299
sensory representation, 295–306
  binding problem in, 303–4, 308, 382
  cognitive maps in, 298, 310; *see also* sensory map of body; spatial maps
  electrophysiological study of, 298–300
  future research on, 423
  reconstructive processes in, 296–98, *297*, 300–305, 308
sensory stimulation, 158–61, 168–70, 309, 313
  in habituation, 160, 166–67, 168–69
  in sensitization, 160, 166, 169–70
serotonin, 223–24, 226, *228*, *229*, 230, 231, *233*, 238, 293, 314, *315*, 371
  in depression, 359–60
  in long-term memory genetics, 254–56, *255*, 262–63, 264, 267, 268, *269*, *271*, 272, 274, *274*, 275
  LSD and, 104–5, 106
  MAOIs and, 359
  receptors for, 104, 105, 325
sexual behavior:
  of *Aplysia*, 188, *190*
  of *Drosophila*, 425
  habituation and, 168
Sharp, Philip, 321
Shaw, E. N., 104–5
Sheinfeld, Izzy, 157, 398
Sheinfeld, Maya, 157, 398
Sheinfeld, Minouche Kandel, 157–58, 181
  birth of, 182
  Nobel Prize award and, 394, 398, *399*, 400, 402
  poem of, *206*, 207
Sheinfeld, Rick, 157, 398, *399*, 402
Shelanski, Michael, 331
Shereshevski, S. V., 266
Sherrington, Charles, 61, *65*, 69–71, *70*, 93, 98, 135, 252, 264
  Nobel Prize of, 79–80
short-term memory, 129, 199, 211, 218, 240, 253, *255*, 267, 268, 292, 401
  of *Aplysia*, 189, 192, 204–5, 213–14, 292
  converting to long-term memory, 128, 129, *130*, 192, 206, 208, 209–10, 213–15, 241, 259, 261–62, 266, 275–76, 314, 328, 422
  duration of, 206
  of H.M., 127, 128
  as primary memory, 210
  storage of, 129, *130*, 212–13, 215, 231, *233*
  storage sites of, 212, 213, 215

short-term memory (*continued*)
see also working memory
short-term memory, molecular biology, 221–36
biochemical signaling pathways in, 225–34
cyclic AMP in, 226–35, *229, 233,* 238, 240–41, 262
dopamine in, 230
first messenger signaling in, 227, *228,* 231
glutamate in, 222, 224, 226, 227, *229,* 230, 232, 240–41
ionotropic receptors in, 226–27, *228*
mediating circuits in, 223, *224*
metabotropic receptors in, 226–28, *228,* 229, 230, 325
modulatory circuits in, 223–24, *224*
modulatory interneurons in, 222–23, *224,* 230
protein kinase A in, 228–30, *229,* 231, *233,* 240–41, 262
S channel in, 231
second-messenger signaling in, 226–28, *228,* 230, 231, 234
serotonin in, 223–24, 226, *228, 229,* 230, 231, *233,* 238
slow synaptic potential in, 222–23, 227, 230, 231
Shumyatsky, Gleb, 348
Si, Kausik, 270–75, 314, 418
Siegelbaum, Steven, 231, 252, 293, 398
Sifneous, Peter, 370
Silva, Alcino, 290, 292
Silvia, Queen of Sweden, 401, 404
Simpson, Eleanor, 353, 357
"Sketch of the Past" (Woolf), 277
Skinner, B. F., 41–42, 132, 135, 186, 427
slips of the tongue, 39–40, 367
smell, sense of, *248,* 251, 309, 405
Snyder, Solomon, 357
social phobias, 342, 369
Society for Neuroscience, 237
Socrates, xiii
sodium ions, 80, 81, *83,* 86–88, *87,* 97–98, 140
Sohlman, Ragnar, 397
somatosensory cortex, 110–14, *112,* 129, 142, 216, 299–300, 344
columns of, 300
receptive field of the neuron in, 300
see also sensory map of body
somatostatin, 320–21
soul 117, 118, 120, 378
"soup versus spark" controversy, 91–98
spatial maps, 281, 282–85, 286, 287, 295, 299, 305, 307–15
allocentric coordinates in, 308
attention and, 311–15
egocentric coordinates in, 308
formation of, 308–9, 310, 312, 424
gender differences in, 315–16
long-term potentiation and, 291, 309–10
long-term stability of, *309,* 310, 312–13, 314–15
spatial memory, 6, 286, 287, 288, 295–96, 328–30, 307–15, 424
attention and, 311, 312, 328
of birds, 306
of London taxi drivers, 306
in mice, 290–94, 295, 302, 306, 308, 310, 314
see also spatial maps
Spencer, Alden, 137–43, *138,* 144, 147, 148, 149, 152, 156, 158, 172, 191, 237, 281, 293, 308
ALS contracted by, 218–19
death of, 183, 219–20, 325
invertebrate research rejected by, 143
memorial of, 220
at NYU Medical School, 183, 185–86, 218–19, 310
personality of, 137–38, 220
spinal reflexes studied by, 143
Spencer, Diane, 138, 183, 219
spinal cord, 44–45, *45,* 66, 75, 76, 93, *95,* 98–100, 135–36, 140, 142
accidental severance of, 341
functions of, 70–71
reflex behaviors mediated by, 44, 70–71, *70*
Spitz, René, 373
Spitzer, Robert, 44
split-brain patients, 379
spreading cortical depression, 114, 136–37
squid:
giant axons of, *83,* 84–85, 86, 88, 107, 108, 422
giant synapses of, 101, 107, 232, 422
Squire, Larry, 131–32, 291
stage fright, 343
stem cells, 334, 360–61
Stern, Fritz, 38, 409
Steward, Oswald, 270
Stockholm, 398–405
awards ceremony in, 395, 398, 400–403
family and friends at, 398–99, *399*
Grand Hotel of, 398
Great Synagogue of, 400
Jewish community of, 398, 400
Karolinska Institute of, 393–94, 399–400, 401
Nobel lectures given in, 399–400
Nobel Medal presented in, 399–400, *400,* 402, *402*
Santa Lucia's Day in, 404
stress, 141, 361, 373
histamine in, 355
striatum, *130,* 132
D2 dopamine receptors in, 357–58, 361
positive emotions associated with, *347,* 349–50
strokes, 121–22, 303, 325, 337
Struhl, Gary, 252
Strumwasser, Felix, 141, 155
Sturm Abteilung (SA), 34

substantia negra, 337
superego, 45, 53–55, *54,* 105, 116, 248
  definition of, 55
surrogate mothers, 373
Sutherland, Earl, 226–27, *228,* 229, 234
Swanson, Robert, 319–21
synapses, *64,* *65,* 67, 90, 93, 100, 101, *101,* 110, 171, 248
  asymmetry of, 69
  change in number of, 213–15, 261, 262
  degeneration of, 327
  of hippocampus, 142
  squid's giant, 101, 107, 232
  see also short-term memory, molecular biology of
synaptic cleft, *65,* 66, 69, 91, 92, 97, 101, *101, 102, 229,* 360
synaptic marking, 267–70
Synaptic Pharmaceuticals, 324–26
  advisory board of, 324–25
synaptic plasticity, 158–62, 171, 185, 186, 187, 198–99, 205, 283, 422
  see also neural analogs of learning
synaptic potentials, 92–93, 96, 97–98, 100, 139, 284
  in *Aplysia,* 148, 200–201, 222–23, 230, 231
  of dendrites, 106
  in neural analogs of learning, 168–69
  slow, 222–23, 227, 230, 231
synaptic strength, 198–99, 200, 201–2, 204–5, 208, 213, 215, 221, 222, 223–24, *224,* 230, 235, 240, 255–56, 279
  long-term facilitation in, 263, 264–66, *265,* 267, 268, 270, 283, 287, 293, 328, 346
  see also long-term potentiation
synaptic transmission, 60, 90–102, *101,* 106–7, 143, 155, 159, 214, *214,* 229
  chemical, 60, 205; *see also* neurotransmitters
  electrical, 100, 205
  "soup versus spark" controversy on, 91–98
synaptic vesicles, 101, *101,* 232, 240–41
syphilis, 67

Talmudic memorist, 266
Tanaka, Akira, 173
Tauc, Ladislav, 145, 146–47, 148, 150, 161, 165–66, *166,* 171, 173, 174, 178–79, 187, 193, 199, 281
Templ, Stephen, 30
temporal lobes, 109, 111, *111, 125,* 126
  medial, 127–33, *128,* 280
Teplist vase, 238–39, *239*
Tessier-Lavigne, Marc, 332
thalamus, 44, 45, 113, 138, 300, 301, 344, *345, 347,* 350
Theis, Martin, 314
Third Man, The, 22
Thompson, Richard, 143
Thorndike, Edward, 40–41, 132, 208, 427
thymine, 243, 244
Tietze, Hans, 409
Tinbergen, Niko, 144
tissue cultures, 108, 249, 253–56, *255,* 259, 262–63, 268
  genes in, 323
  very young cells required for, 253
Tonegawa, Susumu, 290, 291–92
touch, sense of, 109–14, *138,* 209, 299–300, 308, 309
  submodalities of, 300
  see also sensory map of body
touch receptors, 109, 113
tranquilizing drugs, 99, 355–56
transcription, 258–59, 267, 275
transference, 153, 365, 420
  psychotherapy focused on, 370
transformation of information, 242
transgenesis, 288, 289
transmitter-gated ion channels, 97–98, 100
Tritonia, 186
Tsien, Roger, 263
tuberculosis, 359
Tully, Tim, 266
twins, identical, 338

unconscious inference, 209, 305, 424
unconscious memory, *see* implicit memory
unconscious mind, 25, 43, 74, 374–75, 363, 365, 367, 368
  conscious awareness and, 340–42, 344, 376, 384, 385–90, 424–25
  dynamic, 374
  fear in, 340–42, 344
  Freud's view of, 39–40, 53–54, *54,* 55, 56, 133, 388, 424–25
  preconscious, 54, *54,* 126, 365, 374
  psychic determinism and, 39–40, 367, 388
  psychosomatic diseases and, 366
Ungerstadt, Urban, 401–2
University College, London, 86, 94, 97, 172, 385
Unquiet Mind, An (Jamison), 372
Unterbeck, Axel, 326
uracil, 244

vaccines, 322
vagus nerve, 91
Valliant, George, 153
Verrocchio, Andrea del, 418
Vienna, xiv, xv, 3–6, 12–32, 33, 38, 42, 47, 56, 94, 96, 102, 106, 119, 181, 236–37, 391, 394–95, 405–15, 428
  anti-annexation political graffiti in, 15–17, *17*
  anti-Semitism in, 15–18, 20, *21,* 27–31, 152, 412–14; *see also* Kristallnacht
  Art Academy of, 26
  Aryanization of property in, 28

Vienna (*continued*)
beauty of, 20–22
Catholic population of, 15
cultural importance of Jews in, 406–7, 409–10, 412
culture of, 12, 19, 22, 23, 30
Die Neue Frei Presse of, 363
ERK's first return visit to, 150–52, 408
ERK's symposium in, 405, 409–10, 411, 412, 414
Heldenplatz of, 14, *16*
Hitler's entry into, 13–17, *16*
Hofburg palace of, 414
Hotel Sacher of, 414
immigrants in, 22–23
Institute of Molecular Pathology in, 408–9
intellectual environment of, 363–64
Jewish Museum of, 409
Kultusgemeinde agency in, 28, 410–11
new 1867 constitution of, 20, 22
Opera House of, 20, *21*
Orden pour le Mérite in, 412–14
political system of, 19–20
population of, 23
present-day Jewish community of, 405, 409, 410–13, 414
sensuality of, 23–24
vandalism in, 410–11
see also Nazis, Nazism
Vienna, University of, 19, 117, 236, 406–8
Jewish faculty expelled from, 18, 406–7
Nazi faculty of, 30, 407–8, *407*, 414
Vienna and its Jews (Berkley), 17, 20
Vietor, Karl, 38–39, 43
visual cortex, 105, 111, *123*, *129*, 142, 160–61, 296–98, *297*, 302–4
agnosias of, 303
dendrites of, 106–7
face recognition in, 128, 298, 303, 381
memory storage in, 129
orientation-selective cells in, 301
visual hallucinations, 103, 104, 105
visual perception, visual system, 7, 133, 308, 309, 312, 375, 380
emotionally charged, 385–89
illusions and, 297
reconstructive process in, 296–98, *297*, 300–304, 382
retina in, 111, 296, 300–301
vital forces, 75, 83
voltage-gated ion channels, 86–88, 89, 97–98, 100, 140
voluntary attention, 313–14
voluntary movement, 7–8, 59, 313, 389–90
readiness potential in, 389
vom Rath, Ernst, 28
von Frisch, Karl, 144
von Hapsburg, Otto, 406

Walzer, Tina, 30
Watkins, Geoffrey, 283
Watson, James, xi, 162, 377
DNA described by, 242–44
on recombinant DNA, 246
Weinberg, Robert, 248–49
Weisbusch, Claude, 173
Weissman, Charles, 321
Weissman, Myrna, 369–70
Wender, Paul, 153
Wernicke, Carl, 120–24, *121*, 131, 305
Wernicke's area, *121*, 122–23, *123*, 126
"What Is Emotion?" (James), 340
What Is Life? (Schrödinger), 242
Wiesel, Torsten, 180, 299, 301, 308
Wiesenthal, Simon, 28, 408
Wilkins, Maurice, 242
Williams, Tennessee, 281
Winkler, Georg, 414
Wittgenstein, Ludwig, 12, 409
Woolf, Virginia, 277
explicit memories of, 280
Woolley, D. W., 104–5
Woolsey, Clinton, 299
working memory, *111*, 332
in prefrontal cortex, 127, *353–55*, *357*, *358*
World War I, 5, 18, 19, 37, 137
World War II, 28, 29, 57, 86, 94, 95–96, *95*, 151, 176, 364, 400, 405
Bystryn family in, 48–49, 174–75
family disruption in, 373
psychiatric treatment in, 366
see also Nazis, Nazism
worms, 143, 185, 245
Ascaris, 145, 193
C. elegans, 144, 245, 419, 425
Wurtz, Robert, 304–5, 312

Yale University, 420–21
Child Study Center at, 43
Yeshivah of Flatbush, 33, 34, 36, 37, 260
Yin, Jerry, 266
Young, J. Z., 85

Zeilinger, Anton, 414
Zeki, Semir, 304
Zimels, Berman, 28–29, 33
Zimels, Hersch and Dora, 5, 24, 29, 33, 34
Zuckmayer, Carl, 17–18, 37–38